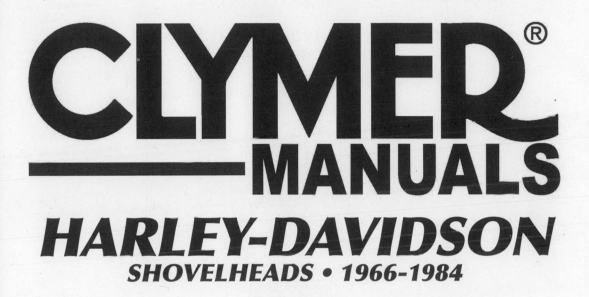

# CLYMER®
## MANUALS
# *HARLEY-DAVIDSON*
### *SHOVELHEADS • 1966-1984*

# WHAT'S IN YOUR TOOLBOX?

# More information available at haynes.com
## Phone: 805-498-6703

**Haynes Group Limited**
Sparkford Nr Yeovil
Somerset BA22 7JJ England

**Haynes North America, Inc.**
2801 Townsgate Road, Suite 340
Thousand Oaks, CA 91361 USA

ISBN-10: 0-89287-566-6
ISBN-13: 978-0-89287-566-5
Library of Congress: 92-70705

*Technical Illustrations:* Steve Amos

*Cover:* Michael Brown Photographic Productions, Los Angeles, California
Harley 74 Low Rider provided by Harley-Davidson West, Venice, California

M420, 7T2, 14-432
ABCDEFGHI

# Common spark plug conditions

## NORMAL

*Symptoms:* Brown to grayish-tan color and slight electrode wear. Correct heat range for engine and operating conditions.

*Recommendation:* When new spark plugs are installed, replace with plugs of the same heat range.

## WORN

*Symptoms:* Rounded electrodes with a small amount of deposits on the firing end. Normal color. Causes hard starting in damp or cold weather and poor fuel economy.

*Recommendation:* Plugs have been left in the engine too long. Replace with new plugs of the same heat range. Follow the recommended maintenance schedule.

## CARBON DEPOSITS

*Symptoms:* Dry sooty deposits indicate a rich mixture or weak ignition. Causes misfiring, hard starting and hesitation.

*Recommendation:* Make sure the plug has the correct heat range. Check for a clogged air filter or problem in the fuel system or engine management system. Also check for ignition system problems.

## ASH DEPOSITS

*Symptoms:* Light brown deposits encrusted on the side or center electrodes or both. Derived from oil and/or fuel additives. Excessive amounts may mask the spark, causing misfiring and hesitation during acceleration.

*Recommendation:* If excessive deposits accumulate over a short time or low mileage, install new valve guide seals to prevent seepage of oil into the combustion chambers. Also try changing gasoline brands.

## OIL DEPOSITS

*Symptoms:* Oily coating caused by poor oil control. Oil is leaking past worn valve guides or piston rings into the combustion chamber. Causes hard starting, misfiring and hesitation.

*Recommendation:* Correct the mechanical condition with necessary repairs and install new plugs.

## GAP BRIDGING

*Symptoms:* Combustion deposits lodge between the electrodes. Heavy deposits accumulate and bridge the electrode gap. The plug ceases to fire, resulting in a dead cylinder.

*Recommendation:* Locate the faulty plug and remove the deposits from between the electrodes.

## TOO HOT

*Symptoms:* Blistered, white insulator, eroded electrode and absence of deposits. Results in shortened plug life.

*Recommendation:* Check for the correct plug heat range, over-advanced ignition timing, lean fuel mixture, intake manifold vacuum leaks, sticking valves and insufficient engine cooling.

## PREIGNITION

*Symptoms:* Melted electrodes. Insulators are white, but may be dirty due to misfiring or flying debris in the combustion chamber. Can lead to engine damage.

*Recommendation:* Check for the correct plug heat range, over-advanced ignition timing, lean fuel mixture, insufficient engine cooling and lack of lubrication.

## HIGH SPEED GLAZING

*Symptoms:* Insulator has yellowish, glazed appearance. Indicates that combustion chamber temperatures have risen suddenly during hard acceleration. Normal deposits melt to form a conductive coating. Causes misfiring at high speeds.

*Recommendation:* Install new plugs. Consider using a colder plug if driving habits warrant.

## DETONATION

*Symptoms:* Insulators may be cracked or chipped. Improper gap setting techniques can also result in a fractured insulator tip. Can lead to piston damage.

*Recommendation:* Make sure the fuel anti-knock values meet engine requirements. Use care when setting the gaps on new plugs. Avoid lugging the engine.

## MECHANICAL DAMAGE

*Symptoms:* May be caused by a foreign object in the combustion chamber or the piston striking an incorrect reach (too long) plug. Causes a dead cylinder and could result in piston damage.

*Recommendation:* Repair the mechanical damage. Remove the foreign object from the engine and/or install the correct reach plug.

# CONTENTS

# QUICK REFERENCE DATA

## MOTORCYCLE INFORMATION

MODEL:_____ YEAR:_____

VIN NUMBER:_____

ENGINE SERIAL NUMBER:_____

CARBURETOR SERIAL NUMBER OR I.D. MARK:_____

## TIRE PRESSURE

| | Front | | Rear | | Sidecar | |
|---|---|---|---|---|---|---|
| | PSI | kg/cm$^2$ | PSI | kg/cm$^2$ | PSI | kg/cm$^2$ |
| **1966-1969** | | | | | | |
| Solo | | | | | | |
| 5.10 × 16 | 20 | 1.41 | 24 | 1.69 | | |
| 5.00 × 16 | 12 | .84 | 18 | 1.26 | | |
| With passenger | | | | | | |
| 5.10 × 16 | 20 | 1.41 | 26 | 1.83 | | |
| 5.00 × 16 | 12 | .84 | 20 | 1.41 | | |
| Rider with sidecar | | | | | | |
| passenger or 150 lb. load | | | 20 | | 1.41 | |
| 5.10 × 16 | 22 | 1.55 | 26 | 1.83 | 20 | 1.41 |
| 5.00 × 16 | 14 | .98 | 20 | 1.41 | 14 | .98 |
| **1970-early 1978** | | | | | | |
| FL series | | | | | | |
| Solo | 20 | 1.41 | 24 | 1.69 | | |
| With passenger | 20 | 1.41 | 26 | 1.83 | | |
| FX series | | | | | | |
| Solo | 24 | 1.69 | 26 | 1.83 | | |
| With passenger | 24 | 1.69 | 28 | 1.97 | | |
| FL series (rider with sidecar | | | | | | |
| passenger or 150 lb. load) | 22 | 1.55 | 26 | 1.83 | 20 | 1.41 |
| **Late 1978-on** | | | | | | |
| FL series | | | | | | |
| Solo | 20 | 1.41 | 24 | 1.69 | | |
| With passenger | 20 | 1.41 | 26 | 1.83 | | |
| FX series (except FXWG) | | | | | | |
| Solo | 24 | 1.69 | 26 | 1.83 | | |
| With passenger | 26 | 1.83 | 28 | 1.97 | | |
| FXWG | | | | | | |
| Solo | 30 | 2.11 | 26 | 1.83 | | |
| With passenger | 30 | 2.11 | 28 | 1.97 | | |
| FL series (rider with sidecar | | | | | | |
| passenger or 140 lb. load) | 22 | 1.55 | 26 | 1.93 | 20 | 1.41 |

**CAUTION:** Maximum pressure for all tires is 36 psi (2.53 kg/cm$^2$).
**NOTE:** These pressures are based on 150 lb. (68.04 kg) rider weight. For each extra 50 lb. (22.68 kg), add 2 psi (0.14 kg/cm$^2$) @ rear, 1 psi (0.07 kg/cm$^2$) @ front and 1 psi (0.07 kg/cm$^2$) @ sidecar tire. Do not exceed maximum pressure of 36 psi (2.53 kg/cm$^2$).

## ENGINE AND TRANSMISSION OIL CAPACITY

| | Quantity |
|---|---|
| Oil tank | 4 quarts (3.78 L, 3.3 Imp. qt.) |
| Transmission | 1 ½ pints (0.71 L, 0.62 Imp. qt.) |

## RECOMMENDED LUBRICANTS AND FLUIDS*

| | |
|---|---|
| **Brake fluid** | |
| Through 1975 | DOT 3 |
| 1976-on | DOT5 |
| **Fork oil** | |
| 1966-1969 | HD Hydra-Glide fork oil |
| 1970-on | HD type B |
| **Battery top up** | Distilled water |
| **Transmission** | HD transmission lubricant |
| **Primary chaincase** | HD primary chaincase |
| **Engine oil** | |
| Below +40 degrees F | HD special light |
| 40-60 degrees F | HD medium heavy |
| +60 degrees F and up | HD regular heavy |
| Severe engine operating | |
| conditions @ +80° F | HD extra heavy grade 60 |
| 10-90 ° F--normal and | |
| severe operating conditions | HD Power Blend Super Premium |

* Lubricants recommended are Harley-Davidson brands.

## BATTERY STATE OF CHARGE

| Specific gravity | State of charge |
|---|---|
| 1.110-1.130 | Discharged |
| 1.140-1.160 | Almost discharged |
| 1.170-1.190 | One-quarter charged |
| 1.200-1.250 | One-half charged |
| 1.230-1.250 | Three-quarters charged |
| 1.260-1.280 | Fully charged |

## SPARK PLUG TYPE AND GAP*

| | |
|---|---|
| **Type** | |
| 1966-1974 | HD No. 3-4 |
| 1975-1976 | HD No. 5-6 |
| 1977-1979 | HD No. 5A6 (standard) or HD No. 5R6 (resistor) |
| 1980-on | HD No. 5R6A or HD No. 5RL |
| **Gap** | |
| 1966-1969 | 0.025-0.030 in. (0.63-0.76 mm) |
| 1970-early 1978 | 0.028-0.033 in. (0.71-0.84 mm) |
| Late 1978-on | 0.038-0.043 in. (0.96-1.09 mm) |

* Spark plugs recommended are Harley-Davidson brands.

## FRONT FORK OIL CAPACITY

| | Quantity |
|---|---|
| **1966-1969** | |
| Wet | 6 1/2 oz. (192.2 cc, 6.76 imp. oz.) |
| Dry | 7 oz. (207 cc, 7.28 imp. oz.) |
| **1970-early 1978** | |
| FX Series | |
| 1970-1972 | |
| Wet | 5 1/2 oz. (162.6 cc, 5.72 imp. oz.) |
| Dry | 6 1/2 oz. (192.2 cc, 6.76 imp. oz.) |
| 1973-early 1978 | |
| Wet | 5 oz. (147.8 cc, 5.2 imp. oz.) |
| Dry | 6 oz. (177.4 cc, 6.24 imp. oz.) |
| FL Series | |
| 1970-early 1977 | |
| Wet | 6 1/2 oz. (192.2 cc, 6.76 imp. oz.) |
| Dry | 7 oz. (207 cc, 7.28 imp. oz.) |
| Early 1977-early 1978 | |
| Wet | 7 3/4 oz. (229.2 cc, 8.06 imp. oz.) |
| Dry | 8 1/2 oz. (251.3 cc, 8.84 imp. oz.) |
| **Late 1978-On** | |
| FL Series | |
| Wet | 7 3/4 oz. (229.2 cc, 8.06 imp. oz.) |
| Dry | 8 1/2 oz. (251.3 cc, 8.84 imp. oz.) |
| FX series (except FXWG) | |
| Wet | 5 oz. (147.8 cc, 5.2 imp. oz.) |
| Dry | 6 oz. (177.4 cc, 6.24 imp. oz.) |
| FXWG | |
| Wet | 9 1/4 oz. (273.5 cc, 9.62 imp. oz.) |
| Dry | 10 oz. (295.7 cc, 10.4 imp. oz.) |

## TUNE-UP SPECIFICATIONS

| | |
|---|---|
| **1966-1969** | |
| Breaker point gap | 0.020 in. (0.51 mm) |
| Dwell reading | 90° @ 2,000 rpm |
| Ignition timing | |
| Retarded | 5° BTDC |
| Automatic | 35° BTDC |
| **1970-Early 1978** | |
| Breaker point gap | 0.020 in. (0.51 mm) |
| Ignition timing | |
| Retarded | 5° BTDC |
| Advance | 34-36° BTDC |
| **Late 1978-on** | |
| Ignition timer air gap | |
| Late 1978-1979 | 0.004-0.006 in. (0.102-0.152 mm) |
| 1980-on | No adjustment |
| Ignition timing | Electronic |
| Compression | |
| Standard | |
| 1966-early 1978 | 90 psi or over (6.33 kg/cm$^2$ or over) |
| Late 1978-1984 | 100 psi or over (7.03 kg/cm$^2$ or over) |
| Acceptable Variance between cylinders | |
| 1966-early 1978 | 29 psi or less (2.04 kg/cm$^2$ or over) |
| Late 1978-1984 | 10 psi or less (0.70 kg/cm$^2$ or over) |

# DECIMAL AND METRIC EQUIVALENTS

| Fractions | Decimal in. | Metric mm | Fractions | Decimal in. | Metric mm |
|-----------|-------------|-----------|-----------|-------------|-----------|
| 11/64 | 0.015625 | 0.39688 | 33/64 | 0.515625 | 13.09687 |
| 1/32 | 0.03125 | 0.79375 | 17/32 | 0.53125 | 13.49375 |
| 3/64 | 0.046875 | 1.19062 | 35/64 | 0.546875 | 13.89062 |
| 1/16 | 0.0625 | 1.5875C | 9/16 | 0.5625 | 14.28750 |
| 5/64 | 0.078125 | 1.98437 | 37/64 | 0.578125 | 14.68437 |
| 3/32 | 0.09375 | 2.38125 | 19/32 | 0.59375 | 15.08125 |
| 7/64 | 0.109375 | 2.77812 | 39/64 | 0.609375 | 15.47812 |
| 1/8 | 0.125 | 3.1750 | 5/8 | 0.625 | 15.87500 |
| 9/64 | 0.140625 | 3.57187 | 41/64 | 0.640625 | 16.27187 |
| 5/32 | 0.15625 | 3.96875 | 21/32 | 0.65625 | 16.66875 |
| 11/64 | 0.171875 | 4.36562 | 43/64 | 0.671875 | 17.06562 |
| 3/16 | 0.1875 | 4.76250 | 11/16 | 0.6875 | 17.46250 |
| 13/64 | 0.2033125 | 5.15937 | 45/64 | 0.703125 | 17.85937 |
| 7/32 | 0.21875 | 5.55625 | 23/32 | 0.71875 | 18.25625 |
| 15/64 | 0.234375 | 5.95312 | 47/64 | 0.734375 | 18.65312 |
| 1/4 | 0.250 | 6.35000 | 3/4 | 0.750 | 19.05000 |
| 17/64 | 0.265625 | 6.74687 | 49/64 | 0.765625 | 19.44687 |
| 9/32 | 0.28125 | 7.14375 | 25/32 | 0.78125 | 19.84375 |
| 19/64 | 0.296875 | 7.54062 | 51/64 | 0.796875 | 20.24062 |
| 5/16 | 0.3125 | 7.93750 | 13/16 | 0.8125 | 20.63750 |
| 21/64 | 0.328125 | 8.33437 | 53/64 | 0.828125 | 21.03437 |
| 11/32 | 0.34375 | 8.73125 | 27/32 | 0.84375 | 21.43125 |
| 23/64 | 0.359375 | 9.12812 | 55/64 | 0.859375 | 21.82812 |
| 3/8 | 0.375 | 9.52500 | 7/8 | 0.875 | 22.22500 |
| 25/64 | 0.390625 | 9.92187 | 57/64 | 0.890625 | 22.62187 |
| 13/32 | 0.40625 | 10.31875 | 29/32 | 0.90625 | 23.01875 |
| 27/64 | 0.421875 | 10.71562 | 59/64 | 0.921875 | 23.41562 |
| 7/16 | 0.4375 | 11.11250 | 15/16 | 0.9375 | 23.81250 |
| 29/64 | 0.453125 | 11.50937 | 61/64 | 0.953125 | 24.20937 |
| 15/32 | 0.46875 | 11.90625 | 31/32 | 0.96875 | 24.60625 |
| 31/64 | 0.484375 | 12.30312 | 63/64 | 0.984375 | 25.00312 |
| 1/2 | 0.500 | 12.70000 | 1 | 1.00 | 25.40000 |

# CLYMER®

# HARLEY-DAVIDSON
## SHOVELHEADS • 1966-1984

# CHAPTER ONE

# GENERAL INFORMATION

Troubleshooting, tune-up, maintenance and repair are not difficult, if you know what tools and equipment to use and what to do. The manual is written simply and clearly enough for owners who have never worked on a motorcycle, but is complete enough for use by experienced mechanics.

It is important to note that because of the parts interchangeability between different models of Harley-Davidson motorcycles, your model, if purchased second hand, may be equipped with parts different from its original equipment.

Some of the procedures require the use of special tools. Using an inferior substitute tool for a special tool is not recommended, as it can be dangerous to you and may damage the part. Where possible, this manual shows suitable special tools that can be fabricated in your garage or by a machinist or purchased at motorcycle or tool stores.

General motorcycle specifications are listed in **Table 1**. General torque specifications are in **Table 3**. **Tables 1-5** are listed at the end of the chapter.

## MANUAL ORGANIZATION

This chapter provides general information and discusses equipment and tools useful both for preventive maintenance and troubleshooting.

Chapter Two provides methods and suggestions for quick and accurate diagnosis and repair of problems. Troubleshooting procedures discuss typical symptoms and logical methods to pinpoint the trouble.

Chapter Three explains all periodic lubrication and routine maintenance necessary to keep your Harley-Davidson operating well. Chapter Three also includes recommended tune-up procedures, eliminating the constant need to consult other chapters on the various assemblies.

Subsequent chapters describe specific systems such as the engine, clutch, primary drive, transmission, fuel, exhaust, suspension, steering and brakes. Each chapter provides disassembly, repair, and assembly procedures in simple step-by-step form. If a repair is impractical for a home mechanic, it is so indicated. It is usually faster and less expensive to take such repairs to a dealer or competent repair shop. Specifications concerning a particular system are included at the end of the appropriate chapter.

## NOTE, CAUTIONS AND WARNINGS

The terms NOTE, CAUTION and WARNING have specific meanings in this manual. A NOTE provides additional information to make a step or procedure easier or clearer. Disregarding a NOTE could cause inconvenience, but would not cause damage or personal injury.

A CAUTION emphasizes areas where equipment damage could occur. Disregarding a CAUTION could cause permanent mechanical damage; however, personal injury is unlikely.

A WARNING emphasizes areas where personal injury or even death could result from negligence. Mechanical damage may also occur. WARNINGS *are to be taken seriously*. In some cases, serious injury and death have resulted from disregarding similar warnings.

## SAFETY FIRST

Professional mechanics can work for years and never sustain a serious injury. If you observe a few rules of common sense and safety, you can enjoy many safe hours servicing your own machine. If you ignore these rules you can hurt yourself or damage the equipment.

1. Never use gasoline as a cleaning solvent.

2. Never smoke or use a torch in the vicinity of flammable liquids, such as cleaning solvent, in open containers.

3. If welding or brazing is required on the machine, remove the fuel tank and rear shocks to a safe distance, at least 50 feet away. Welding on a gas tank requires special safety precautions and must be performed by someone skilled in the process. Do not attempt to weld or braze a leaking gas tank.

4. Use the proper sized wrenches to avoid damage to fasteners and injury to yourself.

5. When loosening a tight or stuck nut, be guided by what would happen if the wrench should slip. Be careful, protect yourself accordingly.

6. When replacing a fastener, make sure to use one with the same measurements and strength as the old one. Incorrect or mismatched fasteners can result in damage to the vehicle and possible personal injury. Beware of fastener kits that are filled with cheap and poorly made nuts, bolts, washers and cotter pins. Refer to Fasteners in this chapter for additional information.

7. Keep all hand and power tools in good condition. Wipe greasy and oily tools after using them. They are difficult to hold and can cause injury. Replace or repair worn or damaged tools.

8. Keep your work area clean and uncluttered.

9. Wear safety goggles during all operations involving drilling, grinding, the use of a cold chisel or anytime you feel unsure about the safety of your eyes. Safety goggles should also be worn anytime compressed air is used to clean a part.

10. Keep an approved fire extinguisher nearby. Be sure it is rated for gasoline (Class B) and electrical (Class C) fires.

11. When drying bearings or other rotating parts with compressed air, never allow the air jet to rotate the bearing or part; the air jet is capable of rotating them at speeds far in excess of those for which they were designed. The bearing or rotating part is very likely to disintegrate and cause serious injury and damage.

## SERVICE HINTS

Most of the service procedures covered are straightforward and can be performed by anyone reasonably handy with tools. It is suggested, however, that you consider your own capabilities carefully before attempting any operation involving major disassembly of the engine or transmission.

1. "Front," as used in this manual, refers to the front of the motorcycle; the front of any component is the end closest to the front of the motorcycle. The "left-" and "right-hand" sides refer to the position of the parts as viewed by a rider sitting on the seat facing forward. For example, the throttle control is on the right-hand side. These rules are simple, but confusion can cause a major inconvenience during service.

2. Whenever servicing the engine or transmission, or when removing a suspension component, the bike should be secured in a safe manner. If the bike is to be parked on its sidestand, check the stand to make sure it is secure and not damaged. Block the front and rear wheels if they remain on the ground. A small hydraulic jack and a block of wood can be used to raise the chassis. If the transmission is not going to be worked on and the drive chain or drive belt is connected to the rear wheel, shift the transmission into first gear.

3. Disconnect the negative battery cable when working on or near the electrical, clutch or starter systems and before disconnecting any wires. On most batteries, the negative terminal will be marked with a minus (–) sign and the positive terminal with a plus (+) sign.

4. When disassembling a part or component, it is a good practice to tag the parts for location and mark all parts which mate together. Small parts, such as

bolts, can be identified by placing them in plastic sandwich bags. Seal the bags and label them with masking tape and a marking pen. When reassembly will take place immediately, an accepted practice is to place nuts and bolts in a cupcake tin or egg carton in the order of disassembly.

5. Finished surfaces should be protected from physical damage or corrosion. Keep gasoline and brake fluid off painted surfaces.

6. Use penetrating oil on frozen or tight bolts, then strike the bolt head a few times with a hammer and punch (use a screwdriver on screws). Avoid the use of heat where possible, as it can warp, melt or affect the temper of parts. Heat also ruins finishes, especially paint and plastics.

7. Keep flames and sparks away from a charging battery or flammable fluids and do not smoke near them. It is a good idea to have a fire extinguisher handy in the work area. Remember that many gas appliances in home garages (water heater, clothes dryer, etc.) have pilot lights.

8. No parts removed or installed (other than bushings and bearings) in the procedures given in this manual should require unusual force during disassembly or assembly. If a part is difficult to remove or install, find out why before proceeding.

9. Cover all openings after removing parts or components to prevent dirt, small tools, etc. from falling in.

10. Read each procedure completely while looking at the actual parts before starting a job. Make sure you *thoroughly* understand what is to be done and then carefully follow the procedure, step by step.

11. Recommendations are occasionally made to refer service or maintenance to a Harley-Davidson dealer or a specialist in a particular field. In these cases, the work will be done more quickly and economically than if you performed the job yourself.

12. In procedural steps, the term "replace" means to discard a defective part and replace it with a new or exchange unit. "Overhaul" means to remove, disassemble, inspect, measure, repair or replace defective parts, reassemble and install major systems or parts.

13. Some operations require the use of a hydraulic press. It would be wiser to have these operations performed by a shop equipped for such work, rather than try to do the job yourself with makeshift equipment that may damage your machine.

14. Repairs go much faster and easier if your machine is clean before you begin to work. There are many special cleaners on the market, like Bel-Ray Degreaser, for washing the engine and related parts. Follow the manufacturer's directions on the container for the best results. Clean all oily or greasy parts with cleaning solvent as you remove them.

*WARNING*
*Never use gasoline as a cleaning agent. It presents an extreme fire hazard. Be sure to work in a well-ventilated area when using cleaning solvent. Keep a fire extinguisher, rated for gasoline fires, handy in any case.*

15. Much of the labor charges for repairs made by dealers are for the labor time involved in removal, disassembly, assembly, and reinstallation of other parts in order to reach the defective part. It is frequently possible to perform the preliminary operations yourself and then take the defective unit to the dealer for repair at considerable savings.

16. If special tools are required, make arrangements to get them before you start. It is frustrating and time-consuming to get partly into a job and then be unable to complete it.

17. Make diagrams (or take a Polaroid picture) wherever similar-appearing parts are found. For instance, crankcase bolts are often not the same length. You may think you can remember where everything came from—but mistakes are costly. There is also the possibility that you may be sidetracked and not return to work for days or even weeks—in which time carefully laid out parts may have become disturbed.

18. When assembling parts, be sure all shims and washers are replaced exactly as they came out.

19. Whenever a rotating part butts against a stationary part, look for a shim or washer. Use new gaskets if there is any doubt about the condition of the old ones. A thin coat of oil on non-pressure type gaskets may help them seal more effectively.

20. If it is necessary to make a gasket, and you do not have a suitable old gasket to use as a guide, apply engine oil to the gasket surface of the part. Then place the part on the new gasket material and press the part slightly. The oil will leave a very accurate outline on the gasket material that can be cut around.

21. Heavy grease can be used to hold small parts in place if they tend to fall out during assembly. However, keep grease and oil away from electrical and brake components.

22. A carburetor is best cleaned by disassembling it and soaking the parts in a commercial carburetor cleaner. Never soak gaskets and rubber parts in these cleaners. Never use wire to clean out jets and air passages unless otherwise instructed to do so in Chapter Six. They are easily damaged. Use compressed air to blow out the carburetor only if the float has been removed first.

23. Take your time and do the job right. Do not forget that a newly rebuilt engine must be broken in just like a new one.

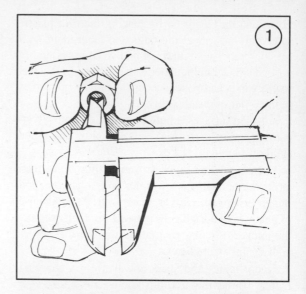

## TORQUE SPECIFICATIONS

Torque specifications throughout this manual are given in foot-pounds (ft.-lb.) and Newton-meters (N·m).

**Table 3** lists general torque specifications for nuts and bolts that are not listed in the respective chapters. To use the table, first determine the size of the nut or bolt. **Figure 1** and **Figure 2** show how this is done.

## FASTENERS

The materials and designs of the various fasteners used on your Harley-Davidson are not arrived at by chance or accident. Fastener design determines the type of tool required to work the fastener. Fastener material is carefully selected to decrease the possibility of physical failure.

### Threads

Nuts, bolts and screws are manufactured in a wide range of thread patterns. To join a nut and bolt, the diameter of the bolt and diameter of the hole in the nut must be the same. It is just as important that the threads on both be properly matched.

The best way to tell if the threads on 2 fasteners are matched is to turn the nut on the bolt (or the bolt into the threaded hole in a piece of equipment) with fingers only. Be sure both pieces are clean. If much force is required, check the thread condition on each fastener. If the thread condition is good but the fasteners jam, the threads are not compatible. A thread pitch gauge can also be used to determine threads per inch.

Four important specifications describe every thread:

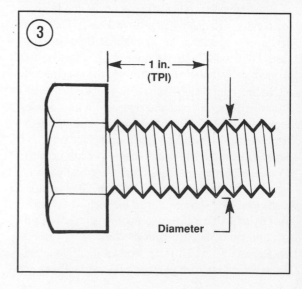

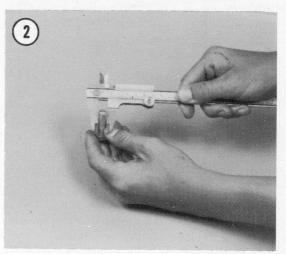

a. Diameter.

b. Threads per inch.

c. Thread pattern.

d. Thread direction.

**Figure 3** shows the first 2 specifications. Thread pattern is more subtle. Italian and British standards exist, but the most commonly used by motorcycle manufacturers are metric standard and American standard. Harley-Davidsons are manufactured with American standard fasteners. The threads are cut differently as shown in **Figure 4**.

Most threads are cut so that the fastener must be turned clockwise to tighten it. These are called right-hand threads. Some fasteners have left-hand threads; they must be turned counterclockwise to be tightened. Left-hand threads are used in locations where normal rotation of the equipment would tend to loosen a right-hand threaded fastener.

### Machine Screws

There are many different types of machine screws. **Figure 5** shows a number of screw heads requiring different types of turning tools. Heads are also designed to protrude above the metal (round) or to be slightly recessed in the metal (flat). See **Figure 6**.

### Bolts

Commonly called bolts, the technical name for these fasteners is cap screw. They are normally described by diameter, threads per inch and length. For example, 1/4-20 × 1 indicates a bolt 1/4 in. in diameter with 20 threads per inch, 1 inch long. The measurement across 2 flats on the head of the bolt indicates the proper wrench size to be used. **Figure 3** shows how to determine bolt diameter.

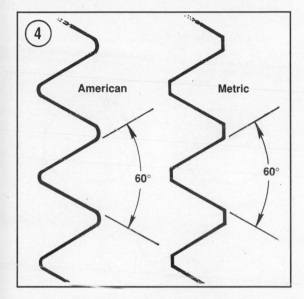

American      Metric

60°      60°

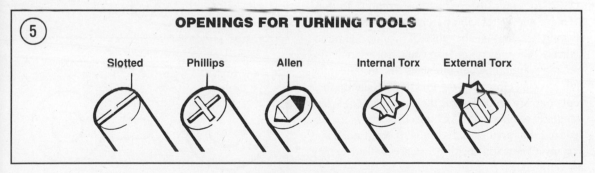

**OPENINGS FOR TURNING TOOLS**

Slotted     Phillips     Allen     Internal Torx     External Torx

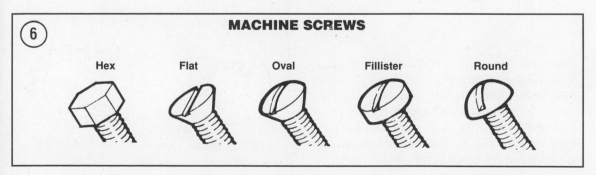

**MACHINE SCREWS**

Hex     Flat     Oval     Fillister     Round

## Nuts

Nuts are manufactured in a variety of types and sizes. Most are hexagonal (6-sided) and fit on bolts, screws and studs with the same diameter and threads per inch.

**Figure 7** shows several types of nuts. The common nut is generally used with a lockwasher. Self-locking nuts have a nylon insert which prevents the nut from loosening; no lockwasher is required. Wing nuts are designed for fast removal by hand. Wing nuts are used for convenience in non-critical locations.

To indicate the size of a nut, manufacturers specify the diameter of the opening and the threads per inch. This is similar to bolt specifications, but without the length dimension. The measurement of the inside bore (**Figure 1**) indicates the diameter of the nut opening.

## Self-Locking Fasteners

Several types of bolts, screws and nuts incorporate a system that develops an interference between the bolt, screw, nut or tapped hole threads. Interference is achieved in various ways: by distorting threads, coating threads with dry adhesive or nylon, distorting the top of an all-metal nut, using a nylon insert in the center or at the top of a nut, etc.

Self-locking fasteners offer greater holding strength and better vibration resistance. Some self-locking fasteners can be reused if in good condition. Others, like the nylon insert nut, form an initial locking condition when the nut is first installed; the nylon forms closely to the bolt thread pattern, thus reducing any tendency for the nut to loosen. When the nut is removed, the locking efficiency is greatly reduced. For greatest safety, it is recommended that you install new self-locking fasteners whenever they are removed.

## Washers

There are 2 basic types of washers: flat washers and lockwashers. Flat washers are simple discs with a hole to fit a screw or bolt. Lockwashers are designed to prevent a fastener from working loose due to vibration, expansion and contraction. **Figure 8** shows several types of washers. Washers are also used in the following functions:

a. As spacers.
b. To prevent galling or damage of the equipment by the fastener.
c. To help distribute fastener load during torquing.
d. As seals.

Note that flat washers are often used between a lockwasher and a fastener to provide a smooth bearing surface. This allows the fastener to be turned easily with a tool.

## Cotter Pins

Cotter pins (**Figure 9**) are used to secure special kinds of fasteners. The threaded stud must have a

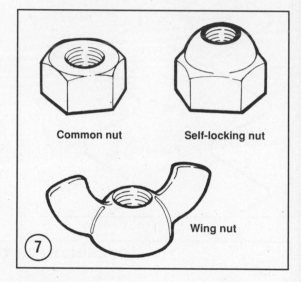

**Common nut**　　　　　　**Self-locking nut**

**Wing nut**

⑦

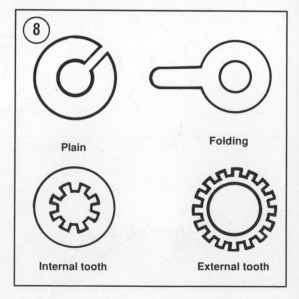

⑧

**Plain**　　　　　　　　　　**Folding**

**Internal tooth**　　　　　　**External tooth**

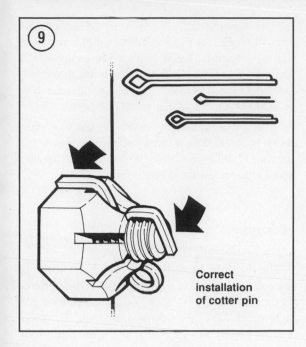

Correct
installation
of cotter pin

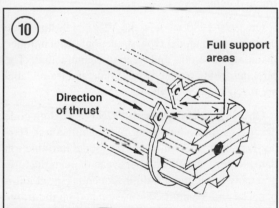

Full support areas

Direction of thrust

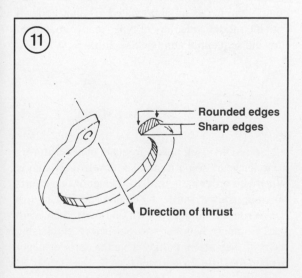

Rounded edges
Sharp edges

Direction of thrust

hole in it; the nut or nut lockpiece has castellations around which the cotter pin ends wrap. Cotter pins should not be reused after removal.

### Snap Rings

Snap rings can be internal or external design. They are used to retain items on shafts (external type) or within tubes (internal type). In some applications, snap rings of varying thicknesses are used to control the end play of parts assemblies. These are often called selective snap rings. Snap rings should be replaced during installation as removal weakens and deforms them.

Two basic styles of snap rings are available: machined and stamped snap rings. Machined snap rings (**Figure 10**) can be installed in either direction (shaft or housing) because both faces are machined, thus creating two sharp edges. Stamped snap rings (**Figure 11**) are manufactured with one sharp edge and one rounded edge. When installing stamped snap rings in a thrust situation (transmission shafts, fork tubes, etc.), the sharp edge must face away from the part producing the thrust. When installing snap rings, observe the following:

    a. Compress or expand snap rings only enough to install them.

    b. After the snap ring is installed, make sure it is completely seated in its groove.

### LUBRICANTS

Periodic lubrication assures long life for any type of equipment. The *type* of lubricant used is just as important as the lubrication service itself, although in an emergency the wrong type of lubricant is better than none at all. The following paragraphs describe the types of lubricants most often used on motorcycle equipment. Be sure to follow the manufacturer's recommendations for lubricant types.

Generally, all liquid lubricants are called "oil." They may be mineral-based (including petroleum bases), natural-based (vegetable and animal bases), synthetic-based or emulsions (mixtures). "Grease" is an oil to which a thickening base has been added so that the end product is semi-solid. Grease is often classified by the type of thickener added; lithium soap is commonly used.

## Engine Oil

Oil for motorcycle and automotive engines is classified by the American Petroleum Institute (API) and the Society of Automotive Engineers (SAE) in several categories. Oil containers display these classifications on the top or label.

API oil grade is indicated by letters; oils for gasoline engines are identified by an "S." The engines covered in this manual require SE or SF graded oil.

Viscosity is an indication of the oil's thickness. The SAE uses numbers to indicate viscosity; thin oils have low numbers while thick oils have high numbers. A "W" after the number indicates that the viscosity testing was done at low temperature to simulate cold-weather operation. Engine oils fall into the 5W-30 and 20W-50 range.

Multi-grade oils (for example 10W-40) are less viscous (thinner) at low temperatures and more viscous (thicker) at high temperatures. This allows the oil to perform efficiently across a wide range of engine operating conditions. The lower the number, the better the engine will start in cold climates. Higher numbers are usually recommended for engine running in hot weather conditions.

## Grease

Greases are graded by the National Lubricating Grease Institute (NLGI). Greases are graded by number according to the consistency of the grease; these range from No. 000 to No. 6, with No. 6 being the most solid. A typical multipurpose grease is NLGI No. 2. For specific applications, equipment manufacturers may require grease with an additive such as molybdenum disulfide (MOS2).

## SERIAL NUMBERS

Three types of serial numbers have been used on 1966-1984 Shovelhead engines.

### 1966-1969

The engine serial number is stamped on a pad on the left-hand crankcase.

From 1966-1969 the engine serial number is a code describing year of manufacture, model type and sequential serial number. The first 2 numbers describe the year of manufacture. The letters following the manufacture year describe the model type (FL, FLH, FLT or FLHP). The last 4 or 5 numbers are the sequential serial number. For example, the serial number 69FLH00000 indicates that the engine is a 1969 model, FLH indicates a Shovelhead engine and 00000 is its sequential serial number.

Most all of the Harley-Davidsons manufactured to 1969 do not have a factory stamped frame serial number. If you have a 1966-1969 model with a frame serial number, it should be suspect.

### 1970-1984

The vehicle identification number (VIN) is stamped on a pad on the right-hand engine crankcase. The frame VIN number is stamped on the right-hand side of the steering head. All 1978-1984 models have a VIN label attached to the right front frame downtube.

On 1970-1980 models, the VIN is a 9-digit code describing the model type, sequential serial number, manufacturer code and year of manufacture. The VIN numbers on the engine, frame and frame label are the same. See **Table 4**.

On 1981-1984 models, the VIN is a 17-digit code describing the country of origin, manufacturer, Harley-Davidson code, model type, model introduction date, factory check number, year of manufacture, assembly plant and a 6-digit sequential serial number. The full 17-digit code is stamped on the frame steering head and printed on the VIN label. The engine number is an abbreviated 10-digit code. It lists the model code, model type, year of manufacture and sequential number. See **Table 5**.

## PARTS REPLACEMENT

Harley-Davidson makes frequent changes during a model year; some minor, some relatively major. When you order parts from the dealer or other parts distributor, always order by engine and frame number. Write the numbers down and carry them with you. Compare new parts to old before purchasing them. If they are not alike, have the parts manager explain the difference to you.

## BASIC HAND TOOLS

Many of the procedures in this manual can be carried out with simple hand tools and test equipment familiar to the average home mechanic. Keep your tools clean and in a tool box. Keep them organized with the sockets and related drives together, the open-end combination wrenches together, etc. After using a tool, wipe off dirt and grease with a clean cloth and return the tool to its correct place.

Top-quality tools are essential; they are also more economical in the long run. If you are now starting to build your tool collection, stay away from the "advertised specials" featured at some parts houses, discount stores and chain drug stores. These are usually a poor grade tool that can be sold cheaply and that is exactly what they are—*cheap*. They are usually made of inferior material, and are thick, heavy and clumsy. Their rough finish makes them difficult to clean and they usually don't last very long. If it is ever your misfortune to use such tools you will probably find out that the wrenches do not fit the heads of bolts and nuts correctly and damage fasteners.

Quality tools are made of alloy steel and are heat treated for greater strength. They are lighter and better balanced than cheap ones. Their surface is smooth, making them a pleasure to work with and easy to clean. The initial cost of good-quality tools may be more, but they are cheaper in the long run. Don't try to buy everything in all sizes in the beginning; do it a little at a time until you have the necessary tools. To sum up tool buying, "...the bitterness of poor quality lingers long after the sweetness of low price has faded."

The following tools are required to perform virtually any repair job. Each tool is described and the recommended size given for starting a tool collection. Additional tools and some duplicates may be added as you become familiar with the vehicle. Harley-Davidson motorcycles are built with American standard fasteners—so if you are starting your collection now, buy American sizes.

### Screwdrivers

The screwdriver is a very basic tool, but if used improperly it will do more damage than good. The slot on a screw has a definite dimension and shape. A screwdriver must be selected to conform with that shape. Use a small screwdriver for small screws and a large one for large screws or the screw head will be damaged.

Two basic types of screwdrivers are required: common (flat-blade) screwdrivers (**Figure 12**) and Phillips screwdrivers (**Figure 13**).

Screwdrivers are available in sets which often include an assortment of common and Phillips blades. If you buy them individually, buy at least the following:

    a. Common screwdriver—5/16 × 6 in. blade.

    b. Common screwdriver—3/8 × 12 in. blade.

    c. Phillips screwdriver—size 2 tip, 6 in. blade.

Use screwdrivers only for driving screws. Never use a screwdriver for prying or chiseling metal. Do not try to remove a Phillips or Allen head screw with a common screwdriver (unless the screw has a combination head that will accept either type); you can damage the head so that the proper tool will be unable to remove it.

Keep screwdrivers in the proper condition and they will last longer and perform better. Always keep the tip of a common screwdriver in good condition. **Figure 14** shows how to grind the tip to the proper shape if it becomes damaged. Note the symmetrical sides of the tip.

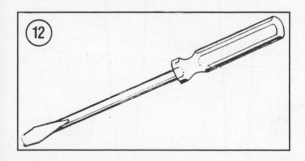

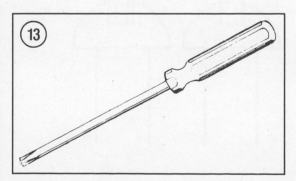

## Pliers

Pliers come in a wide range of types and sizes. Pliers are useful for cutting, bending and crimping. They should never be used to cut hardened objects or turn bolts or nuts. **Figure 15** shows several pliers useful in motorcycle repairs.

Each type of pliers has a specialized function. Gas pliers are general purpose pliers and are used mainly for holding things and for bending. Vise-grips are used as pliers or to hold objects very tightly like a vise. Needlenose pliers are used to hold or bend small objects. Channel-lock pliers can be adjusted to hold various sizes of objects; the jaws remain parallel to grip around objects such as pipe or tubing. There are many more types of pliers.

## Box-end and Open-end Wrenches

Box-end and open-end wrenches are available in sets or separately in a variety of sizes. The size number stamped near the end refers to the distance between 2 parallel flats on the hex head bolt or nut.

Box-end wrenches (**Figure 16**) are usually superior to open-end wrenches (**Figure 17**). Open-end wrenches grip the nut on only 2 flats. Unless a wrench fits well, it may slip and round off the points on the nut. The box-end wrench grips on all 6 flats. Both 6-point and 12-point openings on box-end

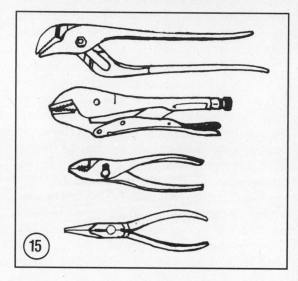

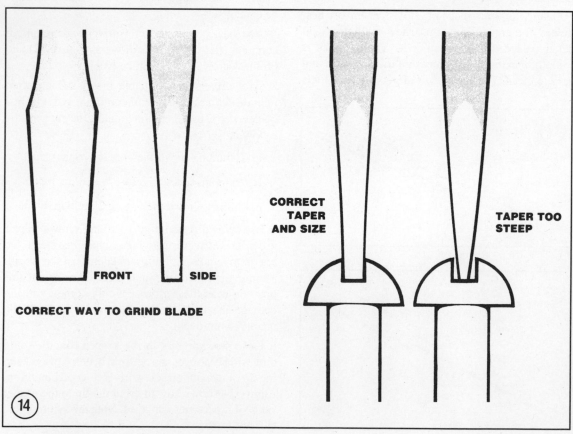

CORRECT
TAPER
AND SIZE

TAPER TOO
STEEP

FRONT    SIDE

**CORRECT WAY TO GRIND BLADE**

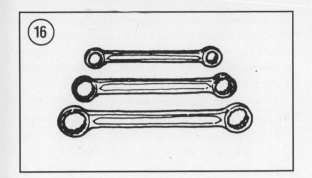

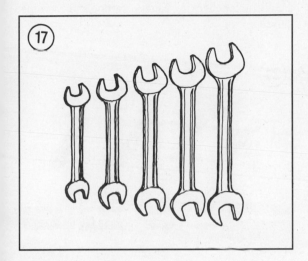

wrenches are available. The 6-point gives superior holding power; the 12-point allows a shorter swing.

Combination wrenches which are open on one side and boxed on the other are also available. Both ends are the same size.

### Adjustable Wrenches

An adjustable wrench can be adjusted to fit a variety of nuts or bolt heads (**Figure 18**). However, it can loosen and slip, causing damage to the nut and injury to your knuckles. Use an adjustable wrench only when other wrenches are not available.

Adjustable wrenches come in sizes ranging from 4-18 in. overall. A 6 or 8 in. wrench is recommended as an all-purpose wrench.

### Socket Wrenches

This type is undoubtedly the fastest, safest and most convenient to use. Sockets which attach to a ratchet handle (**Figure 19**) are available with 6-point or 12-point openings and 1/4, 3/8, 1/2 and 3/4 inch drives. The drive size indicates the size of the square hole which mates with the ratchet handle.

### Torque Wrench

A torque wrench (**Figure 20**) is used with a socket to measure how tightly a nut or bolt is installed. They come in a wide price range and with either 3/8 or 1/2 square drive. The drive size indicates the size of the square drive which mates with the socket. Purchase one that measures 0-150 ft.-lb. (0-207 N•m).

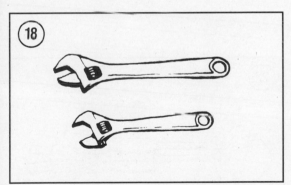

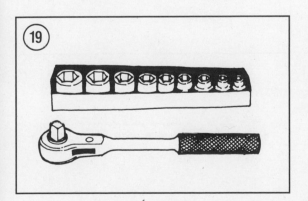

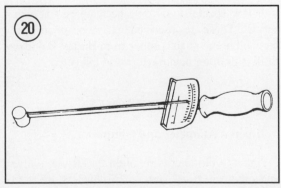

## Impact Driver

This tool makes removal of tight fasteners easy and eliminates damage to bolts and screw slots. Impact drivers and interchangeable bits (**Figure 21**) are available at most large hardware and motorcycle dealers. Sockets can also be used with a hand impact driver. However, make sure the socket is designed for impact use. Do not use regular hand type sockets as they may shatter.

## Hammers

The correct hammer is necessary for repairs. Use only a hammer with a face (or head) of rubber or plastic or the soft-faced type that is filled with buckshot. These are sometimes necessary in engine teardowns. *Never* use a metal-faced hammer as severe damage will result in most cases. You can always produce the same amount of force with a soft-faced hammer.

## Feeler Gauge

This tool has both flat and wire measuring gauges and is used to measure spark plug gap. See **Figure 22**. Wire gauges are used to measure spark plug gap; flat gauges are used for all other measurements.

## Vernier Caliper

This tool is invaluable when reading inside, outside and depth measurements to close precision. The vernier caliper can be purchased from large dealers or mail order houses. See **Figure 23**.

## Special Tools

A few special tools may be required for major service. These are described in the appropriate chapters and are available either from Harley-Davidson dealers or other manufacturers as indicated.

## TEST EQUIPMENT

### Voltmeter, Ammeter and Ohmmeter

A good voltmeter is required for testing ignition and other electrical systems. Voltmeters are avail-able with analog meter scales or digital readouts. An instrument covering 0-20 volts is satisfactory. It should also have a 0-2 volt scale for testing points or individual contacts where voltage drops are much smaller. Accuracy should be ±1/2 volt.

An ohmmeter measures electrical resistance. This instrument is useful in checking continuity (for open and short circuits) and testing lights. A self-powered 12-volt test light can often be used in its place.

The ammeter measures electrical current. These are useful for checking battery starting and charging currents.

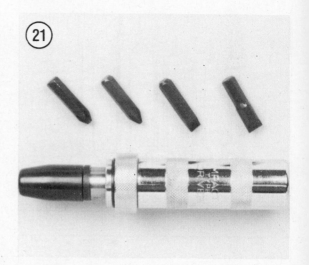

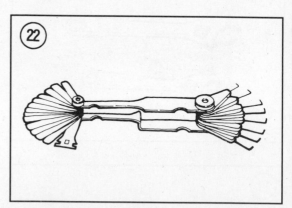

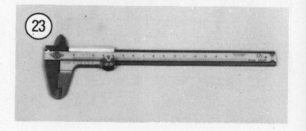

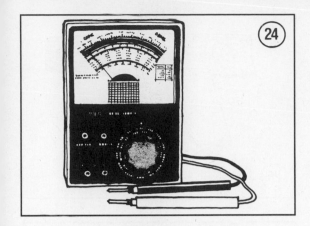

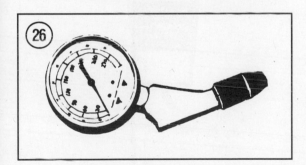

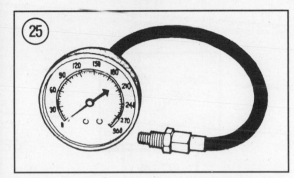

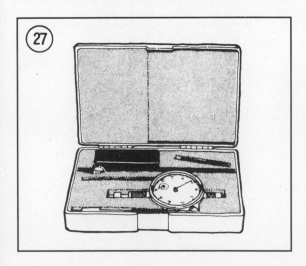

Some manufacturers combine the 3 instruments into one unit called a multimeter or VOM. See **Figure 24**.

### Compression Gauge

An engine with low compression cannot be properly tuned and will not develop full power. A compression gauge measures the amount of pressure present in the engine's combustion chambers during the compression stroke. This indicates general engine condition.

The easiest type to use has screw-in adaptors that fit into the spark plug holes (**Figure 25**). Press-in rubber-tipped types (**Figure 26**) are also available.

### Dial Indicator

Dial indicators (**Figure 27**) are precision tools used to check dimension variations on machined parts such as transmission shafts and axles and to check crankshaft and axle shaft end play. Dial indicators are available with various dial types for different measuring requirements.

### Strobe Timing Light

This instrument is necessary for checking ignition timing. By flashing a light at the precise instant the spark plug fires, the position of the timing mark can be seen. The flashing light makes a moving mark appear to stand still opposite a stationary mark.

Suitable lights range from inexpensive neon bulb types to powerful xenon strobe lights. See **Figure 28**. A light with an inductive pickup is recommended to eliminate any possible damage to ignition wiring.

### Portable Tachometer

A portable tachometer is necessary for tuning. See **Figure 29**. Ignition timing and carburetor adjustments must be performed at the specified idle speed. The best instrument for this purpose is one with a low range of 0-1,000 or 0-2,000 rpm and a high range of 0-4,000 rpm. Extended range (0-6,000 or 0-8,000 rpm) instruments lack accuracy at lower speeds. The instrument should be capable of detecting changes of 25 rpm on the low range.

### Expendable Supplies

Certain expendable supplies are also required. These include grease, oil, gasket cement, shop rags and cleaning solvent. Ask your dealer for the special locking compounds, silicone lubricants and lube products which make vehicle maintenance simpler and easier. Cleaning solvent is available at some service stations.

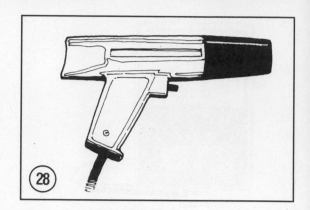

## MECHANIC'S TIPS

### Removing Frozen Nuts and Screws

When a fastener rusts and cannot be removed, several methods may be used to loosen it. First, apply penetrating oil such as Liquid Wrench or WD-40 (available at hardware or auto supply stores). Apply it liberally and let it penetrate for 10-15 minutes. Rap the fastener several times with a small hammer; do not hit it hard enough to cause damage. Reapply the penetrating oil if necessary.

For frozen screws, apply penetrating oil as described, then insert a screwdriver in the slot and rap the top of the screwdriver with a hammer. This loosens the rust so the screw can be removed in the normal way. If the screw head is too chewed up to use this method, grip the head with Vise-grips pliers and twist the screw out.

Avoid applying heat unless specifically instructed as it may melt, warp or remove the temper from parts.

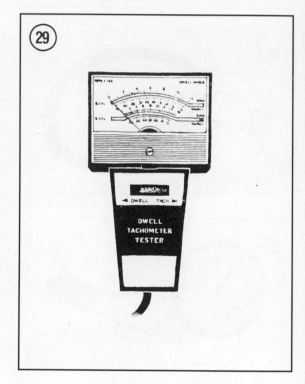

### Remedying Stripped Threads

Occasionally, threads are stripped through carelessness or impact damage. Often the threads can be cleaned up by running a tap (for internal threads on nuts) or die (for external threads on bolts) through the threads. See **Figure 30**. To clean or repair spark plug threads, a spark plug tap can be used (**Figure 31**).

### Removing Broken Screws or Bolts

When the head breaks off a screw or bolt, several methods are available for removing the remaining portion.

If a large portion of the remainder projects out, try gripping it with Vise-grips. If the projecting portion

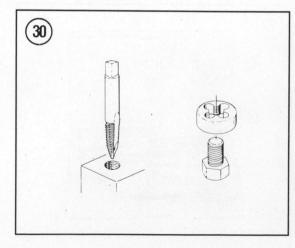

is too small, file it to fit a wrench or cut a slot in it to fit a screwdriver. See **Figure 32**.

If the head breaks off flush, use a screw extractor. To do this, centerpunch the exact center of the re- maining portion of the screw or bolt. Drill a small hole in the screw and tap the extractor into the hole. Back the screw out with a wrench on the extractor. See **Figure 33**.

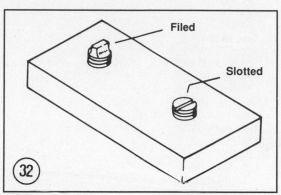

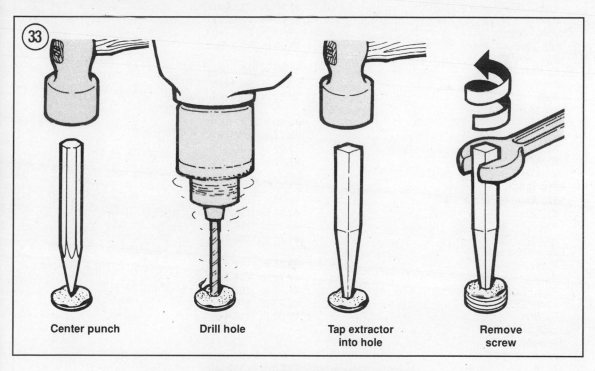

| Center punch | Drill hole | Tap extractor into hole | Remove screw |

**Tables 1-5 are on the following pages.**

## Table 1  GENERAL SPECIFICATIONS

| | |
|---|---|
| **Wheelbase** | |
| 1966-1969 | 60.00 in. (152.4 cm) |
| 1970-early 1978 | |
| FL/FLH | 61.00 in. (154.9 cm) |
| FX | 63.00 in. (160 cm) |
| FXS | 63.50 in. (161.3 cm) |
| Late 1978-1984 | |
| FLH/FLHS | 61.12 in. (155.2 cm) |
| FXB/FXSB/FXS | 63.50 in. (161.3 cm) |
| FXE/FXEF | 63.00 in. (160 cm) |
| FXWG | 65.00 in. (165.1 cm) |
| **Overall length** | |
| 1966-1969 | 92.00 in. (233.7 cm) |
| 1970-early 1978 | |
| FL/FLH | 89.00 in. (226.1 cm) |
| FX/FXS | 92.00 in. (233.7 cm) |
| Late 1978-1984 | |
| FLH/FLHS | 92.88 in. (235.9 cm) |
| FXB/FXSB/FXS | 92.00 in. (233.7 cm) |
| FXE/FXEF | 91.50 in. (232.4 cm) |
| FXWG | 93.00 in. (236.2 cm) |
| **Overall width** | |
| 1966-1969 | 35.00 in. (88.9 cm) |
| 1970-early 1978 | |
| FL/FLH | 39.00 in. (99.1 cm) |
| FX | 34.00 in. (86.4 cm) |
| FXS | 29.00 in. (73.7 cm) |
| Late 1978-1984 | |
| FLH | 42.50 in. (107.9 cm) |
| FLHS/FXE/FXEF | 33.75 in. (85.7 cm) |
| FXB/FXSB/FXS | 29.00 in. (73.7 cm) |
| FXWG | 27.50 in. (69.8 cm) |
| **Overall height** | |
| 1966-1969 | * |
| 1970-early 1978 | |
| FL/FLH | 43.50 in. (110.5 cm) |
| FX | 45.75 in. (116.2 cm) |
| FXS | 41.75 in. (106 cm) |
| Late 1978-1984 | |
| FLH | 63.25 in. (160.6 cm) |
| FLHS | 48.12 in. (122.2 cm) |
| FXB/FXSB/FXS | 41.75 in. (106 cm) |
| FXE/FXEF | 45.75 in. (116.2 cm) |
| FXWG | 47.00 in. (119.4 cm) |
| **Vehicle weight** | |
| 1966-early 1978 | * |
| Late 1978-1984 | |
| FLH | 722 lbs. (327.5 kg) |
| FLHS | 626 lbs. (283.9 kg) |
| FX | 572 lbs. (259.4 kg) |
| **Horsepower** | |
| 1966-1969 | |
| FL | 57.0 hp @ 5,200 rpm |
| FLH | 66.0 hp @ 5,600 rpm |
| 1970-early 1978 | |
| FL | 57.0 hp @ 5,200 rpm |
| FLH | 62.0 hp @ 5,400 rpm |
| Late 1978-1984 | |
| 1200 cc | 60.0 hp @ 5,200 rpm |

(continued)

## Table 1 GENERAL SPECIFICATIONS (continued)

| | |
|---|---|
| Late 1978-1984 (continued) | |
| 1340 cc (1978-1980) | 60.0 hp @ 4,800 rpm |
| 1340 cc (1981-1984) | 65.0 hp @ 5,400 rpm |
| Fuel tank capacity | |
| 1966-1969 | |
| Total | 3.5 or 5.0 gal. (13.2 or 18.9 L, 2.91 or 4.16 imp. gal.) |
| Reserve | 1 or 1.2 gal. (3.78 or 4.54 L, 0.83 or 1.0 imp. gal.) |
| 1970-early 1978 | |
| FL/FLH | |
| Total | 3.5 or 5.0 gal. (13.2 or 18.9 L, 2.91 or 4.16 imp. gal.) |
| Reserve | 1 or 1.2 gal. (3.78 or 4.54 L, 0.83 or 1.0 imp. gal.) |
| FX/FXE | |
| 1973-1974 | |
| Total | 3.6 gal. (13.6 L, 2.99 imp. gal.) |
| Reserve | 0.7 gal. (2.65 L, 0.58 imp. gal.) |
| 1975-early 1978 | |
| FX/FXE | |
| Total | 3.6 gal. (13.6 L, 2.99 imp. gal.) |
| Reserve | 0.6 gal. (2.27 L, 0.49 imp. gal.) |
| FXS | |
| Total | 3.5 gal. (13.2 L, 2.91 imp. gal.) |
| Reserve | 1.0 gal. (3.78 L, 0.83 imp. gal.) |
| Late 1978-1984 | |
| FLH/FLHS/FXWG | |
| Total | 5.0 gal. (18.9 L, 4.16 imp. gal.) |
| Reserve | 1.2 gal. (4.54 L, 1.0 imp. gal.) |
| FXE | |
| Total | 3.2 gal. (12.1 L, 2.66 imp. gal.) |
| Reserve | 0.6 gal. (2.27 L, 0.5 imp. gal.) |
| FXS/FXEF/FXB/FXSB | |
| Total | 3.5 gal. (13.2 L, 2.91 imp. gal.) |
| Reserve | 1.0 gal. (3.78 L, 0.83 imp. gal.) |

* Not specified.

## Table 2 DECIMAL AND METRIC EQUIVALENTS

| Fractions | Decimal in. | Metric mm | Fractions | Decimal in. | Metric mm |
|---|---|---|---|---|---|
| 1/64 | 0.015625 | 0.39688 | 9/32 | 0.28125 | 7.14375 |
| 1/32 | 0.03125 | 0.79375 | 19/64 | 0.296875 | 7.54062 |
| 3/64 | 0.046875 | 1.19062 | 5/16 | 0.3125 | 7.93750 |
| 1/16 | 0.0625 | 1.58750 | 21/64 | 0.328125 | 8.33437 |
| 5/64 | 0.078125 | 1.98437 | 11/32 | 0.34375 | 8.73125 |
| 3/32 | 0.09375 | 2.38125 | 23/64 | 0.359375 | 9.1289 |
| 7/64 | 0.109375 | 2.77812 | 3/8 | 0.375 | 9.52500 |
| 1/8 | 0.125 | 3.1750 | 25/64 | 0.390625 | 9.92187 |
| 9/64 | 0.140625 | 3.57187 | 13/32 | 0.40625 | 10.31875 |
| 5/32 | 0.15625 | 3.96875 | 27/64 | 0.421875 | 10.71562 |
| 11/64 | 0.171875 | 4.36562 | 7/16 | 0.4375 | 11.11250 |
| 3/16 | 0.1875 | 4.76250 | 29/64 | 0.453125 | 11.50937 |
| 13/64 | 0.2033125 | 5.15937 | 15/32 | 0.46875 | 11.90625 |
| 7/32 | 0.21875 | 5.556 | 31/64 | 0.484375 | 12.30312 |
| 15/64 | 0.234375 | 5.95312 | 1/2 | 0.500 | 12.70000 |
| 1/4 | 0.250 | 6.35000 | 33/64 | 0.515625 | 13.09687 |
| 17/64 | 0.265625 | 6.74687 | 17/32 | 0.53125 | 13.49375 |

(continued)

### Table 2 DECIMAL AND METRIC EQUIVALENTS (continued)

| Fractions | Decimal in. | Metric mm | Fractions | Decimal in. | Metric mm |
|---|---|---|---|---|---|
| 35/64 | 0.546875 | 13.89062 | 25/32 | 0.78125 | 19.84375 |
| 9/16 | 0.5625 | 14.28750 | 51/64 | 0.796875 | 20.24062 |
| 37/64 | 0.578125 | 14.68437 | 13/16 | 0.8125 | 20.63750 |
| 19/32 | 0.59375 | 15.08125 | 53/64 | 0.828125 | 21.03437 |
| 39/64 | 0.609375 | 15.47812 | 27/32 | 0.84375 | 21.43125 |
| 5/8 | 0.625 | 15.87500 | 55/64 | 0.859375 | 21.82812 |
| 41/64 | 0.640625 | 16.27187 | 7/8 | 0.875 | 22.22500 |
| 21/32 | 0.65625 | 16.66875 | 57/64 | 0.890625 | 22.62187 |
| 43/64 | 0.671875 | 17.06562 | 29/32 | 0.90625 | 23.01875 |
| 11/16 | 0.6875 | 17.46250 | 59/64 | 0.921875 | 23.41562 |
| 45/64 | 0.703125 | 17.85937 | 15/16 | 0.9375 | 23.81250 |
| 23/32 | 0.71875 | 18.25625 | 61/64 | 0.953125 | 24.20937 |
| 47/64 | 0.734375 | 18.65312 | 31/32 | 0.96875 | 24.60625 |
| 3/4 | 0.750 | 19.05000 | 63/64 | 0.984375 | 25.00312 |
| 49/64 | 0.765625 | 19.44687 | 1 | 1.00 | 25.40000 |

### Table 3 GENERAL TORQUE SPECIFICATIONS (FT.-LB.)*

| Type** | Body Size or Outside Diameter | | | | | | | | | |
|---|---|---|---|---|---|---|---|---|---|---|
| | 1/4 | 5/16 | 3/8 | 7/16 | 1/2 | 9/16 | 5/8 | 3/4 | 7/8 | 1 |
| SAE 2 | 6 | 12 | 20 | 32 | 47 | 69 | 96 | 155 | 206 | 310 |
| SAE 5 | 10 | 19 | 33 | 54 | 78 | 114 | 154 | 257 | 382 | 587 |
| SAE 7 | 13 | 25 | 44 | 71 | 110 | 154 | 215 | 360 | 570 | 840 |
| SAE 8 | 14 | 29 | 47 | 78 | 119 | 169 | 230 | 380 | 600 | 700 |

* Convert ft.-lb. specification to N•m by multiplying by 1.38.

** Fastener strength of SAE bolts can be determined by the bolt head "grade markings." Unmarked bolt heads and cap screws are usually to be mild steel. More "grade markings" indicate higher fastener quality.

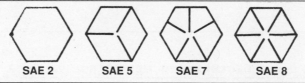

SAE 2          SAE 5          SAE 7          SAE 8

### Table 4 VEHICLE IDENTIFICATION NUMBER CHART--1970-1980

|  |  |  |  |  |  |
|---|---|---|---|---|---|
| | VIN number: | 3G | 62152 | H | 9 |

Model code
Sequential number
Manufacturer code
Model year

**MODEL CODE:**

| 1A FL-1200 | |
|---|---|
| 2A FLH-1200 | 6H FL-80 |
| 7E FLHS-1200 | 7H FLH-1200 Police |
| 3G FLH-80 | 8H FLH-1200 Shrine |
| 3G 1979 Classic | 9H FLH-80 Police |
| 3H 1980 Classic | 1K FLH-80 Shrine |
| 5H FLHS-80 | 2H CLE |

**SEQUENTIAL NUMBER:**

10000 and up: 1970 to Dec. 31, 1977
60000 and up: Jan. 1, 1978 through 1980
All CLE numbers start at 80000 and up

**MANUFACTURER CODE:**

H: 1970-1979
J: 1980

**MODEL YEAR:**

| | 1970* | | |
|---|---|---|---|
| 1 | 1971 | 6 | 1976 |
| 2 | 1972 | 7 | 1977 |
| 3 | 1973 | 8 | 1978 |
| 4 | 1974 | 9 | 1979 |
| 5 | 1975 | 0 | 1980 |

* Unavailable

**SAMPLE VIN NUMBER:** 1979 FLH-80 (3G 62152 H9)

## Table 5 VEHICLE IDENTIFICATION NUMBER CHART--1981-1984

VIN number:  1 HD 1 AA K 1 ** E Y 110000

- Country of origin
- Harley-Davidson
- HD code
- Model code
- Engine displacement
- Introduction date
- Factory check number
- Model year
- Assembly plant
- Sequential number

### HD CODE
1: Heavyweight motorcycle

### MODEL CODE
AA FLH-80
AB FLH-Police-Std.
AC FLH-Shrine-Chain
AD FLH Classic
AG FLH Classic with sidecar
AH FLH-Police-Dlx.
AJ FLH-80 Heritage
AK FLHS-80
AL FLH-Shrine-Belt
SA Sidecar not used with Classic
SD Sidecar sold with Classic
SE Sidecar sold separately for
   uoo with Classic

### ENGINE DISPLACEMENT
K: 80 cu. in.

### MODEL YEAR
B: 1981
C: 1982
D: 1983
E: 1984

### INTRODUCTION DATE
1: Regular introduction date
2: Mid-year introduction
3: Special-Early 1984 models

### FACTORY CHECK NUMBER
Can be 0-9 or X

### SAMPLE VIN NUMBER ON STEERING HEAD:
1 HD 1AAK1 1 EY110000

# CHAPTER TWO

# TROUBLESHOOTING

Every motorcycle engine requires an uninterrupted supply of fuel and air, proper ignition and adequate compression. If any of these are lacking, the engine will not run.

Diagnosing mechanical problems is relatively simple if you use orderly procedures and keep a few basic principles in mind.

The troubleshooting procedures in this chapter analyze typical symptoms and show logical methods of isolating causes. These are not the only methods. There may be several ways to solve a problem, but only a systematic approach can guarantee success.

Never assume anything. Do not overlook the obvious. If you are riding along and the bike suddenly quits, check the easiest, most accessible problem spots first. Is there gasoline in the tank? Has a spark plug wire fallen off?

If nothing obvious turns up in a quick check, look a little further. Learning to recognize and describe symptoms will make repairs easier for you or a mechanic at the shop. Describe problems accurately and fully. Saying that "it won't run" isn't the same thing as saying "it quit at high speed and won't start," or, "it sat in my garage for 3 months and then wouldn't start."

Gather as many symptoms as possible to aid in diagnosis. Note whether the engine lost power gradually or all at once. Remember that the more complicated a machine is, the easier it is to troubleshoot because symptoms point to specific problems.

After the symptoms are defined, areas which could cause problems are tested and analyzed. Guessing at the cause of a problem may provide the solution, but it can easily lead to frustration, wasted time and a series of expensive, unnecessary parts replacements.

You do not need fancy equipment or complicated test gear to determine whether repairs can be attempted at home. A few simple checks could save a large repair bill and lost time while the bike sits in a dealer's service department. On the other hand, be realistic and do not attempt repairs beyond your abilities. Service departments tend to charge heavily for putting together a disassembled engine that may have been abused. Some won't even take on such a job—so use common sense and don't get in over your head.

## OPERATING REQUIREMENTS

An engine needs 3 basics to run properly: correct fuel/air mixture, compression and a spark at the correct time. If one or more are missing, the engine will not run. Four-stroke engine operating principles are described in Chapter Four under *Engine Princi-*

*ples*. The electrical system is the weakest link of the 3 basics. More problems result from electrical breakdowns than from any other source. Keep that in mind before you begin tampering with carburetor adjustments and the like.

If the machine has been sitting for any length of time and refuses to start, check and clean the spark plugs and then look to the gasoline delivery system. This includes the fuel tank, fuel shutoff valve and fuel line to the carburetor. Gasoline deposits may have formed and gummed up the carburetor jets and air passages. Gasoline tends to lose its potency after standing for long periods. Condensation may contaminate the fuel with water. Drain the old fuel (fuel tank, fuel lines and carburetor) and try starting with a fresh tankful.

## TROUBLESHOOTING INSTRUMENTS

Chapter One lists the instruments needed and gives instruction on their use.

## EMERGENCY TROUBLESHOOTING

When the bike is difficult to start, or won't start at all, it doesn't help to wear down the battery using the starter or your leg and foot on kickstart models.

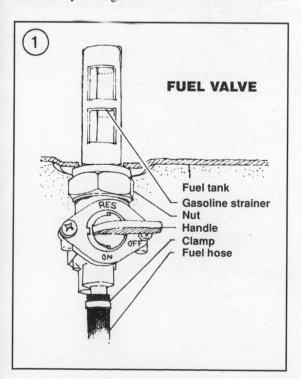

**FUEL VALVE**

- Fuel tank
- Gasoline strainer
- Nut
- Handle
- Clamp
- Fuel hose

Check for obvious problems even before getting out your tools. Go down the following list step by step. Do each one; you may be embarrassed to find the kill switch off, but that is better than wearing down the battery. If the bike still will not start, refer to the appropriate troubleshooting procedures which follow in this chapter.

1. Is there fuel in the tank? Open the filler cap and rock the bike. Listen for fuel sloshing around.

> *WARNING*
> *Do not use an open flame to check in the tank. A serious explosion is certain to result.*

2. Is the fuel supply valve in the ON position? Turn the valve to the reserve position to be sure you get the last remaining gas. See **Figure 1**.

3. Make sure the kill switch is not stuck in the OFF position or that the wire is not broken and shorting out.

4. Are the spark plug wires on tight? Push both spark plug wires on and slightly rotate them to clean the electrical connection between the plug and the spark plug wire connector.

5. Is the choke in the right position?

## ENGINE STARTING

An engine that refuses to start or is difficult to start is very frustrating. More often than not, the problem is very minor and can be found with a simple and logical troubleshooting approach.

The following items will help isolate engine starting problems.

### Engine Fails to Start

Perform the following spark test to determine if the ignition system is operating properly.

1. Remove one of the spark plugs.

2. Connect the spark plug wire and connector to the spark plug and touch the spark plug base to a good ground like the engine cylinder head. Position the spark plug so you can see the electrodes.

> *WARNING*
> *During the next step, do not hold the spark plug, wire or connector with fingers or a serious electrical shock may result. If necessary, use a pair of insu-*

*lated pliers to hold the spark plug or wire. The high voltage generated by the ignition system could produce serious or fatal shocks.*

3. Crank the engine over with the starter. A fat blue spark should be evident across the spark plug electrodes.

4. If the spark is good, check for one or more of the following possible malfunctions:

    a. Obstructed fuel line or fuel filter.

    b. Leaking head gasket(s).

    c. Low compression.

5. If the spark is not good, check for one or more of the following:

    a. Loose electrical connections.

    b. Dirty electrical connections.

    c. Loose or broken ignition coil ground wire.

    d. Broken or shorted high tension lead to the spark plug.

    e. Discharged battery.

    f. Disconnected or damaged battery connection.

    g. Breaker points oxidized (1966-early 1978).

### Engine is Difficult to Start

Check for one or more of the following possible malfunctions:

    a. Fouled spark plug(s).

    b. Improperly adjusted choke.

    c. Intake manifold air leak.

    d. Contaminated fuel system.

    e. Improperly adjusted carburetor.

    f. Weak ignition unit.

    g. Weak ignition coil(s).

    h. Poor compression.

    i. Timing advance weight sticking in advanced position (1966-1979 models only).

    j. Engine and transmission oil too heavy.

### Engine Will Not Crank

Check for one or more of the following possible malfunctions:

    a. Blown fuse or damaged circuit breaker.

    b. Discharged battery.

    c. Defective starter motor.

    d. Seized piston(s).

    e. Seized crankshaft bearings.

    f. Broken connecting rod.

## ENGINE PERFORMANCE

In the following checklist, it is assumed that the engine runs, but is not operating at peak performance. This will serve as a starting point from which to isolate a performance malfunction.

The possible causes for each malfunction are listed in a logical sequence and in order of probability.

### Engine Will Not Idle

    a. Carburetor incorrectly adjusted.

    b. Fouled or improperly gapped spark plug(s).

    c. Leaking head gasket(s).

    d. Obstructed fuel line or fuel shutoff valve.

    e. Obstructed fuel filter.

    f. Ignition timing incorrect due to defective ignition component(s).

    g. Pushrod clearance incorrect.

### Engine Misses at High Speed

    a. Fouled or improperly gapped spark plugs.

    b. Improper carburetor main jet selection.

    c. Ignition timing incorrect due to defective ignition component(s).

    d. Weak ignition coil(s).

    e. Obstructed fuel line or fuel shutoff valve.

    f. Obstructed fuel filter.

    g. Clogged carburetor jets.

### Engine Overheating

    a. Incorrect carburetor adjustment or jet selection.

    b. Ignition timing retarded due to improper adjustment or defective ignition component(s).

    c. 1966-1979: Low power—circuit breaker cam sticking in retarded position.

    d. Improper spark plug heat range.

    e. Damaged or blocked cooling fins.

    f. Oil level low.

    g. Oil not circulating properly.

    h. Valves leaking.

    i. Heavy engine carbon deposits.

## Smoky Exhaust and Engine Runs Roughly

a. Clogged air filter element.

b. Carburetor adjustment incorrect—mixture too rich.

c. Choke not operating correctly.

d. Water or other contaminants in fuel.

e. Clogged fuel line.

f. Spark plugs fouled.

g. Ignition coil defective.

h. Breaker point models: Points worn, dirty or out of adjustment. Loose condenser connections. Defective condenser.

i. Electronic ignition models: Ignition module or sensor defective.

j. Loose or defective ignition circuit wire.

k. Short circuit from damaged wire insulation.

l. Loose battery cable connection.

m. Cam timing incorrect.

n. Intake manifold or air cleaner air leak.

## Engine Loses Power at Normal Riding Speed

a. Carburetor incorrectly adjusted.

b. Engine overheating.

c. Ignition timing incorrect due to defective ignition component(s). Circuit breaker cam sticking on 1966-1979 models.

d. Incorrectly gapped spark plugs.

e. Obstructed muffler.

f. Dragging brakes(s).

## Engine Lacks Acceleration

a. Carburetor mixture too lean.

b. Clogged fuel line.

c. Ignition timing incorrect due to faulty ignition component(s).

d. Dragging brakes(s).

## ENGINE NOISES

Often the first evidence of an internal engine problem is a strange noise. That knocking, clicking or tapping sound which you never heard before may be warning you of impending trouble.

While engine noises can indicate problems, they are difficult to interpret correctly; inexperienced mechanics can be seriously misled by them.

Professional mechanics often use a special stethoscope (which looks like a doctor's stethoscope) for isolating engine noises. You can do nearly as well with a "sounding stick" which can be an ordinary piece of doweling, a length of broom handle or a section of small hose. By placing one end in contact with the area to which you want to listen and the other end near your ear, you can hear sounds emanating from that area. The first time you do this, you may be horrified at the strange sounds coming from even a normal engine. If possible, have an experienced friend or mechanic help you sort out the noises.

Consider the following when troubleshooting engine noises:

1. *Knocking or pinging during acceleration*—Caused by using a lower octane fuel than recommended. May also be caused by poor fuel. Pinging can also be caused by a spark plug of the wrong heat range. Refer to *Correct Spark Plug Heat Range* in Chapter Three.

2. *Slapping or rattling noises at low speed or during acceleration*—May be caused by piston slap, i.e., excessive piston-cylinder wall clearance.

3. *Knocking or rapping while decelerating*—Usually caused by excessive rod bearing clearance.

4. *Persistent knocking and vibration*—Usually caused by worn main bearing(s).

5. *Rapid on-off squeal*—Compression leak around cylinder head gasket(s) or spark plugs.

6. *Valve train noise*—Check for the following:

    a. Bent pushrod(s).

    b. Hydraulic tappets adjusted incorrectly.

    c. Defective hydraulic tappets.

    d. Valve sticking in guide.

    e. Worn cam gears and/or cam.

    f. Low oil pressure—probably caused by obstructed oil screen. Also check oil feed pump operation.

    g. Damaged rocker arm or shaft. Rocker arm may be binding on shaft.

## ENGINE LUBRICATION

A too low oil tank level or a plugged or damaged lubricating system will quickly lead to engine damage. The engine oil tank should be checked weekly and the tank refilled, as described in Chapter Three. Oil pump service is covered in Chapter Four.

## Oil Light

The oil light should come on when the ignition switch is turned to ON before starting the engine. After the engine is started, the oil light should go off when the engine is above idle.

If the oil light does not come on when the ignition is turned to ON and the engine is not running, check for a burned out oil light bulb. If the bulb is okay, check the oil pressure switch.

If the oil light remains on when the engine speed is above idle, turn the engine off and check the oil level in the oil tank. If the oil level is satisfactory, check the following:
  a. Check for a plugged tappet screen. Remove and service the screen as described in Chapter Three.
  b. Oil may not be returning to the tank from the return line. Check for a clogged or damaged return line or a damaged oil pump.
  c. If you are operating your Harley in conditions where the ambient temperature is below freezing, ice and sludge may be blocking the oil feed pipe. Oil cannot flow to the engine and engine damage will result.

*NOTE*
*Because water is formed during combustion, it can collect in the engine if the engine is not run hard enough to vaporize the water; a condition inherent with winter riding conditions. Water buildup will form sludge deposits as it mixes with the oil. Sludge is a thick, creamy substance that will clog oil feed and return systems (oil filter, tappet screen, lines, etc.). Sludge will accelerate engine wear and result in engine failure. In addition, when the ambient temperature falls below freezing, water in the tank and oil lines will freeze, thus preventing proper oil circulation and lubrication. To prevent engine damage from sludge buildup, note the following:*

1. When operating your Harley in cold/freezing weather, note how long you actually run the engine—the engine is colder in winter months and will take longer to warm up. Run the engine for longer periods so that it can reach maximum operating temperature. The water will then vaporize and be blown out through the engine breather.
2. Change the engine oil more frequently. The oil change intervals specified in Chapter Three should be reduced when you are operating your Harley during cold weather. This step can interrupt sludge accumulation and prevent engine damage.
3. Flush the oil tank at each oil change. Refer to Chapter Three for oil change and tank flushing procedures.

## Oil Consumption High or Engine Smokes Excessively

Check engine compression and perform a cylinder leakage test as described in Chapter Three. Causes can be one or more of the following:
  a. Worn valve guides.
  b. Worn valve guide seals.
  c. Worn or damaged piston rings.
  d. Breather valve damaged or timed incorrectly.
  e. Restricted oil tank return line.
  f. Oil tank overfilled.
  g. Restricted oil filter.
  h. Leaking cylinder head surfaces.
  i. Insufficient primary chain case vacuum on dry clutch models only (early 1984). Perform the *Primary Housing Vacuum Check (Early 1984 With Dry Clutch)* as described in Chapter Five.

## Oil Fails to Return to Oil Tank

  a. Oil lines or fittings restricted or damaged.
  b. Oil pump damaged or operating incorrectly.
  c. Empty oil tank.
  d. Restricted oil filter.

## Excessive Engine Oil Leaks

  a. Clogged air cleaner breather hose.
  b. Restricted or damaged oil return line to oil tank.
  c. Loose engine parts.
  d. Damaged gasket sealing surfaces.
  e. Restricted air cleaner breather hose.
  f. Oil tank overfilled.

## CLUTCH

The three basic clutch troubles are:

a. Clutch noise.

b. Clutch slipping.

c. Improper clutch disengagement or dragging.

All clutch troubles, except adjustments, require partial clutch disassembly to identify and cure the problem. The troubleshooting chart in **Figure 2** lists clutch troubles and checks to make. Refer to Chapter Five for clutch service procedures.

## TRANSMISSION

The basic transmission troubles are:

a. Excessive gear noise.

b. Difficult shifting.

c. Gears pop out of mesh.

d. Incorrect shift lever operation.

Transmission symptoms are sometimes hard to distinguish from clutch symptoms. The troubleshooting chart in **Figure 3** lists transmission troubles and checks to make. Refer to Chapter Five for transmission service procedures. Be sure that the clutch is not causing the trouble before working on the transmission.

## CHARGING SYSTEM

Charging system testing procedures are described in Chapter Seven.

## STARTING SYSTEM

The basic starter-related troubles are:

a. The starter does not crank.

b. The starter cranks, but the engine does not start.

### Testing

Starting system problems are relatively easy to find. In most cases, the trouble is a loose or dirty electrical connection. Use the troubleshooting chart in **Figure 4** with the following tests.

### *Starter does not crank*

1. Turn on the headlight and push the starter button. Check for one of the following conditions.

2. *Starter does not crank and headlight does not come on:* The battery is dead or there is a loose battery connection. Check the battery charge as described in Chapter Three. If the battery is okay, check

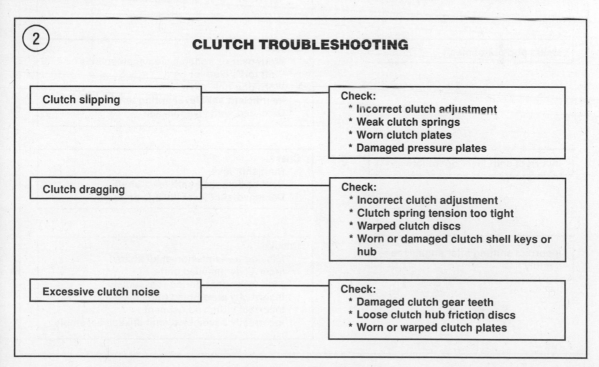

**CLUTCH TROUBLESHOOTING**

**Clutch slipping**

Check:
* Incorrect clutch adjustment
* Weak clutch springs
* Worn clutch plates
* Damaged pressure plates

**Clutch dragging**

Check:
* Incorrect clutch adjustment
* Clutch spring tension too tight
* Warped clutch discs
* Worn or damaged clutch shell keys or hub

**Excessive clutch noise**

Check:
* Damaged clutch gear teeth
* Loose clutch hub friction discs
* Worn or warped clutch plates

the starter connections at the battery, solenoid and at the starter switch.

3. *Headlight comes on, but goes out when the starter button is pushed:* There may be a bad connection at the battery. Wiggle the battery terminals and re-check. If the starter starts cranking, you've found the problem. Remove and clean the battery terminal clamps. Clean the battery posts also. Reinstall the terminal clamps and tighten securely.

4. *Headlight comes on, but dims slightly when the starter button is pushed:* The problem is probably in

the starter. Remove and test the starter as described in Chapter Seven.

5. *Headlight comes on, but dims severely when the starter button is pushed:* Either the battery is nearly dead or the starter or engine is partially seized. Check the battery as described in Chapter Three. Check the starter as described in Chapter Seven before checking for partial engine seizure.

6. *Headlight comes on and stays bright when the starter button is pushed:* The problem is in the starter button-to-solenoid wiring or in the starter itself. Check the starter switch, kill switch, starter relay and

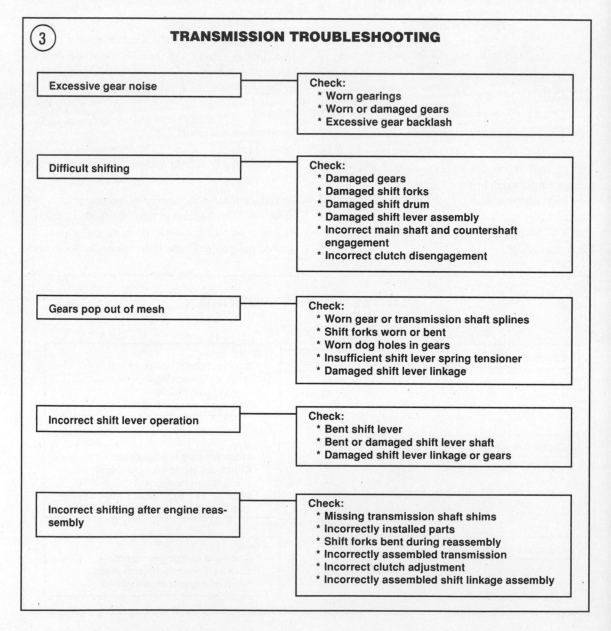

③ **TRANSMISSION TROUBLESHOOTING**

**Excessive gear noise**

Check:
* Worn gearings
* Worn or damaged gears
* Excessive gear backlash

**Difficult shifting**

Check:
* Damaged gears
* Damaged shift forks
* Damaged shift drum
* Damaged shift lever assembly
* Incorrect main shaft and countershaft engagement
* Incorrect clutch disengagement

**Gears pop out of mesh**

Check:
* Worn gear or transmission shaft splines
* Shift forks worn or bent
* Worn dog holes in gears
* Insufficient shift lever spring tensioner
* Damaged shift lever linkage

**Incorrect shift lever operation**

Check:
* Bent shift lever
* Bent or damaged shift lever shaft
* Damaged shift lever linkage or gears

**Incorrect shifting after engine reassembly**

Check:
* Missing transmission shaft shims
* Incorrectly installed parts
* Shift forks bent during reassembly
* Incorrectly assembled transmission
* Incorrect clutch adjustment
* Incorrectly assembled shift linkage assembly

the solenoid switch. Check each switch by bypass-ing it with a jumper wire. Check the starter as de-scribed in Chapter Seven.

### Starter spins but engine does not crank

If the starter spins at normal or high speed but the engine fails to crank, the problem is in the starter drive mechanism.

> *NOTE*
> *Depending upon battery condition, the battery will eventually run down as the starter button is continually pressed. Remember that if the starter cranks nor-mally, but the engine fails to start, the starter is working properly. It's time to start checking other engine systems. Don't wear the battery down.*

## ELECTRICAL PROBLEMS

If bulbs burn out frequently, the cause may be excessive vibration, loose connections that permit sudden current surges or the installation of the wrong type of bulb.

Most light and ignition problems are caused by loose or corroded ground connections. Check these prior to replacing a bulb or electrical component.

## IGNITION SYSTEM

The ignition system may be either a breaker point or breakerless type. See Chapter Seven. Most prob-lems involving failure to start, poor driveability or rough running are caused by trouble in the ignition system, particularly in breaker point systems.

Note the following symptoms:

a. Engine misses.

b. Stumbles on acceleration (misfiring).

c. Loss of power at high speed (misfiring).

d. Hard starting (or failure to start).

e. Rough idle.

Most of the symptoms can also be caused by a carburetor that is worn or improperly adjusted. But considering the law of averages, the odds are far better that the source of the problem will be found in the ignition system rather than the fuel system.

## BREAKER POINT IGNITION TROUBLESHOOTING

The following basic tests are designed to pinpoint and isolate quickly problems in the primary circuit of a breaker point ignition. Before starting, make sure that the battery is in good condition. See Chap-ter Three. If the primary circuit checks out satisfac-

---

**(4) STARTER TROUBLESHOOTING**

| Symptom | Probable Cause | Remedy |
|---|---|---|
| Starter does not work | Low battery | Recharge battery |
| | Worn brushes | Replace brushes |
| | Defective relay | Repair or replace |
| | Defective switch | Repair or replace |
| | Defective wiring connection | Repair wire or clean connection |
| | Internal short circuit | Repair or replace defective component |
| Starter action is weak | Low battery | Recharge battery |
| | Pitted relay contacts | Clean or replace |
| | Worn brushes | Replace brushes |
| | Defective connection | Clean and tighten |
| | Short circuit in commutator | Replace armature |
| Starter runs continuously | Stuck relay | Replace relay |
| Starter turns; does not turn engine | Defective starter drive | Repair or replace |

torily, refer to *Tune-up* in Chapter Three and check the circuit breaker, spark plug wires and spark plugs.

*NOTE*
*Before starting, connect a voltmeter between the battery terminals (positive to positive; negative to negative) and write down the reading. This is battery voltage.*

1. Remove the circuit breaker point cover (**Figure 5**).
2. Rotate the engine until the points are closed (D, **Figure 6**).
3. Disconnect the high-voltage lead from one of the spark plugs.
4. Insert a metal adapter in the plug boot and hold it about 3/16 in. from a clean engine ground with insulated pliers (**Figure 7**).
5. Turn the ignition switch ON.
6. Open the points with an insulated tool made of wood. A fat, blue-white spark should jump from the spark plug lead to the engine. If the spark is good, the ignition system is good, but ignition timing may be incorrect. Check as described in Chapter Three. If there is no spark, or if it is thin, yellowish, or weak, continue with Step 7.
7. Turn the ignition switch OFF and close the points.
8. Connect the positive lead of a voltmeter to the wire on the points (A, **Figure 6**) and the negative lead to a good ground. Turn the ignition switch ON. If the voltmeter indicates more than 0.125 volts, the points are defective. Replace them as described in Chapter Three.
9. Open the points with the same tool used in Step 6. The voltmeter should indicate battery voltage. If not, the following may be the problem:
    a. Shorted points.
    b. Shorted condenser.
    c. Open coil primary winding.
    Perform Steps 10-12 to isolate the problem.
10. With the ignition switch ON, disconnect the condenser and the wire from the points (A, **Figure 6**). Connect the voltmeter positive lead to the wire which was connected to the points. Connect the voltmeter negative lead to a good ground (bare metal). The voltmeter should indicate battery voltage. If not, current is not reaching the points.
    a. Check the ignition primary circuit wiring for breaks or bad connections. Repair as needed and retest.

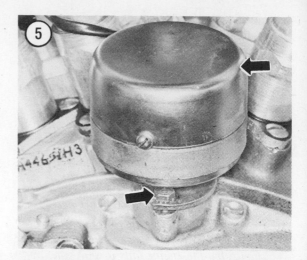

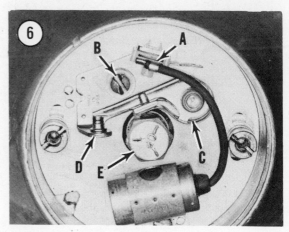

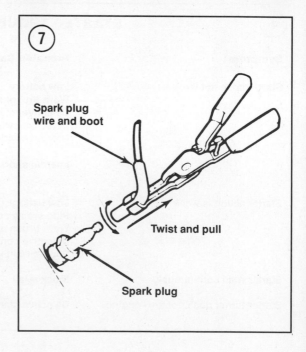

Spark plug wire and boot

Twist and pull

Spark plug

b. If current does not reach the points with the wiring in good condition, the primary winding inside the ignition coil may be defective. Substitute a known good ignition coil and retest.

11. If the voltmeter indicated battery voltage in Step 9, the coil primary circuit and primary winding are okay. Connect the positive voltmeter lead to the wire which goes from the coil to the points. Block the points open as in Step 6. Connect the negative voltmeter lead to the movable point. If the voltmeter indicates any voltage, the points are shorted and must be replaced.

12. If the preceding checks are satisfactory, the problem is in the coil or condenser. Replace each of these separately with a known good one to determine which is defective.

### Ignition Coil

Ignition coil testing for 1970 and later models is described in Chapter Seven. On 1966-1969 bikes, test by substituting a known good coil.

## BREAKERLESS IGNITION TROUBLESHOOTING

The following basic tests are designed to pinpoint and isolate problems quickly in the primary circuit of the breakerless inductive discharge ignition system. The procedure requires a voltmeter, non-magnetic feeler gauge (late 1978-1979) and an ignition test adapter (1980-on).

### Spark Test

Perform the following test to determine if the ignition system is operating properly.

1. Remove one of the spark plugs.
2. Connect the spark plug wire and connector to the spark plug and touch the spark plug base to a good ground like the engine cylinder head. Position the spark plug so you can see the electrodes.

*WARNING*
*During the next step, do not hold the spark plug, wire or connector with fingers or a serious electrical shock may result. If necessary, use a pair of insulated pliers (Figure 7) to hold the spark plug or wire. The high voltage gener-*

*ated by the ignition system could produce serious or fatal shocks.*

3. Crank the engine over with the starter. A fat blue spark should be evident across the spark plug electrodes.

4A. If a spark is obtained in Step 3, the problem is not in the breakerless ignition or coil. Check the fuel system and spark plugs. On late 1978-1979 models, check the ignition advance mechanism as described in Chapter Seven.

4B. If no spark is obtained, proceed with the following tests for your model.

### Late 1978-1979 models

1. Check the battery charge as described in Chapter Three. If battery is okay, proceed to Step 2.
2. Remove the ignition timer cover (2, **Figure 8**).

*NOTE*
*The ignition module (3, Figure 8) is fastened to the back of the cover.*

3. Check the control module black wire (ground) connection at the timer plate (11, **Figure 8**). Make sure the connection is tight and that the wire is not chafed or damaged.

4. Check the sensor air gap as described in Chapter Three.

5. If the air gap cannot be adjusted for both rotor lobes, the trigger rotor (7, **Figure 8**) and/or the timer mechanism shaft on the advance assembly base is bent or damaged. Replace the damaged part as described in Chapter Seven.

*NOTE*
*After adjusting the sensor air gap, check and adjust the timing as described in Chapter Three.*

6. Turn the engine over so that the sensor is located between the 2 trigger rotor lobes (**Figure 9**).

7. Connect a voltmeter between the ignition coil positive terminal (white wires) and a good engine ground. **Figure 10** shows the ignition circuit and wire color codes.

*NOTE*
*The ignition and engine stop switches must be ON when performing Steps 8-11.*

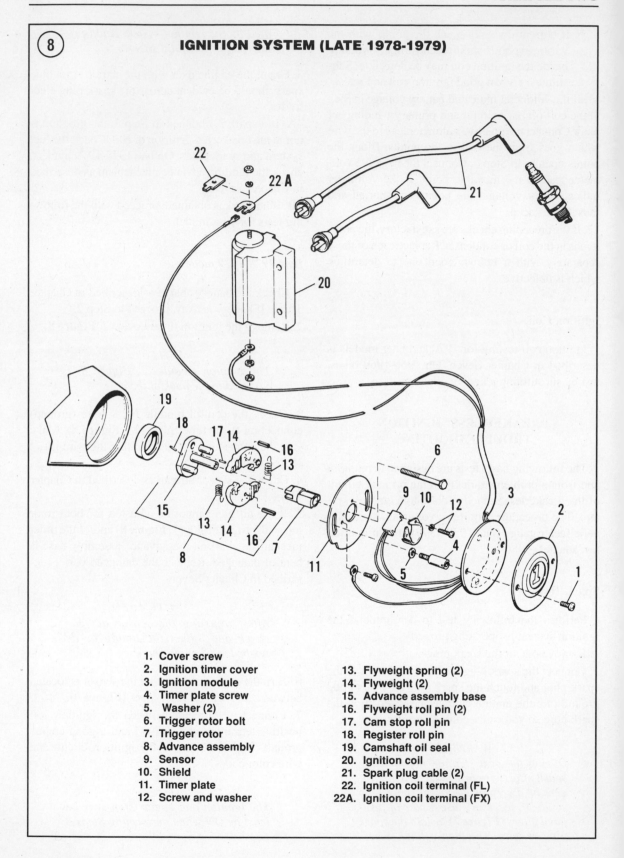

**IGNITION SYSTEM (LATE 1978-1979)**

1. Cover screw
2. Ignition timer cover
3. Ignition module
4. Timer plate screw
5. Washer (2)
6. Trigger rotor bolt
7. Trigger rotor
8. Advance assembly
9. Sensor
10. Shield
11. Timer plate
12. Screw and washer
13. Flyweight spring (2)
14. Flyweight (2)
15. Advance assembly base
16. Flyweight roll pin (2)
17. Cam stop roll pin
18. Register roll pin
19. Camshaft oil seal
20. Ignition coil
21. Spark plug cable (2)
22. Ignition coil terminal (FL)
22A. Ignition coil terminal (FX)

8. Turn the ignition and engine stop switches ON. The voltmeter should read $11.5 \pm 0.5$ volts. Interpret results as follows:

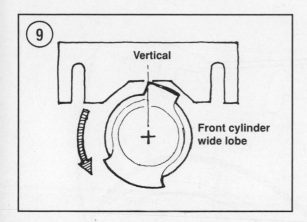

Vertical

Front cylinder wide lobe

a. If the voltage is incorrect, the problem is in the battery-to-ignition coil circuit. Check the wiring connectors at the circuit breakers and at the ignition switch. See **Figure 10**.

b. If the voltage is correct, proceed to Step 9.

9. Disconnect the ignition coil negative (blue) connector (**Figure 10**). Connect a voltmeter between the coil negative terminal and a good engine ground. Turn the ignition and engine stop switches ON. The voltmeter should read $11.5 \pm 0.5$ volts. Interpret results as follows:

a. If the voltage is incorrect, the problem is in the ignition coil primary winding. Replace the ignition coil and perform the *Spark Test* as described in this chapter.

b. If the voltage is correct, proceed to Step 10.

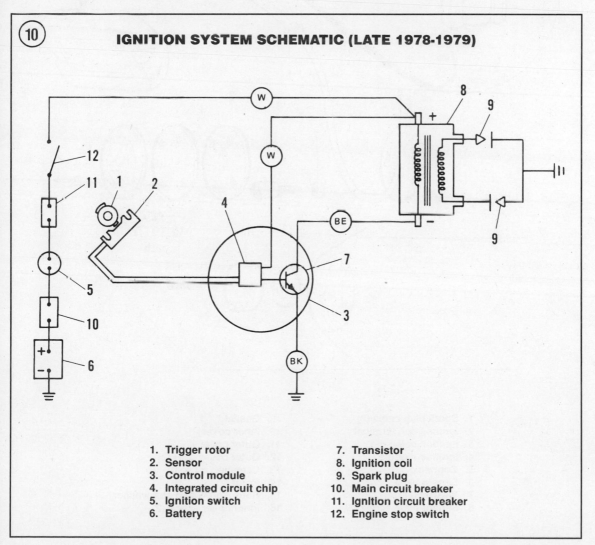

**IGNITION SYSTEM SCHEMATIC (LATE 1978-1979)**

1. Trigger rotor
2. Sensor
3. Control module
4. Integrated circuit chip
5. Ignition switch
6. Battery
7. Transistor
8. Ignition coil
9. Spark plug
10. Main circuit breaker
11. Ignition circuit breaker
12. Engine stop switch

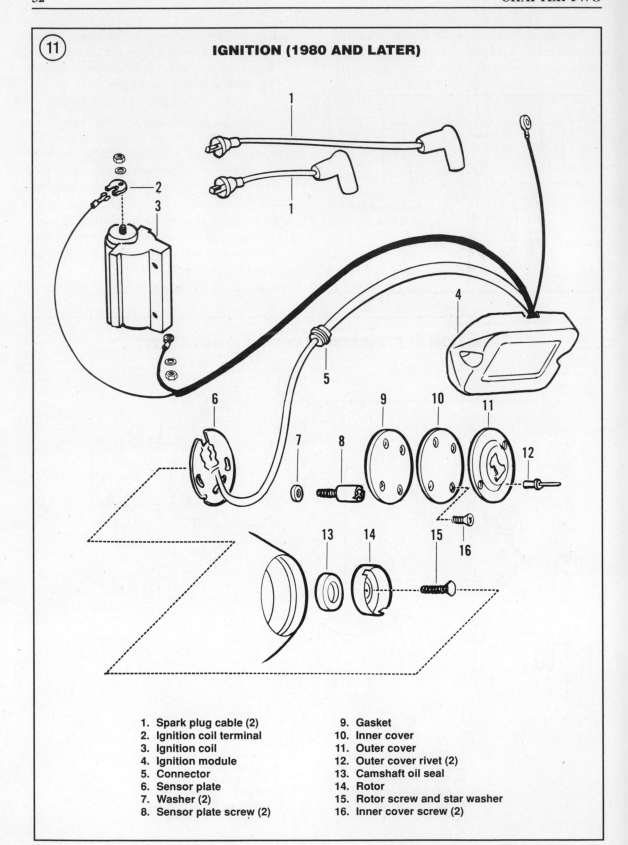

**IGNITION (1980 AND LATER)**

1. Spark plug cable (2)
2. Ignition coil terminal
3. Ignition coil
4. Ignition module
5. Connector
6. Sensor plate
7. Washer (2)
8. Sensor plate screw (2)
9. Gasket
10. Inner cover
11. Outer cover
12. Outer cover rivet (2)
13. Camshaft oil seal
14. Rotor
15. Rotor screw and star washer
16. Inner cover screw (2)

10. Reconnect the ignition coil negative connector (blue).

11. Connect a voltmeter between the ignition coil negative terminal (**Figure 10**) and a good engine ground. The voltmeter should read 1-2 volts. Next place a screwdriver against the face of the sensor. The voltmeter should read 11.5-13 volts. Interpret results as follows:

   a. If the voltage is incorrect in one or both tests, the ignition module is defective. Replace it as described in Chapter Seven.

   b. If the voltage is correct, proceed to Step 12.

12. Disconnect one of the spark plug caps. Insert an adapter in the spark plug cap in place of the spark plug. Position the adapter so that it is approximately 3/16 in. away from the cylinder head surface. Hold the spark plug cap with insulated pliers (**Figure 7**). Place a screwdriver against the face of the sensor and at the same time watch the end of the adapter. A spark should occur each time the screwdriver touches the sensor face. If sparks do not occur, the ignition coil secondary winding is defective. Replace the ignition coil and retest.

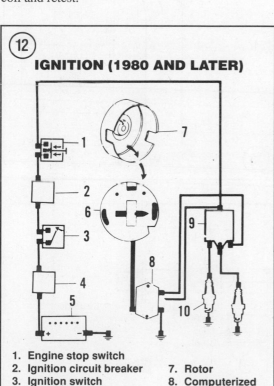

**(12)**

## IGNITION (1980 AND LATER)

1. Engine stop switch
2. Ignition circuit breaker
3. Ignition switch
4. Main circuit breaker
5. Battery
6. Sensor plate
7. Rotor
8. Computerized control module
9. Ignition coil
10. Spark plug

### *1980-on*

1. Check the battery charge as described in Chapter Three. If battery is okay, proceed to Step 2.

2. Remove the timer case cover (11, **Figure 11**).

3. Check that the black lead to the ignition module (located on frame above regulator, 8, **Figure 12**) is fastened securely. Check also that the battery negative cable is fastened and in good condition.

*NOTE*
*When performing the following tests, a voltmeter with an input resistance of 20,000 ohms/volts or higher is required. A lower-resistance meter may give a faulty reading.*

4. Turn the engine over so that the sensor is located between the 2 rotor slots (**Figure 9**).

*NOTE*
*The ignition switch and kill switch must be turned ON when performing Step 5 and Step 6.*

5. Connect a voltmeter between the ignition coil positive terminal (white wires) and a good engine ground. The voltmeter should read $11.5 \pm 0.5$ volts. Interpret results as follows:

   a. If the voltage is incorrect, the problem is in the battery-to-ignition coil circuit. Check the wiring connectors at the circuit breakers (4, **Figure 12**) and at the ignition switch.

   b. If the voltage is correct, proceed to Step 6.

6. Disconnect the ignition coil negative (blue) connector. Connect a voltmeter between the coil negative winding and a good engine ground. The voltmeter should read $11.5 \pm 0.5$ volts. Interpret results as follows:

   a. If the voltage is incorrect, the problem is in the ignition coil primary winding. Replace the ignition coil and perform the *Spark Test* as described in this chapter.

   b. If the voltage is correct, check the control module and sensor, beginning with Step 7.

   c. Turn the ignition switch OFF.

   d. Reconnect the ignition coil negative connection (blue).

7. Disconnect the sensor plate from the control module at the connector (6, **Figure 12**).

8. Install an ignition test adapter (HD-94465-81) and a voltmeter as shown in **Figure 13**.

*NOTE*
*Figure 14 shows how to make a test adapter.*

*CAUTION*
*When using the ignition test adapter, make sure the exposed terminal connector does not touch another connector or ground (bare metal) as this will damage the ignition module.*

9. Turn the ignition switch and the kill switch ON and measure the voltage between the No. 1 pin (red wire) and the No. 2 pin (black wire) as shown in **Figure 13**.

10. The meter should read 5.4-5.5 volts. If it does not, the control module is defective and should be replaced as described in Chapter Seven.

11. To test the sensor operation, connect the voltmeter to the ignition test adapter as shown in **Figure 15**.

12. If the ignition switch and the kill switch were turned OFF, turn them back ON.

13. Turn the engine so that the rotor slots are not aligned with the sensor (**Figure 16**). Measure the voltage reading. It should be 4.5-5.5 volts.

14. Turn the engine so that the center of the sensor is between the 2 rotor slots. The voltmeter should read 0-1 volts.

15. If the voltage readings in Step 13 and Step 14 are incorrect and the control module passes the previous test, the sensor is faulty and should be replaced as described in Chapter Seven.

16. Turn the ignition switch OFF. Remove the ignition test adapter and reconnect the module connectors.

## EXCESSIVE VIBRATION

Usually this is caused by loose engine mounting hardware. If not, it can be difficult to find without disassembling the engine. High speed vibration may be due to a bent axle shaft or loose or faulty suspension components. Vibration can also be caused by the following conditions:

a. Broken frame.
b. Severely worn primary chain.
c. Primary chain links tight due to improper lubrication.
d. Loose transmission mounting bolts.
e. Loose transmission sub-mounting bolts.

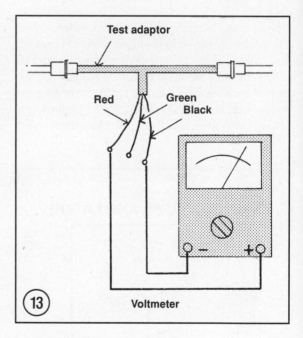

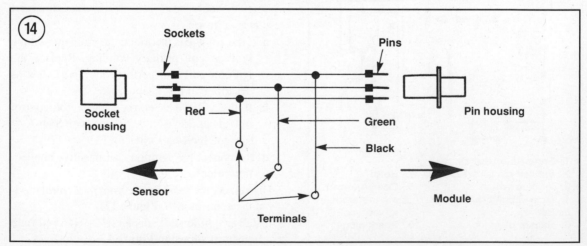

f. Improperly balanced wheels.

g. Defective or damaged wheels.

h. Defective or damaged tires.

i. Internal engine wear or damage.

## CARBURETOR TROUBLESHOOTING

Troubleshooting procedures unique to HD carburetors are found in **Figure 17**. **Figure 18** lists troubleshooting procedures for Keihin carburetors. For other carburetors, use the HD carburetor troubleshooting procedures in **Figure 17** as a guide.

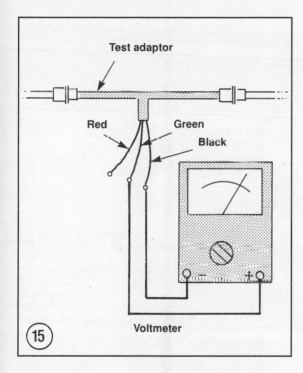

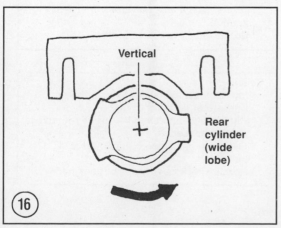

## FRONT SUSPENSION AND STEERING

Poor handling may be caused by improper tire pressure, a damaged or bent frame or front steering components, worn wheel bearings or dragging brakes. Possible causes of suspension and steering malfunctions are listed below.

### Irregular or Wobbly Steering

a. Loose wheel axle nuts.

b. Loose or worn steering head bearings.

c. Excessive wheel hub bearing play.

d. Damaged cast wheel.

e. Spoke wheel out of alignment.

f. Unbalanced wheel assembly.

g. Worn hub bearings.

h. Incorrect wheel alignment.

i. Bent or damaged steering stem or frame (at steering neck).

j. Tire incorrectly seated on rim.

k. Excessive front end loading from non-standard equipment.

### Stiff Steering

a. Low front tire air pressure.

b. Bent or damaged steering stem or frame (at steering neck).

c. Loose or worn steering head bearings.

d. Steering damper too tight (FL models with adjustable fork).

### Stiff or Heavy Fork Operation

a. Incorrect fork springs.

b. Incorrect fork oil viscosity.

c. Excessive amount of fork oil.

d. Bent fork tubes.

### Poor Fork Operation

a. Worn or damaged fork tubes.

b. Fork oil level low due to leaking fork seals.

c. Bent or damaged fork tubes.

d. Contaminated fork oil.

e. Incorrect fork springs.

f. Heavy front end loading from non-standard equipment.

## Poor Rear Shock Absorber Operation

a. Weak or worn springs.

b. Damper unit leaking.

c. Shock shaft worn or bent.

d. Incorrect rear shock springs.

e. Rear shock adjusted incorrectly.

f. Heavy rear end loading from non-standard equipment.

g. Incorrect loading.

## BRAKE PROBLEMS

Sticking disc brakes may be caused by a stuck piston(s) in a caliper assembly, warped pad shim(s) or improper rear brake adjustment. See **Figure 19** for disc brake troubles and checks to make.

A sticking drum brake may be caused by worn or weak return springs, dry pivot and cam bushings or improper adjustment. Grabbing brakes may be caused by greasy linings which must be replaced. Brake grab may also be due to an out-of-round drum. Glazed linings will cause loss of stopping power. See **Figure 20** for drum brake troubles and checks to make.

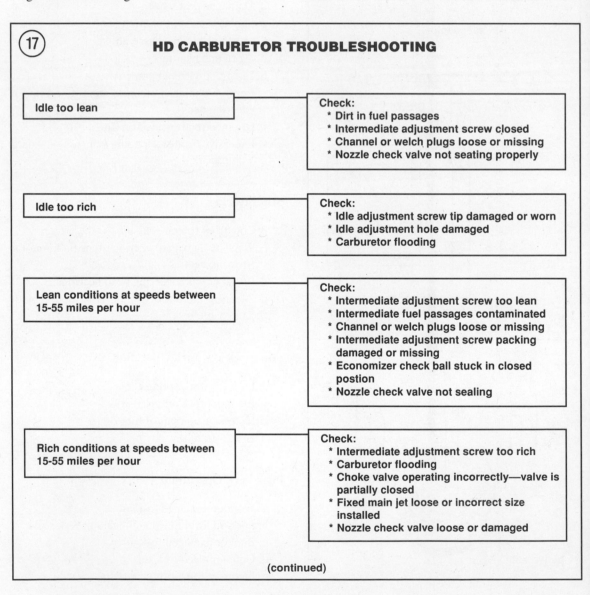

(17) **HD CARBURETOR TROUBLESHOOTING**

| | |
|---|---|
| **Idle too lean** | **Check:**<br>* **Dirt in fuel passages**<br>* **Intermediate adjustment screw closed**<br>* **Channel or welch plugs loose or missing**<br>* **Nozzle check valve not seating properly** |
| **Idle too rich** | **Check:**<br>* **Idle adjustment screw tip damaged or worn**<br>* **Idle adjustment hole damaged**<br>* **Carburetor flooding** |
| **Lean conditions at speeds between 15-55 miles per hour** | **Check:**<br>* **Intermediate adjustment screw too lean**<br>* **Intermediate fuel passages contaminated**<br>* **Channel or welch plugs loose or missing**<br>* **Intermediate adjustment screw packing damaged or missing**<br>* **Economizer check ball stuck in closed position**<br>* **Nozzle check valve not sealing** |
| **Rich conditions at speeds between 15-55 miles per hour** | **Check:**<br>* **Intermediate adjustment screw too rich**<br>* **Carburetor flooding**<br>* **Choke valve operating incorrectly—valve is partially closed**<br>* **Fixed main jet loose or incorrect size installed**<br>* **Nozzle check valve loose or damaged** |

(continued)

(17) (continued)

Lean conditions at highway speeds

Check:
* Main fuel jet too small
* Loose main fuel jet causing air leak
* Nozzle system contaminated
* Nozzle check valve loose or damaged

Rich conditions at highway speeds

Check:
* Main fuel jet too large
* Economizer check ball not seating
* Carburetor flooding

Accelerating pump inoperative causing a lean acceleration

Check:
* Carburetor adjustment incorrect—adjust intermediate and idle adjustment screws
* Diaphragm cover plug screw loose or missing
* Diaphragm flap check valves worn or damaged
* Economizer check valve stuck closed
* Acceleration fuel ports and channels contaminated
* Accelerator pump worn or damaged

Lean conditions at all speeds

Check:
* Plugged or dirty filter screens
* Loose diaphragm cover plate screws
* Stretched or damaged inlet tension spring
* Inlet control lever adjusted incorrectly
* Air leak in metering system

Rich conditions at all speeds

Check:
* Carburetor flooding
* Inlet control lever adjusted incorrectly
* Choke valve not staying open due to worn or damaged choke friction spring and friction ball

Carburetor flooding

Check:
* Inlet needle and seat assembly contaminated with dirt
* Inlet seat gasket worn or damaged
* Diaphragm installed incorrectly
* Diaphragm worn or damaged
* Inlet needle and seat assembly worn
* Inlet control lever tight on lever pin

2

# KEIHIN CARBURETOR TROUBLESHOOTING

**Hard starting**

Check:
* Choke not operating correctly
* Idle mixture misadjusted (early models idle mixture adjustment only)
* Air leak at carburetor mounting
* Fuel overflow

**Fuel overflows**

Check:
* Worn float needle valve or dirty seat
* Incorrect float lever
* Damaged flaot bowl o-ring or loose float bowl mounting screws
* Damaged float pin or loose locking screw
* Damaged float

**Poor idling**

Check:
* Idle misadjusted
* Worn idle mixture screw
* Blocked jet or port in carburetor bore
* Air leak at carburetor mounting
* Accelerator pump rod too long or misadjusted

**Poor acceleration**

Check:
* Clogged accelerator pump
* Worn accelerator pump diaphragm
* Idle mixture misadjusted (early models with idle mixture adjustment)
* Clogged slow jet
* Float level too high

**Low power at all speeds**

Check:
* Dirty or plugged carburetor passages
* Clogged fuel lines
* Clogged fuel strainer in tank
* Air leak at carburetor mount
* Dirty air filter
* Loose carburetor jets

**Poor power at high speeds**

Check:
* Loose or clogged main jet
* Incorrect float lever
* Dirty or plugged carburetor passages

**Poor fuel economy**

Check:
* Float level too low
* Loose jets
* Clogged bleed tubes of jets
* Choke not opening fully
* Dirty air filter

**Fuel starvation**

Check:
* Clogged fuel line
* Carburetor dirty
* Fuel tank strainers clogged or dirty
* Accelerating pump not operating correctly
* Fuel tank dirty

(18)

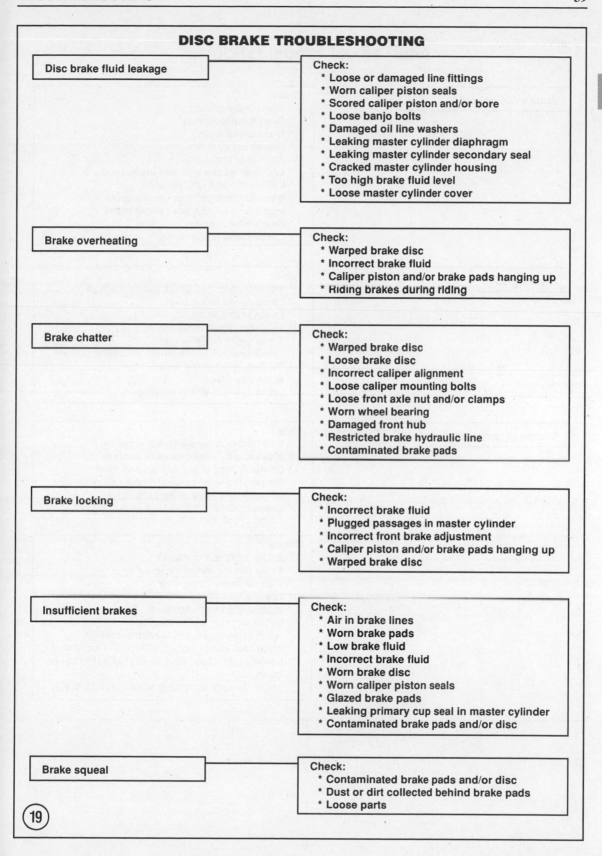

# DISC BRAKE TROUBLESHOOTING

**Disc brake fluid leakage**

Check:
* Loose or damaged line fittings
* Worn caliper piston seals
* Scored caliper piston and/or bore
* Loose banjo bolts
* Damaged oil line washers
* Leaking master cylinder diaphragm
* Leaking master cylinder secondary seal
* Cracked master cylinder housing
* Too high brake fluid level
* Loose master cylinder cover

**Brake overheating**

Check:
* Warped brake disc
* Incorrect brake fluid
* Caliper piston and/or brake pads hanging up
* Riding brakes during riding

**Brake chatter**

Check:
* Warped brake disc
* Loose brake disc
* Incorrect caliper alignment
* Loose caliper mounting bolts
* Loose front axle nut and/or clamps
* Worn wheel bearing
* Damaged front hub
* Restricted brake hydraulic line
* Contaminated brake pads

**Brake locking**

Check:
* Incorrect brake fluid
* Plugged passages in master cylinder
* Incorrect front brake adjustment
* Caliper piston and/or brake pads hanging up
* Warped brake disc

**Insufficient brakes**

Check:
* Air in brake lines
* Worn brake pads
* Low brake fluid
* Incorrect brake fluid
* Worn brake disc
* Worn caliper piston seals
* Glazed brake pads
* Leaking primary cup seal in master cylinder
* Contaminated brake pads and/or disc

**Brake squeal**

Check:
* Contaminated brake pads and/or disc
* Dust or dirt collected behind brake pads
* Loose parts

19

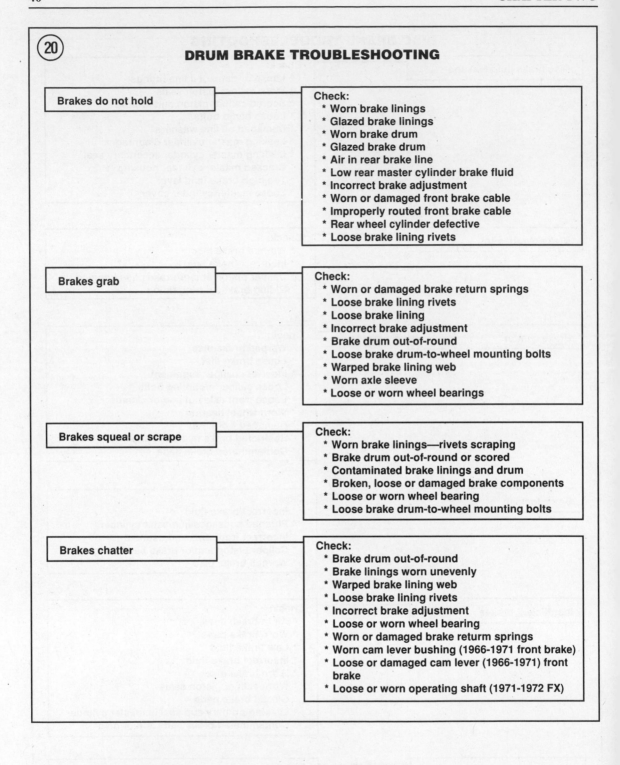

**(20)**

# DRUM BRAKE TROUBLESHOOTING

| Brakes do not hold |
|---|

Check:
* Worn brake linings
* Glazed brake linings
* Worn brake drum
* Glazed brake drum
* Air in rear brake line
* Low rear master cylinder brake fluid
* Incorrect brake adjustment
* Worn or damaged front brake cable
* Improperly routed front brake cable
* Rear wheel cylinder defective
* Loose brake lining rivets

| Brakes grab |
|---|

Check:
* Worn or damaged brake return springs
* Loose brake lining rivets
* Loose brake lining
* Incorrect brake adjustment
* Brake drum out-of-round
* Loose brake drum-to-wheel mounting bolts
* Warped brake lining web
* Worn axle sleeve
* Loose or worn wheel bearings

| Brakes squeal or scrape |
|---|

Check:
* Worn brake linings—rivets scraping
* Brake drum out-of-round or scored
* Contaminated brake linings and drum
* Broken, loose or damaged brake components
* Loose or worn wheel bearing
* Loose brake drum-to-wheel mounting bolts

| Brakes chatter |
|---|

Check:
* Brake drum out-of-round
* Brake linings worn unevenly
* Warped brake lining web
* Loose brake lining rivets
* Incorrect brake adjustment
* Loose or worn wheel bearing
* Worn or damaged brake return springs
* Worn cam lever bushing (1966-1971 front brake)
* Loose or damaged cam lever (1966-1971) front brake
* Loose or worn operating shaft (1971-1972 FX)

# CHAPTER THREE

# PERIODIC LUBRICATION, MAINTENANCE AND TUNE-UP

Your bike can be cared for by two methods: preventive or corrective maintenance. Because a motorcycle is subjected to tremendous heat, stress and vibration (even in normal use), preventive maintenance prevents costly and unexpected corrective maintenance. A careful program of lubrication, preventive maintenance and regular tune-ups will result in longer engine and vehicle life, ensuring good performance, dependability and safety. It will also pay dividends in fewer and less expensive repair bills. Such a program is especially important if the bike is used in remote areas or on heavily traveled freeways where breakdowns are not only inconvenient but dangerous. Breakdowns are much less likely to occur if the bike has been well maintained.

Certain maintenance tasks and checks should be performed weekly. Others should be performed at certain time or mileage intervals. Still others should be done whenever certain symptoms appear. Some maintenance procedures are included under *Tune-up* at the end of this chapter. Detailed instructions will be found there. Other steps are described in the following chapters. Chapter references are included with these steps.

Scheduled maintenance requirements are provided in **Table 1** (1966-1969), **Table 2** (1970-early 1978) and **Table 3** (late 1978-1984). **Tables 1-11** are at the end of the chapter.

## ROUTINE CHECKS

The following simple checks should be carried out at each fuel stop.

### Engine Oil Tank Level

Refer to *Periodic Lubrication* in this chapter.

### General Inspection

1. Examine the engine for signs of oil or fuel leakage.
2. Check the tires for embedded stones.
3. Make sure all lights work.

> *NOTE*
> *At least check the brake light. It can burn out anytime. Motorists can't stop as quickly as you and need all the warning you can give.*

### Tire Pressure

Tire pressure must be checked with the tires cold. Correct tire pressure depends on the load you are carrying. See **Table 4**.

## Lights and Horn

With the engine running, check the following:

1. Pull the front brake lever and check that the brake light comes on.

2. Push the rear brake pedal and check that the brake light comes on soon after you have begun depressing the pedal.

3. With the engine running, check to see that the headlight and taillight are on.

4. Move the dimmer switch up and down between the high and low positions, and check to see that both headlight elements are working.

5. Push the turn signal switch to the left position and the right position and check that all 4 turn signal lights are working.

6. Operate the horn and note that the horn blows loudly.

7. If during the test the brake light did not come on, check the taillight bulb and replace it as required. If the bulb is okay, but it did not light when either the front brake lever or rear brake pedal was operated, replace the defective brake light switch as described in Chapter Seven.

8. If the horn or any light failed to work properly, refer to Chapter Seven.

## MAINTENANCE INTERVALS

The services and intervals shown in **Tables 1-3** are recommended by the factory. Strict adherence to these recommendations will go a long way toward insuring long service from your Harley-Davidson. If the bike is run in an area of high humidity, the lubrication service must be done more frequently to prevent possible rust damage.

For convenient maintenance of your motorcycle, most of the services shown in **Tables 1-3** are described in this chapter. Those procedures which require more than minor disassembly or adjustment are covered elsewhere in the appropriate chapter. The *Table of Contents* and *Index* can help you locate a particular service procedure.

## TIRES AND WHEELS

### Tire Pressure

Tire pressure should be checked and adjusted to accommodate rider and luggage weight. A simple,

accurate gauge (**Figure 1**) can be purchased for a few dollars and should be carried in your motorcycle tool kit. The appropriate tire pressures are shown in **Table 4**.

> *NOTE*
> *After checking and adjusting the air pressure, make sure to reinstall the air valve cap (**Figure 2**). The cap prevents small pebbles and/or dirt from collecting in the valve stem; these could allow air leakage or result in incorrect tire pressure readings.*

### Inspection

Check tire tread for excessive wear, deep cuts, embedded objects such as stones, nails, etc. If you find a nail in a tire, mark its location with a light crayon before pulling it out. This will help to locate the hole. Refer to *Tire Changing* in Chapter Eight.

Check local traffic regulations concerning minimum tread depth. Measure with a tread depth gauge (**Figure 3**) or small ruler. As a guideline, replace tires when the tread depth is 5/16 in. (7.94 mm) or less.

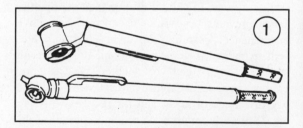

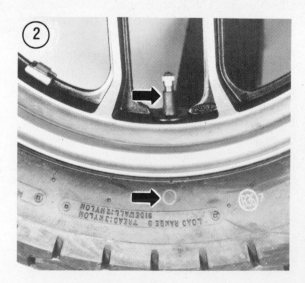

Tread depth indicators appear across the tire when tread reaches minimum safe depth. Replace the tire at this point.

## Wheel Spoke Tension

On spoked wheels, tap each spoke with a screwdriver (**Figure 4**). The higher the pitch of sound it

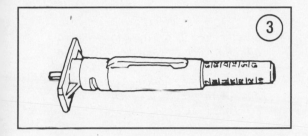

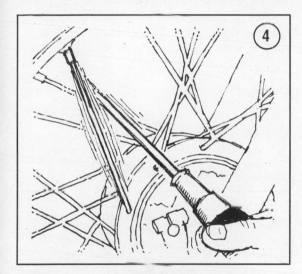

makes, the tighter the spoke. The lower the sound frequency, the looser the spoke. A "ping" is good; a "clunk" says the spoke is loose.

If one or more spokes are loose, tighten them as described in Chapter Eight.

## Rim Inspection

Frequently inspect the wheel rims. If a rim has been damaged it might have been knocked out of alignment. Improper wheel alignment can cause severe vibration and result in an unsafe riding condition. If the rim portion of an alloy wheel is damaged the wheel must be replaced as it cannot be repaired. Refer to Chapter Eight for rim service.

## BATTERY

*CAUTION*
*If it becomes necessary to remove the battery breather tube when performing any of the following procedures, make sure to route the tube correctly during installation to prevent acid from spilling on parts.*

## Checking Electrolyte Level

The battery is the heart of the electrical system. It should be checked and serviced as indicated. Most electrical system troubles can be attributed to neglect of this vital component.

In order to service the electrolyte level correctly, it is necessary to remove the battery from the frame. On batteries with transparent cases, the electrolyte level should be maintained between the two marks on the battery case. See **Figure 5** and **Figure 6**. On other batteries, remove the battery caps and observe the level in the individual battery cells. If the electrolyte level is low, it's a good idea to remove the battery completely so that it can be thoroughly cleaned, serviced and checked.

Refer to **Figure 7** (typical), for this procedure.

1. Remove all necessary components to gain access to the battery terminals.

2. Disconnect the negative cable from the battery.

3. Disconnect the battery positive cable from the battery.

4. Slide the battery out of the box.

5. Remove the caps from the battery cells and add distilled water to correct the level. Never add electrolyte (acid) or tap water to correct the level.

6. After the level has been corrected and the battery allowed to stand for a few minutes, check the specific gravity of the electrolyte in each cell with a hydrometer (**Figure 8**). Follow the manufacturer's instructions for reading the instrument. See *Battery Testing* in this chapter.

7. After the battery has been refilled, recharged or replaced, install it by reversing these removal steps.

## Cleaning

After the battery has been removed from the bike, check it for corrosion or excessive dirt. The top of the battery in particular should be kept clean. Acid film and dirt will permit current to flow between the terminals, causing the battery to discharge slowly.

For best results when cleaning, first rinse off the top of the battery with plenty of clean water (avoid letting water enter the cells). Then carefully wash the case, both terminals and the battery box with a solution of baking soda and tap water (about 2 tablespoons of baking soda in a 1-lb. coffee can of water is a good ratio). Keep the cells sealed tight with the filler plugs so that none of the cleaning solution enters a cell as this would neutralize the cell's electrolyte and seriously damage the battery. Brush the solution on liberally with a stiff bristle parts cleaning brush. Using a strong spray from a garden hose, clean all the residue from the solution off the battery and all painted surfaces.

## Testing

Hydrometer testing is the best way to check battery condition. Use a hydrometer with numbered graduations from 1.100 to 1.300 rather than one with just color-coded bands. To use the hydrometer, squeeze the rubber ball, insert the tip into the cell and release the ball. Draw enough electrolyte to float the weighted float inside the hydrometer. Note the number in line with the electrolyte surface; this is the specific gravity for this cell. Return the electrolyte to the cell from which it came. The specific gravity of the electrolyte in each battery cell is an excellent indication of that cell's condition. A fully charged cell will read 1.260-1.280 while a cell in good condition reads from 1.230-1.250 and anything below 1.220 is weak.

If the cells test in the poor range, the battery requires recharging. The hydrometer is useful for checking the progress of the charging operation. **Table 5** shows approximate state of charge.

## Charging

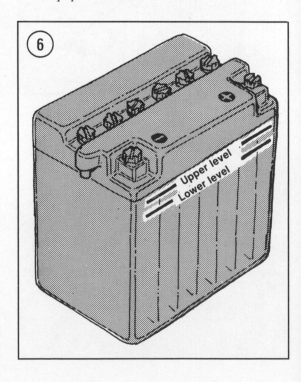

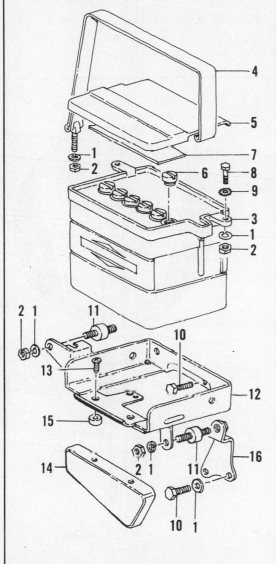

**BATTERY AND CARRIER (TYPICAL)**

1. Lockwasher
2. Nut
3. 12-volt battery
4. Battery strap
5. Cover
6. Battery plug
7. Battery cover cushion
8. Bolt
9. Lockwasher
10. Bolt
11. Rubber mount
12. Battery carrier
13. Screw
14. Cover
15. Damper
16. Support bracket

*WARNING*
*During charging, highly explosive hydrogen gas is released from the battery. The battery should be charged only in a well-ventilated area, and open flames and any other source of ignition should be kept away. Never check the charge of the battery by arcing across the terminals; the resulting spark can ignite the hydrogen gas.*

1. Connect the positive (+) charger lead to the positive battery terminal and the negative (–) charger lead to the negative battery terminal.

2. Remove all vent caps from the battery, set the charger at 12 volts (the same voltage as the battery), and switch it on. If the output of the charger is variable, it is best to select a low setting—1 1/2 to 2 amps.

3. After battery has been charged for about 8 hours, turn the charger off, disconnect the leads and check the specific gravity (**Figure 8**). It should be within the limits specified in **Table 5**. If it is, and remains stable for one hour, the battery is charged.

4. To ensure good electrical contact, cables must be clean and tight on the battery's terminals. If the cable terminals are badly corroded, even after performing the above cleaning procedures, the cables should be disconnected, removed from the bike and cleaned separately with a wire brush and a baking soda solution. After cleaning, apply a very thin coating of petroleum jelly (Vaseline) to the battery terminals before reattaching the cables. After connecting the cables, apply a light coating to the connections also—this will delay future corrosion.

**New Battery Installation**

When replacing the old battery with a new one, be sure to charge it completely (specific gravity, 1.260-1.280) before installing it in the bike.

Failure to do so, or using the battery with a low electrolyte level, will permanently damage the battery.

## PERIODIC LUBRICATION

All lubrication checks and adjustments should be performed at the intervals specified in **Tables 1-3**.

## Oil Tank Level Check

The engine oil runs cooler when the oil level is kept high in the tank. Refer to **Figure 9** for this procedure.

1. Have an assistant support the bike so that it stands straight up. If the bike is supported on the sidestand, an incorrect reading will be obtained.

2. Remove oil tank cap (**Figure 10**) and dipstick. Oil level should be above "Refill" mark. If not, add the recommended oil (**Table 6**) to bring it within one inch from top of oil tank. Do not overfill beyond this point as an air space is required.

*NOTE*
*On all models, 2 quarts can be added when oil level is at "Refill" dipstick mark.*

3. Reinstall dipstick and oil tank cap.

## Engine Oil and Filter Change

The factory-recommended oil and filter change interval is specified in **Tables 1-3**. This assumes that the motorcycle is operated in moderate climates. The time interval is more important than the mileage interval because combustion acids, formed by gasoline and water vapor, will contaminate the oil even if the motorcycle is not run for several months. If a motorcycle is operated under dusty conditions, the oil will get dirty more quickly and should be changed more frequently than recommended.

Use only a detergent oil with an API classification of SE or SF. The classification is stamped on top of the can or on the label (**Figure 11**). Try always to use the same brand of oil. Use of oil additives is not recommended. Refer to **Table 6** for correct viscosity of oil to use under different temperatures.

To change the engine oil and filter you will need the following:

   a. Drain pan.
   b. Funnel.
   c. Can opener or pour spout.
   d. Wrench set.
   e. Engine oil (see **Table 7** for quantity).
   f. Oil filter element.

There are a number of ways to discard the used oil safely. The easiest way is to pour it from the drain pan into a gallon plastic bleach, juice or milk container for proper disposal.

*NOTE*
*Never dispose of motor oil in the trash, on the ground, or down a storm drain. Many service stations accept used motor oil and waste haulers provide curbside used motor oil collection. Do not combine other fluids with motor oil to be recycled. To locate a recycler, contact the American Petroleum Institute (API) at **www.recycleoil.org**.*

1. Place the motorcycle on its sidestand.

2. Start the engine and run it until it is at normal operating temperature, then turn it off.

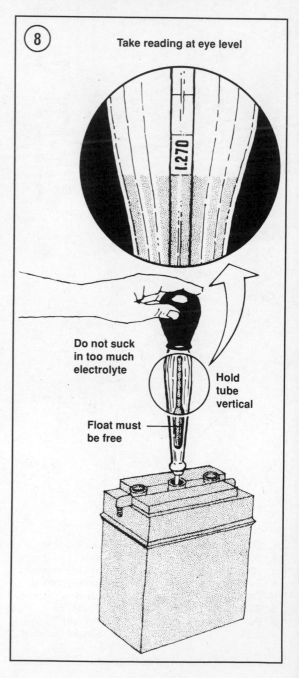

**8**

Take reading at eye level

1.270

Do not suck in too much electrolyte

Hold tube vertical

Float must be free

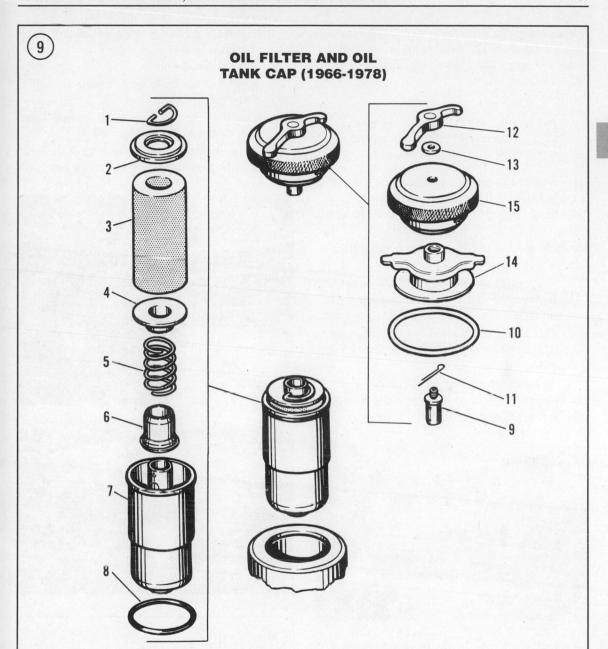

**OIL FILTER AND OIL
TANK CAP (1966-1978)**

1. Filter clip
2. Cap seal washer
3. Filter element
4. Filter lower retainer
5. Cup spring
6. Cup seal
7. Cup
8. O-ring
9. Valve assembly
10. Cap gasket
11. Cap cotter pin
12. Cap screw
13. Cap washer
14. Cap nut
15. Cap top

3. Place a drip pan under the oil tank and remove the drain plug (**Figure 12**). Remove the oil tank cap and dipstick (**Figure 10**); this will speed up the flow of oil.

> *NOTE*
> *Before removing the oil tank cap, thor-*
> *oughly clean off all dirt and oil around*
> *it.*

4. If so equipped, remove the primary case drain plug (**Figure 13**) located underneath the clutch cover and drain the primary case. Reinstall the drain plug and washer.

5A. *1966-1981:* Remove and service the oil filter as follows. See Figure 9.

    a. Remove the oil filter cup from the oil tank.

    b. Remove the clip and washer and remove the oil filter.

    c. Thoroughly clean the oil filter in solvent every 2,000 miles and replace the filter every 5,000 miles.

    d. Reinstall the oil filter into the cup.

5B. *1982-on:* These models use an automotive type spin-on filter. Service the oil filter as follows:

a. Remove the filter (**Figure 14**) with a filter wrench.

b. Discard the oil filter.

c. Wipe the crankcase gasket surface with a clean, lint-free cloth.

d. Coat the neoprene gasket on the new filter with clean oil (**Figure 15**).

e. Screw the filter onto the crankcase *by hand* until the filter gasket just touches the base, i.e.,

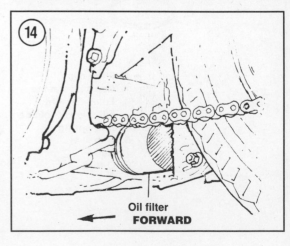

Oil filter
**FORWARD**

until you feel the slightest resistance when turning the filter. Then tighten the filter *by hand* 2/3 more turn.

> *CAUTION*
> *Do not overtighten and do not use a filter wrench or the filter may leak.*

6. At the first 500 miles (after rebuilding engine), and at every second oil change thereafter, flush the oil tank as described in this chapter. During cold weather, the oil tank should be flushed at each oil change.

7. Fill the oil tank with the correct viscosity (**Table 6**) and quantity (**Table 7**) of oil.

8. Reinstall the oil tank cap and dipstick securely.

## Oil Tank Flushing

At the first 500 miles (after rebuilding engine), and at every second oil change during warm riding weather, flush the oil tank before refilling it with new oil. During cold weather, the oil tank should be flushed at each oil change.

1. Drain the oil tank as previously described.

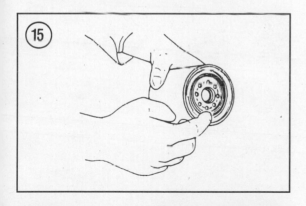

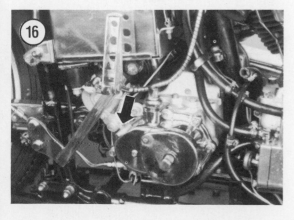

2. ID all oil hoses at the oil tank so you don't mix them up during reassembly. Then remove the oil tank fasteners and remove the oil tank from the bike.

> *CAUTION*
> *Total flushing of the oil tank while it is mounted on the bike is difficult. Sludge broken loose from the tank and not removed during flushing will pass through the main oil feed line and into the oil pump; there it will clog oil passages and cause engine seizure.*

3. Reinstall the oil tank drain plug and plug all tank hose openings.

4. Fill the oil tank 3/4 full with kerosene.

5. Vigorously swish the tank from side to side to break loose sludge and sediment accumulation in the tank.

6. Remove the dipstick from the top of the tank and drain the tank. Using a small flashlight, check the tank for sludge and sediment deposits that did not flush out. Repeat until *all* of these deposits are removed. If necessary, break hard deposits loose with a wooden dowel or similar tool inserted into the tank.

7. When the tank is clean, pour some clean engine oil into the tank and shake the tank once again to cover the tank walls with the oil. Then drain and discard the oil.

8. Clean the filler cap/dipstick assembly before installing it back into the tank.

9. Remove the plugs from the oil tank hoses and reinstall the oil tank. Tighten all tank mounting bolts or nuts securely. Wipe the oil hoses off before reconnecting them onto the tank. Reconnect the oil hoses, following your ID marks made prior to removing the hoses.

## Transmission Oil Check

Remove the filler cap. The oil should be level with the filler cap opening (**Figure 16**) or the level plug opening. If not, add oil until the level is correct. Reinstall the filler cap and gasket.

## Transmission Oil Change

Change the transmission oil at the intervals specified in **Tables 1-3**.

1. Set the bike level with both wheels on the ground.
2. Remove the oil filler plug from the top of the side cover (**Figure 16**).
3. Place a drip pan underneath the transmission.
4. Remove the drain plug and gasket at the bottom of the transmission (**Figure 17**). Allow the oil to drain for 10 minutes.
5. Clean the drain plug and install it and its gasket in the cover.
6. Refill the transmission through the side cover hole (**Figure 16**) with the recommended type (**Table 6**) and quantity (**Table 7**) of transmission oil.

## Front Fork Oil change

The fork oil should be changed at the intervals specified in **Tables 1-3**.
1. On 1966-1978 models, remove the front wheel as described in Chapter Eight.
2. Place a drip pan under the fork and remove the fork slider drain screw (**Figure 18**). Apply the front brake lever and push down on the forks and release. Repeat this procedure until all of the fork oil is drained.

> *CAUTION*
> *Do not allow the fork oil to come in contact with any of the brake components.*

3. Install the slider drain screw (**Figure 18**).
4A. On FL models with adjustable forks, perform the following:
　　a. Remove the front fork side cover panels or headlight housing as required.
　　b. Remove the filler screw (27, **Figure 19**) from the fork cap (26, **Figure 19**).
4B. On all models except FL with adjustable forks, remove the fork cap (**Figure 20**).
5. Fill the fork tube with the correct amount (**Table 8**) and viscosity (**Table 6**) fork oil. On FL models with adjustable forks, an oil filler container can be constructed with a 1 quart tin can as shown in **Figure 21** to facilitate filling the fork tubes through the filler plug hole. **Figure 22** shows how to use the filler container.

> *NOTE*
> *If fork has been disassembled, refill with "dry" quantity; otherwise refill with "wet" quantity.*

> *NOTE*
> *In order to measure the correct amount of fluid, use a plastic baby bottle. These bottles have measurements in fluid ounces (oz.) and cubic centimeters (cc) imprinted on the side.*

6. Check the condition of the fork cap O-ring (if so equipped), and replace it if necessary.
7. Install the fork cap or filler screw.
8. Repeat Steps 2-7 for the opposite side.
9. Road test the bike and check for oil leaks.

## Control Cables

The control cables should be lubricated at the intervals specified in **Tables 1-3**. At this time, they should also be inspected for fraying, and the cable sheath should be checked for chafing. The cables are relatively inexpensive and should be replaced when found to be faulty.

3

## ⑲ STEERING ASSEMBLY (FL ADJUSTABLE FORK)

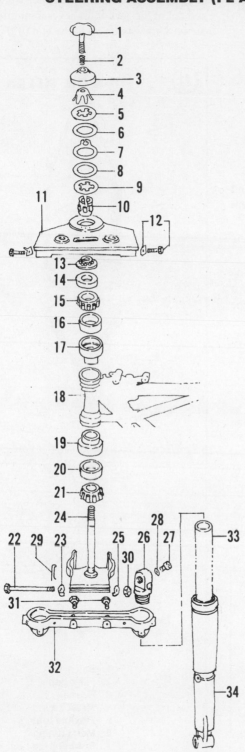

1. Steering damper adjusting screw
2. Spring
3. Spider spring cover
4. Spider spring
5. Pressure disc
6. Friction washer
7. Anchor plate
8. Friction washer
9. Pressure disc
10. Fork stem nut
11. Upper bracket
12. Bolt and washer
13. Head bearing nut
14. Dust shield
15. Bearing cone
16. Bearing race
17. Bearing cup
18. Frame
19. Bearing cup
20. Bearing race
21. Bearing cone
22. Bracket bolt
23. Bracket bolt washer
24. Steering stem and lower bracket
25. Bracket bolt washer
26. Fork cap
27. Filler screw
28. O-ring
29. Cotter pin
30. Nut
31. Bracket clamp bolt
32. Lower bracket
33. Fork tube
34. Slider

They can be lubricated with any of the popular cable lubricants and a cable lubricator. The second method requires more time and complete lubrication of the entire cable is less certain.

> *NOTE*
> *The main cause of cable breakage or cable stiffness is lack of lubrication. Maintaining the cables as described in this section will assure long service life.*

### Lubricator method

1. Disconnect the clutch and cable from the left-hand side handlebar. Disconnect the throttle cable(s) from the throttle grip as described in Chapter Eight.
2. Attach a lubricator to the cable following the manufacturer's instructions.
3. Insert the nozzle of the lubricant can into the lubricator, press the button on the can and hold it down until the lubricant begins to flow out of the other end of the cable. See **Figure 23**.

> *NOTE*
> *Place a shop cloth at the end of the cable(s) to catch all excess lubricant that will flow out.*

4. Remove the lubricator, reconnect and adjust the cable(s) as described in this chapter.

### Oil method

1. Disconnect the cables as previously described.
2. Make a cone of stiff paper and tape it to the end of the cable sheath (**Figure 24**).

3. Hold the cable upright and pour a small amount of light machine oil into the cone. Work the cable in and out of the sheath for several minutes to help the oil work its way down to the end of the cable.

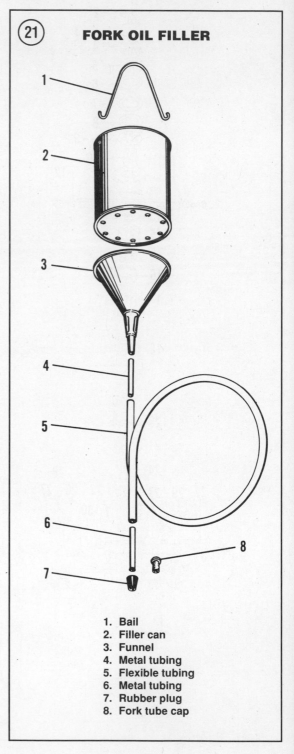

**FORK OIL FILLER**

1. Bail
2. Filler can
3. Funnel
4. Metal tubing
5. Flexible tubing
6. Metal tubing
7. Rubber plug
8. Fork tube cap

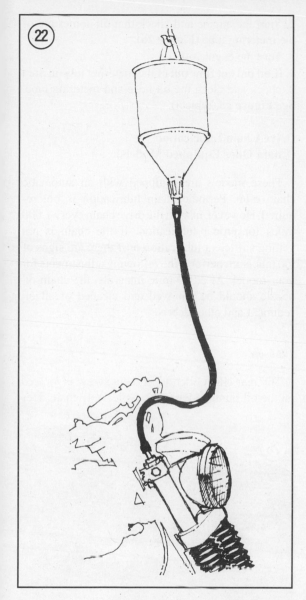

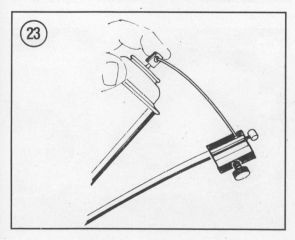

4. Remove the cone, reconnect the cable and adjust the cable(s) as described in this chapter.

### Swing Arm Bearings

Repack the rear swing arm bearings with a lithium-base, waterproof wheel bearing grease.

Refer to *Swing Arm, Removal/Installation* in Chapter Nine for complete details.

### Circuit Breaker Cam

The circuit breaker cam (1966-1979) should be lightly lubricated with special breaker cam grease (**Figure 25**) at the intervals specified in **Tables 1-3**. Refer to Chapter Seven for procedures on removing and installing ignition components.

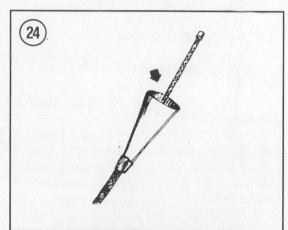

3

## Brake Cam Lubrication

Lubricate the brake cam whenever the rear wheel is removed (if required).

1. Remove the front or rear wheel as described in Chapter Nine.

2. Wipe away the old grease, being careful not to get any on the brake shoes.

3. Sparingly apply high-temperature grease to the camming surfaces of the camshaft, the camshaft groove, the brake shoe pivots and the ends of the springs. *Do not* get any grease on the brake shoes.

> ### WARNING
> *Make sure to use only a high-temperature grease. Because of the temperatures created by braking conditions, other types of grease may thin considerably and run onto the brake shoes, causing a loss of braking.*

4. Reassemble the rear wheel and install it. See Chapter Eight or Chapter Nine.

## Speedometer/Tachometer Cable Lubrication

Lubricate the cable every year or whenever needle operation is erratic.

1. Remove the cable from the instrument cluster.

2. Pull the cable from the sheath.

3. If the grease is contaminated, thoroughly clean off all old grease.

4. Thoroughly coat the cable with a good grade of multi-purpose grease and reinstall into the sheath.

5. Make sure the cable is correctly seated into the drive unit. If not, it will be necessary to disconnect the cable at its lower connection and reattach.

## Primary Chain Lubrication

### 1966-on

The primary chain is lubricated through a metering tube (**Figure 26**) connected to an oil line attached to the oil pump. Oil flow is controlled by a fixed metering orifice; adjustment of the oil flow is not possible. Excess oil that collects in the primary cover is drawn back into the engine through the gearcase breather. Whenever the primary chain is adjusted, check to see that oil drops out of the metering tube as follows.

1. Remove the adjustment cover (**Figure 27**).

2. Start the engine and check that oil comes out of the metering tube (**Figure 26**).

3. Turn the engine off.

4. If oil did not flow out of the metering tube in Step 2, check and clean the oil hose and metering tube. See **Figure 28** (typical).

## Drive Chain Lubrication (Chain Oiler Equipped Models)

These models are equipped with an automatic chain oiler. Periodic chain lubrication is not required. However, inspect the drive chain every 1,000 miles for proper lubrication. If the chain is not getting sufficient lubrication or if there are signs of too much oil, perform the following adjustments for your model. At 2,000-mile intervals, the chain oil nozzle should be removed and cleaned of all oil sediment and other debris.

### 1966-on

The rear chain oiler adjustment screw is located on the oil pump (**Figure 29**, typical). If chain does

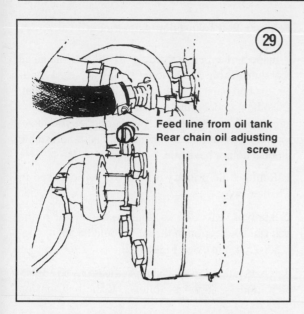

Feed line from oil tank
Rear chain oil adjusting screw

not appear to be receiving enough oil, clean and adjust chain oiler as follows.

1. Loosen the rear chain oiler adjustment screw.

2. Turn adjustment screw in until it seats lightly. Count and record number of turns required.

3. Remove adjusting screw by turning it out completely.

4. Blow out the chain oiler orifice with compressed air.

5. Reinstall the adjusting screw, then turn it in until it seats lightly.

6. Back out adjusting screw amount recorded in Step 2.

7. Oil should flow at approximately 2 to 3 drops per minute. This flow rate should occur when screw is backed out about 1/4 turn.

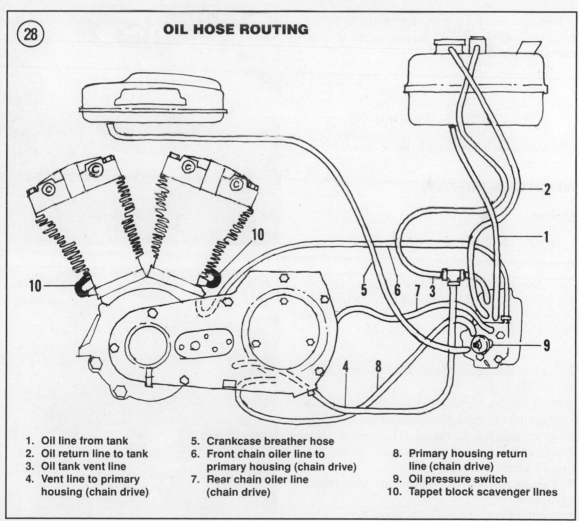

## OIL HOSE ROUTING

1. Oil line from tank
2. Oil return line to tank
3. Oil tank vent line
4. Vent line to primary housing (chain drive)
5. Crankcase breather hose
6. Front chain oiler line to primary housing (chain drive)
7. Rear chain oiler line (chain drive)
8. Primary housing return line (chain drive)
9. Oil pressure switch
10. Tappet block scavenger lines

## Drive Chain Lubrication (Non-Automatic Chain Oiler Models)

On models not equipped with an automatic rear chain oiler, oil the drive chain every 300 miles or sooner if it becomes dry. A properly maintained chain will provide maximum service life and reliability.

1. Support the bike so that the rear wheel clears the ground.

2. Shift the transmission to NEUTRAL.

2. Oil the bottom run of the chain with a commercial chain lubricant. Concentrate on getting the lubricant down between the side plates, pins, bushings and rollers of each chain link. Rotate the wheel and oil the entire chain.

## Miscellaneous Lubrication Points

Lubricate the clutch lever, front brake lever, rear brake lever, sidestand pivot and footrest pivot points. Use SAE 10W-30 motor oil.

## PERIODIC MAINTENANCE

Maintenance intervals are listed in **Tables 1-3**.

## Primary Chain Adjustment

*1966-on*

1. Remove the inspection cover (**Figure 27**).

> *NOTE*
> *The primary chain can be adjusted through the inspection hole (**Figure 30**). However, this procedure shows the adjustment with the primary cover removed for clarity.*

2. Loosen the chain adjuster shoe center bolt (A, **Figure 31**).

3. Move the shoe support (B, **Figure 31**) up or down to adjust tension chain play. Free vertical movement in upper chain run (C, **Figure 31**) should be 5/8 to 7/8 in. (15.8-22.2 mm) (cold engine) or 3/8 to 5/8 in. (9.5-15.8 mm) (hot engine).

4. Tighten the bolt (A, **Figure 31**) and recheck adjustment.

5. Install the inspection cover (**Figure 29**).

## Primary Belt Inspection/Adjustment

Inspect the belt according to the maintenance schedule (**Table 3**). On some 1980-early 1981 models, the belt must be replaced when slack is too great. On other models, the primary belt tension can be adjusted. If in doubt about which model you have, look for an adjuster mechanism. Non-adjustable models do not have one.

1. Remove the primary drive cover as described in Chapter Five.

2. Apply a 10 lb. (4.5 kg) force to the middle of the belt run. Deflection of the belt should be 5/8 to 1 in. (15.8-25.4 mm). See **Figure 32**.

3. If belt deflection exceeds specifications on adjustable models, adjust as follows. If the belt tension

cannot be adjusted on your model, replace the belt as described in Chapter Five.

4. Remove the safety wire from the 2 bolts securing the primary drive housing to the crankcase (**Figure 33**). Then loosen the 4 primary case-to-crankcase bolts.

5. Loosen the 4 bottom nuts securing the transmission to the transmission mounting plate.

6. Loosen the 2 starter motor and bracket mounting nuts on the right end of the starter, at the top.

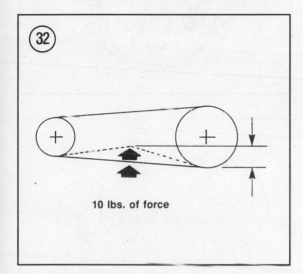

**10 lbs. of force**

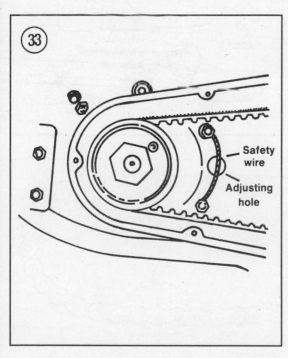

Safety wire

Adjusting hole

7. To tighten the belt, pry against the crankcase with a screwdriver inserted through the hole in the primary drive housing (**Figure 33**).

> *NOTE*
> *Some late 1981 models have no hole drilled in the primary housing. On these models, pry carefully between the alternator rotor and the primary housing.*

8. When belt tension is correct, tighten the 2 inner crankcase mounting bolts lightly. Recheck belt tension and readjust if necessary. Torque the 4 crankcase bolts to 19 ft.-lb. (26.2 N•m)

9. Install new safety wire through the 2 inner crankcase bolts.

> *CAUTION*
> *If you do not install safety wire through these bolts, a bolt could back out unnoticed and cause severe engine damage.*

10. Torque the 4 bottom transmission/plate nuts and the one transmission/frame nut to 20 ft.-lb. (27.6 N•m).

11. Tighten the 2 starter mounting nuts.

12. After adjusting primary belt tension, check and adjust the following as described in this chapter.

   a. Final drive belt.

   b. Rear brake pedal free play.

   c. Clutch.

13. Adjust the shift linkage as described in Chapter Five.

## Drive Chain Inspection/Adjustment

> *NOTE*
> *As drive chains stretch and wear in use, the chain will become tighter at one point. The chain must be checked and adjusted at this point.*

1. Turn the rear wheel by pushing the bike and check the chain for its tightest point. Mark this spot and turn the wheel so that the mark is located on the chain's lower run, midway between the drive sprockets. Check and adjust the drive chain as follows.

2. Have a rider mounted on the seat.

3. Push the chain up midway between the sprockets on the lower chain run and check the free play (**Figure 34**); it should be 1/2 in. (12.7 mm).

4. If the chain adjustment is incorrect, adjust it as follows.

5A. *1966-1972*—Referring to **Figure 35**, perform the following:

    a. Loosen the axle nut (1, **Figure 35**).

    b. Loosen the axle adjuster locknut (2) on both sides of the wheel.

    c. Loosen the brake anchor stud nut.

    d. Turn each adjustment stud (3) clockwise to take up slack in the chain. To loosen the chain, turn each adjustment stud counterclockwise. Be sure to turn each adjusting stud equally to maintain rear wheel alignment. Adjust the chain until the lower chain run moves 1/2 in. (12.7 mm) up and down (**Figure 34**).

    e. Check rear wheel alignment by sighting along chain as it runs over the rear sprocket. It should not appear to bend sideways.

    f. Tighten all nuts securely.

5B. *1973-on:*

    a. Loosen the rear axle nut (**Figure 36**).

    b. Loosen the anchor nut.

    c. Loosen the axle adjuster locknuts.

    d. Turn each axle adjuster in or out as required, in equal amounts to maintain rear wheel alignment. The correct amount of chain free play is 1/2 in. (12.7 mm).

    e. Tighten all nuts.

6. If you cannot adjust the drive chain within the limits of the chain adjusters, it is excessively worn and stretched and must be replaced. If the chain can be adjusted, but you feel it may be worn, pull the chain away from the rear sprocket as shown in **Figure 37**. If more than 1/2 of a sprocket tooth is exposed, the chain is worn. Always replace both sprockets when replacing the drive chain; never

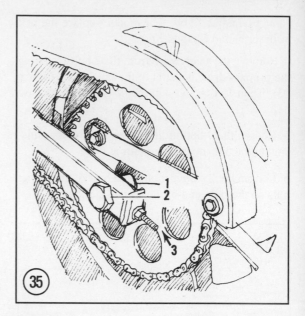

Washer
Lockwasher
Axle nut

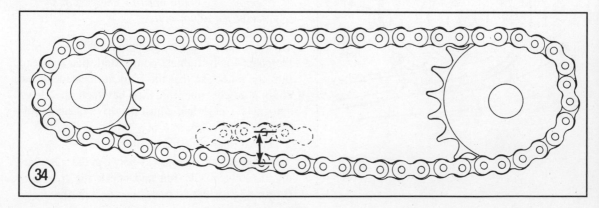

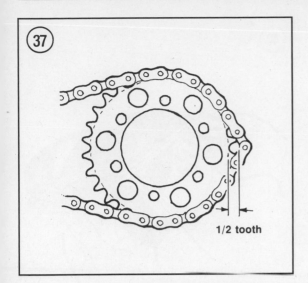

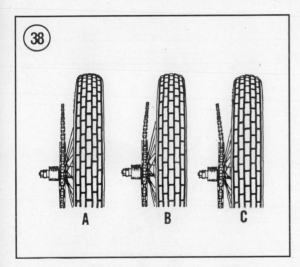

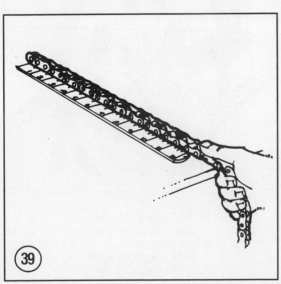

install a new chain over worn sprockets. Replace the drive chain as described in Chapter Nine.

> *WARNING*
> *Excessive free play or a worn chain can result in chain breakage; this could cause a serious accident.*

7. Sight along the top of the drive chain from the rear sprocket to see that it is correctly aligned. It should leave the top of the rear sprocket in a straight line. If it is cocked to one side, the rear wheel must be realigned. See **Figure 38**.

8. After the drive chain has been adjusted, the rear brake pedal free play must be adjusted as described in this chapter.

## Drive Chain Cleaning, Inspection and Lubrication

The drive chain should be cleaned and inspected every 3,000 miles or more frequently if ridden in dusty conditions.

1. Remove the drive chain as described in Chapter Nine.

2. Immerse the chain in a pan of cleaning solvent and allow it to soak for about a half hour. Move it around and flex it during this period so that dirt between the pins and rollers may work its way out.

3. Scrub the rollers and side plates with a still brush and rinse away loosened grit. Rinse the chain a couple of times to make sure all dirt is washed out. Hang up the chain and allow it to dry thoroughly.

4. After cleaning the chain, examine it for the following conditions:

    a. Excessive wear.

    b. Loose pins.

    c. Damaged rollers.

    d. Damaged plates.

    e. Dry plates.

If any of these conditions are visible, replace the chain.

5. Lay the chain on a bench and push all links together so chain is shorter. Measure and record its length. See **Figure 39**.

6. Stretch chain to remove all slack between links. Measure chain again (**Figure 39**).

7. If there is one inch or more between measurements obtained in Step 5 and Step 6, replace the chain.

*CAUTION*
*Always check both sprockets every time the drive chain is removed. If any wear is visible on the teeth, replace the sprockets. Never install a new chain over worn sprockets or a worn chain over new sprockets.*

## Final Drive Belt Inspection/Adjustment

The final drive belt stretches very little after the first 500 miles of operation, but it should be inspected for tension and alignment according to the maintenance schedule (**Table 3**).

1. Remove the belt guard bolts and remove the belt guard.

2. On FLH models, remove the left-hand saddlebag.

3. Have a normal weight rider sit on the seat with the wheels on the ground.

4. Apply 10 lb. (4.5 kg) of force to the middle of the belt run. Deflection of the belt should be 5/8 to 3/4 in. (15.8-19 mm) (**Figure 32**).

5. Check the sprocket alignment. Place a straightedge along the side of the rear sprocket near the top (**Figure 40**). There should be an equal space between the belt and the straightedge all along the belt.

6. If the belt tension or alignment is out of specification, adjust as follows.

*WARNING*
*When adjusting the final drive belt, rear wheel and sprocket alignment must be maintained. A misaligned rear wheel will drastically shorten belt life and it may cause poor handling and pulling to one side or the other. Once the alignment is set correctly, if both adjusters are moved an equal amount the rear wheel will be aligned correctly.*

7. Loosen the axle nut (3, **Figure 41**).

8. Remove the brake anchor cotter pin and loosen the anchor nut(4).

9. If the belt was too tight, back out both adjuster nuts (1) an equal amount and kick the rear wheel forward or tap on the edges of the axle until the belt is looser than desired.

10. Turn both adjuster nuts in inward equal amount gradually until the belt tension free play is 5/8 to 3/4 in. (15.8-19 mm). If sprocket realignment is required, turn one adjuster more than the other.

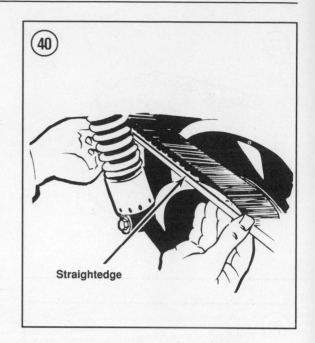

Straightedge

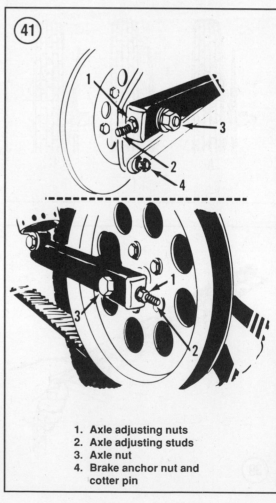

1. Axle adjusting nuts
2. Axle adjusting studs
3. Axle nut
4. Brake anchor nut and cotter pin

11. Torque the axle nut to 50 ft.-lb. (69 N•m) and recheck belt tension and alignment.

12. Tighten the brake anchor nut and install a new cotter pin. Spread the ends of the cotter pin to lock it.

13. Install the belt guard and left-hand saddlebag on FLH models.

14. If the condition of the final drive belt is questionable, refer to Chapter Nine for additional information.

### Disc Brake Inspection

The hydraulic brake fluid in the disc brake master cylinder should be checked every month. The disc brake pads should be checked at the intervals specified in **Tables 1-3**. Replacement is described in Chapter Ten.

### Disc Brake Fluid Level

1A. *Front brake:* The fluid level in the reservoir should be level with the gasket surface (**Figure 42**). To check, level the master cylinder assembly by turning the handlebar assembly to the left.

1B. *Rear brake:* The fluid level in the reservoir should be 1/8 in. (3.2 mm) below the gasket surface. To check, level the bike.

2. Wipe the master cylinder cover with a clean shop cloth.

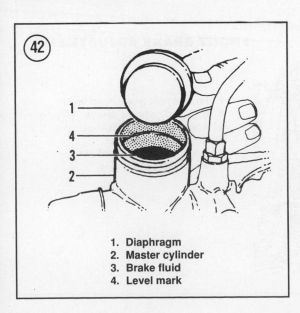

1. Diaphragm
2. Master cylinder
3. Brake fluid
4. Level mark

3. Remove the cover screws and cover and lift the diaphragm out of the housing. If necessary, correct the level by adding fresh brake fluid.

> **WARNING**
> *Harley-Davidson specifies DOT 3 brake fluid for models produced prior to September, 1976 and DOT 5 for later models. Mixing the two types of brake fluids can cause brake failure. If you own a 1976 or 1977 model take your frame number to a Harley-Davidson dealer to find out your bike's production date.*

> **CAUTION**
> *Be careful not to spill DOT 3 brake fluid on painted or plated surfaces as it will destroy the surface. Wash immediately with soapy water and thoroughly rinse it off.*

4. Reinstall all parts.

> **WARNING**
> *If the brake fluid was so low as to allow air into the hydraulic system, the brakes will have to be bled. Refer to **Bleeding The System** in Chapter Ten. A level this low indicates badly worn pads, a fluid leak or both. Correct the situation immediately.*

### Disc Brake Lines and Seals

Check brake lines between the master cylinder and the brake caliper. If there is any leakage, tighten the connections and bleed the brakes as described in Chapter Ten. If this does not stop the leak or if a line is obviously damaged, cracked or chafed, replace the line and seals and bleed the brake.

### Disc Brake Pad Wear (Front and Rear)

Brake pads must be inspected for excessive or uneven wear, scoring and oil or grease on the friction surface. Because the disc brake units are not equipped with a wear indicator, the caliper must be partially disassembled to measure the brake pad thickness. See Chapter Ten.

*NOTE*
*A low fluid level in the master cylinder*
*may indicate pad wear.*

### Disc Brake Fluid Change

Every time you remove the reservoir cap a small amount of dirt and moisture enters the brake fluid. The same thing happens if a leak occurs or when any part of the hydraulic system is loosened or disconnected. Dirt can clog the system and cause unnecessary wear. Water in the fluid vaporizes at high temperatures, impairing the hydraulic action and reducing brake performance.

To change brake fluid, refer to *Bleeding the System* in Chapter Ten. Continue adding new fluid to the master cylinder and bleeding at the calipers until the fluid leaving the calipers is clean and free of contaminants and air bubbles.

*WARNING*
*Harley-Davidson specifies DOT 3 brake fluid for models produced prior to September, 1976 and DOT 5 for later models. Mixing the two types of brake fluids can cause brake failure. If you own a 1976 or 1977 model take your frame number to a Harley-Davidson dealer to find out your bike's production date.*

### Drum Brake Lining (Front and Rear)

The front and rear brake pads must be inspected for excessive or uneven wear, wear down to the rivet heads, scoring and oil or grease on the friction surface. Install new brake linings or replace the brake shoes and linings as described in Chapter Ten.

Because the drum brake units are not equipped with a wear indicator, the wheels must first be removed to inspect the brake linings. See Chapter Eight or Chapter Nine.

### Front Brake Adjustment
### (1966-1972 Drum Brake)

The front brake on these models requires periodic adjustment. See **Figure 43** for this procedure.

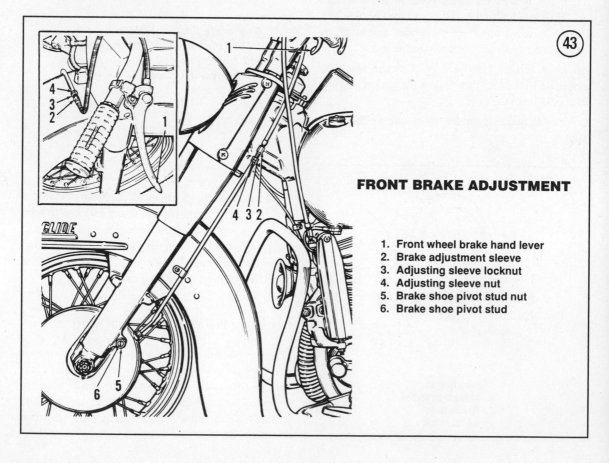

**FRONT BRAKE ADJUSTMENT**

1. Front wheel brake hand lever
2. Brake adjustment sleeve
3. Adjusting sleeve locknut
4. Adjusting sleeve nut
5. Brake shoe pivot stud nut
6. Brake shoe pivot stud

1. Support bike so that front wheel clears ground.

2. Loosen the adjusting sleeve locknut (3, **Figure 43**).

3. Turn the adjusting sleeve nut (4) clockwise toward the cable support to tighten brake or turn sleeve nut (4) counterclockwise to loosen brake. Hand lever should move approximately 1/4 of its total travel before brake starts to operate.

4. Tighten the adjusting sleeve locknut (3). Recheck the adjustment.

5. Spin the wheel to make sure the brake does not drag. If the brake drags with the proper adjustment, proceed with Step 6.

6. Loosen the brake shoe pivot stud (6) and the front axle nut. Do not remove them.

7. Spin the wheel. While wheel is spinning, apply front brake, then tighten pivot stud (6) and axle nut.

8. Recheck brake adjustment as specified in Steps 2-3.

### Front Disc Brake Adjustment

The front disc brake does not require periodic adjustment.

### Rear Drum Brake Adjustment

1. Support motorcycle so that rear wheel clears the ground.

2. Referring to **Figure 44**, locate adjusting cam nuts on outside of brake side cover.

3. Turn forward cam nut counterclockwise until wheel has noticeable drag.

4. Turn the wheel forward and backward. This will center the forward brake shoe.

5. Turn the front cam nut clockwise until the wheel turns freely.

6. Turn rear cam nut clockwise until wheel has noticeable drag.

7. Turn the wheel forward and backward. This will center the rear brake shoe.

8. Turn the rear cam nut counterclockwise until the wheel turns freely.

9. Brake adjustment is now complete.

### Rear Brake Pedal Adjustment

The rear brake pedal should be adjusted anytime the rear brake pedal is removed or when the brake linings or shoes are replaced.

#### *1966-1969*

1. Place the motorcycle on the sidestand.

2. Check that the brake pedal is in the at-rest position.

3. Work the brake pedal by hand. When the brake pedal is adjusted correctly, the pedal will move 1 1/2 in. (38.1 mm) before the brake takes effect.

4. If necessary, adjust as follows.

5. Slide the rubber boot (**Figure 45**) away from the master cylinder housing.

6. Loosen the locknut (**Figure 45**) and turn the plunger to lengthen or shorten the piston rod as

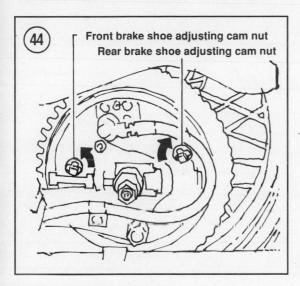

(44) Front brake shoe adjusting cam nut
Rear brake shoe adjusting cam nut

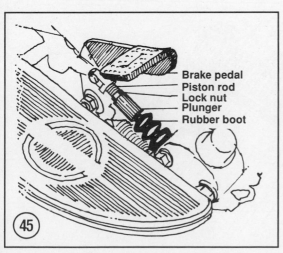

(45) Brake pedal
Piston rod
Lock nut
Plunger
Rubber boot

required to obtain the correct pedal movement (Step 3).

7. Tighten the locknut and recheck the adjustment.

### *1970-on FL*

1. Place the motorcycle on the sidestand.

2. Check that the brake pedal is in the at-rest position.

3. Work the brake pedal by hand. When the brake pedal is adjusted correctly, the pushrod will move approximately 1/16 in. (1.58 mm) before it contacts the master cylinder piston.

4. If necessary, adjust as follows.

5A. *1970-early 1979:*

  a. Loosen the rear master cylinder bolt and the brake pedal stop bolt (**Figure 46**).

  b. Move the pedal stop plate (**Figure 46**) up to increase free play or down to decrease it.

  c. Tighten all bolts securely.

5B. *Late 1979-on:*

  a. Loosen the stop bolt locknut (**Figure 47**) at the brake pedal.

  b. Turn the stop bolt (**Figure 47**) to obtain 1/16 in. (1.58 mm) master cylinder plunger free play.

  c. Tighten the locknut.

### *1972-on FX*

1. Place the motorcycle on the sidestand.

2. Check that the brake pedal is in the at-rest position.

3. Work the brake pedal by hand. When the brake pedal is adjusted correctly, the pushrod will move approximately 1/16 in. (1.58 mm) before it contacts the master cylinder piston.

4. If necessary, adjust as follows.

5. Loosen the locknut (**Figure 48**) and turn the brake rod forward to increase free play or rearward to decrease free play.

6. Tighten the locknut.

### *1983 and earlier FXWG*

1. Place the motorcycle on the sidestand.

2. Check that the brake pedal is in the at-rest position.

3. Work the brake pedal by hand. When the brake pedal is adjusted correctly, the pushrod will move

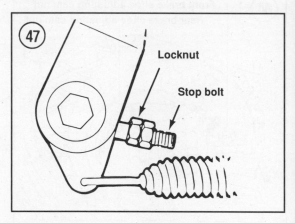

**47**

Locknut

Stop bolt

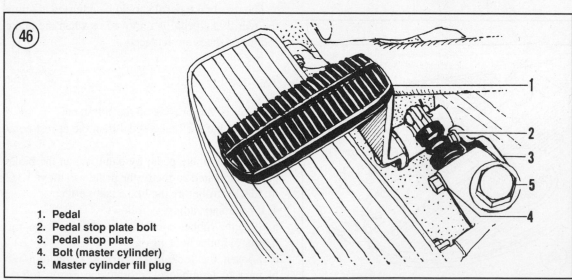

**46**

1. Pedal
2. Pedal stop plate bolt
3. Pedal stop plate
4. Bolt (master cylinder)
5. Master cylinder fill plug

approximately 1/16 in. (1.58 mm) before it contacts the master cylinder piston.

4. If necessary, adjust as follows.

5. Loosen the locknut and adjust the stop bolt (**Figure 48**) counterclockwise to increase free play or clockwise to decrease it.

6. Tighten the locknut.

### *1984 FXWG and FXDG*

1. Place the motorcycle on the sidestand.

2. Check that the brake pedal is in the at-rest position.

3. Work the brake pedal by hand. When the brake pedal is adjusted correctly, the pushrod will move

approximately 1/16 in. (1.58 mm) before it contacts the master cylinder piston.

4. If necessary, adjust as follows.

5. Loosen the brake pedal jam nut (**Figure 49**).

6. Turn the stop bolt to obtain a suitable brake pedal height. Check height while sitting on bike with your right foot in position. Tighten the jam nut.

7. Loosen the locknut and turn the clevis rod (**Figure 49**) counterclockwise to increase free play or clockwise to decrease it.

### Foot Clutch Control Adjustment (**1966-early 1978**)

Refer to **Figure 50** for this procedure.

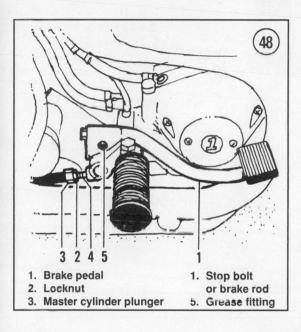

48

1. Brake pedal
2. Locknut
3. Master cylinder plunger

1. Stop bolt or brake rod
5. Grease fitting

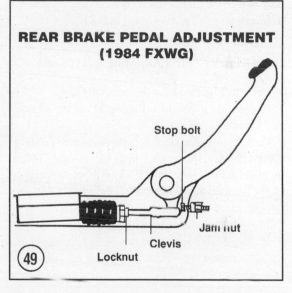

**REAR BRAKE PEDAL ADJUSTMENT (1984 FXWG)**

Stop bolt

Jam nut

Clevis

Locknut

49

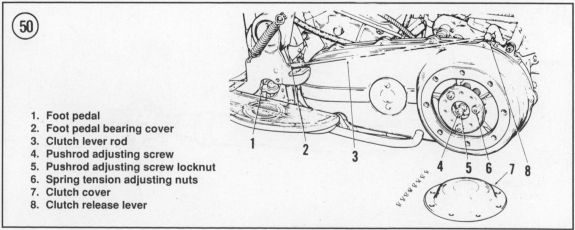

50

1. Foot pedal
2. Foot pedal bearing cover
3. Clutch lever rod
4. Pushrod adjusting screw
5. Pushrod adjusting screw locknut
6. Spring tension adjusting nuts
7. Clutch cover
8. Clutch release lever

1. Disengage the clutch by pushing the foot pedal (1, **Figure 50**) all the way down (heel down). Check that the clutch lever (8) just touches the transmission cover.

2. Loosen the clutch lever rod (3) locknuts and turn the lever rod adjuster so that the rod just clears the foot pedal bearing cover (2). The rod should not be bent down by the bearing cover. Tighten the locknut and recheck the clutch lever rod position.

3. Remove the screws securing the clutch cover (7) and remove the cover.

4. Move the foot pedal (1) to the fully engaged position (toe down).

5. Loosen the pushrod adjusting screw locknut (5). Next, turn the pushrod adjusting screw (4) so that the end of the clutch lever has approximately 1/4 in. (6.3 mm) (1966-1969) or 1/8 in. (3.2 mm) (1970-on) free play before the clutch disengages.

6. Install all parts previously removed.

## Hand Clutch Control Adjustment (1968-on)

Refer to **Figure 51** for this procedure.

1. Loosen the locknut (2, **Figure 51**) at the engine. See **Figure 52**.

2. Turn the threaded sleeve (1) as required to produce approximately 1/4 in. (6.3 mm) (1968-early 1978) or 1/16 in. (1.6 mm) (late 1978-on) free movement of the clutch hand lever before the clutch starts to release.

3. Tighten the locknut.

4. If the sleeve adjustment has been taken up, perform the following adjustment.

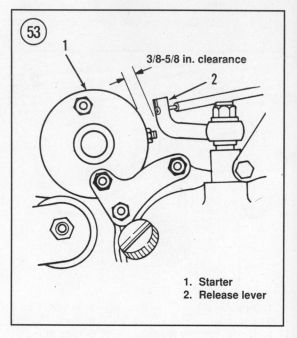

3/8-5/8 in. clearance

1. Starter
2. Release lever

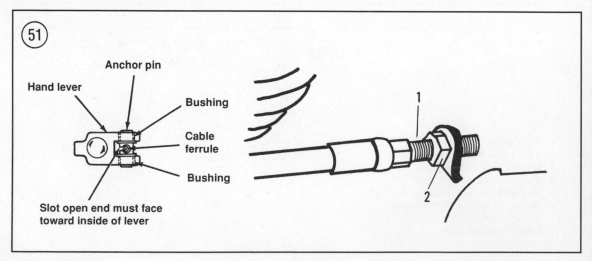

Hand lever
Anchor pin
Bushing
Cable ferrule
Bushing
Slot open end must face toward inside of lever

5. Move the release lever on transmission as far forward as possible. See **Figure 53** (1968-early 1979) or **Figure 54** (late 1979-on).

6. Measure the clearance at the clutch release lever as indicated in **Figure 53** or **Figure 54**. Clearance should be as follows:

    a. 1968-early 1979: 3/8 to 5/8 in. (9.5-15.8 mm)

    b. Late 1979-on: 13/16 in. (20.6 mm)

If the clearance is incorrect, perform the following.

7. Loosen the locknut (2, **Figure 51**) and turn the sleeve (1, **Figure 51**) all the way into its bracket.

8. Remove the clutch adjustment cover (**Figure 55**).

9. Loosen the pushrod locknut (1, **Figure 56**).

10. Turn the clutch adjusting screw (2, **Figure 56**) counterclockwise to remove tension on the pushrod.

11. Turn the adjusting sleeve (1, **Figure 51**) until the proper clearance specified in **Figure 53** or **Figure 54** is obtained. Tighten the locknut (2, **Figure 51**).

12. Turn the clutch adjusting screw (2, **Figure 56**) clockwise until it contacts the pushrod, then back it out 1/8 turn. Tighten the locknut (1, **Figure 56**).

13. Install the clutch adjustment cover.

14. Perform Steps 1-3 to adjust the clutch cable.

### Hand Clutch Booster Adjustment (1969 and Earlier Models)

Check the booster adjustment when the clutch hand lever operation becomes difficult, or if improper clutch operation is indicated. Refer to **Figure 57** for this procedure.

1. Loosen the locknut (7, **Figure 57**).

2. Adjust the clutch lever rod (5) so that the clutch release lever (8, **Figure 50**) has approximately 1/2 in. free movement. Move lever forward until all slack in mechanism is taken up.

3

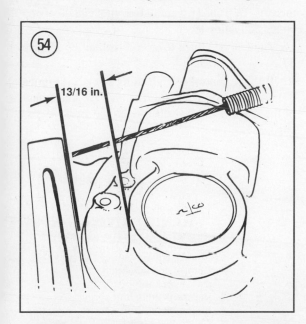

(54)

13/16 in.

(55)

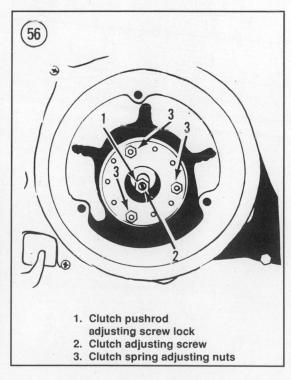

(56)

1. Clutch pushrod adjusting screw lock
2. Clutch adjusting screw
3. Clutch spring adjusting nuts

3. The distance between the chain housing and clutch lever rod should be 1/4 in. (6.3 mm). If specifications are incorrect, perform Steps 4-20.

4. Remove the clutch inspection cover (7, **Figure 50**) or the chain cover, as required.

5. Loosen the pushrod screw locknut (5, **Figure 50**).

6. Turn the pushrod adjusting screw (4, **Figure 50**) as required to obtain the clearance specified in Step 3 for your model.

7. Tighten the locknut (5, **Figure 50**).

8. Loosen the control coil adjusting sleeve locknut (2, **Figure 57**).

9. Turn the adjusting sleeve (2, **Figure 57**) until there is approximately one inch free play at the clutch hand lever.

10. Loosen the bellcrank adjusting screw locknut (4, **Figure 57**).

11. Turn the bellcrank adjusting screw (3, **Figure 57**) until the bellcrank (8, **Figure 57**) does not cross dead center when moved back and forth by hand.

*NOTE*
*Dead center is indicated in* **Figure 57**.

12. Loosen the upper adjuster nut (13, **Figure 57**) as far as it will go.

13. Turn the bellcrank adjusting screw (3, **Figure 57**) a little at a time until the bellcrank moves over dead center and remains in that position when released. Do not move bellcrank with clutch lever; move it by hand. Bellcrank should lock in position approximately 1/8 in. (3.2 mm) past dead center.

14. Tighten the bellcrank adjusting screw locknut (4, **Figure 57**).

15. Adjust the clutch lever rod (5, **Figure 57**) so that there is 1/16 in. (1.6 mm) free movement of the clutch activating lever.

16. Tighten the clutch lever rod locknut (7, **Figure 57**).

17. Turn the control coil adjusting sleeve (1, **Figure 57**) until there is 1/2 in. (12.7 mm) free movement of the clutch hand lever before clutch starts to disengage.

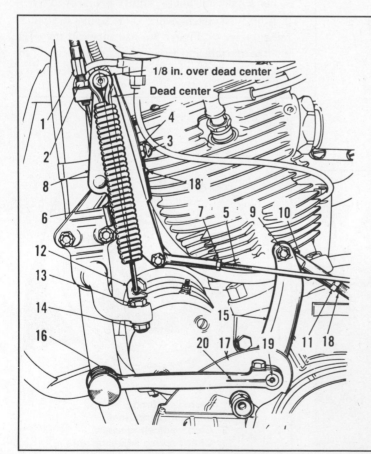

(57)

### HAND CLUTCH BOOSTER ADJUSTMENT

1. Control coil adjusting sleeve
2. Control coil adjusting sleeve locknut
3. Bellcrank adjusting screw
4. Bellcrank adjusting screw locknut
5. Clutch lever rod
6. Clutch control booster spring
7. Clutch lever rod locknut
8. Clutch control booster bellcrank
9. Shifter rod end bolt
10. Shifter rod end
11. Shifter rod end locknut
12. Clutch booster spring tension adjuster
13. Clutch booster spring tension upper adjuster nut
14. Clutch booster spring tension lower adjuster nut
15. Gear shifter lever
16. Gear shifter foot lever and rubber pedal
17. Grease fittings
18. Shifter rod
19. Foot lever positioning mark
20. Foot lever clamping slot

18. Tighten the control coil adjusting sleeve locknut (2, **Figure 57**).

19. Pull the clutch hand lever in fully. Then tighten the adjuster nut (14, **Figure 57**) until the lever remains in. Next, slowly loosen the adjuster nut (14, **Figure 57**) until the clutch hand lever returns to its released position.

20. Tighten the adjuster nut (13, **Figure 57**).

### Throttle Cable(s)

Check the throttle cable(s) from grip to carburetor(s). Make sure they are not kinked or chafed. Replace if necessary.

Make sure that the throttle grip rotates smoothly from fully closed to fully open. Check at center, full left and full right position of steering.

*WARNING*
*If idle speed increases when the handlebar is turned to right or left, check throttle cable routing. Do not ride the motorcycle in this unsafe condition.*

### Throttle Cable Adjustment (1966-1974)

Refer to *Handlebar* in Chapter Eight for adjustment procedures.

### Throttle Adjustment (1975-1980)

Refer to **Figure 58** for this procedure.

1. When the throttle grip is operated and then released, it must return to its closed or idle position. If not, loosen the adjusting screw (12, **Figure 58**) a little at a time until the grip returns correctly. If the

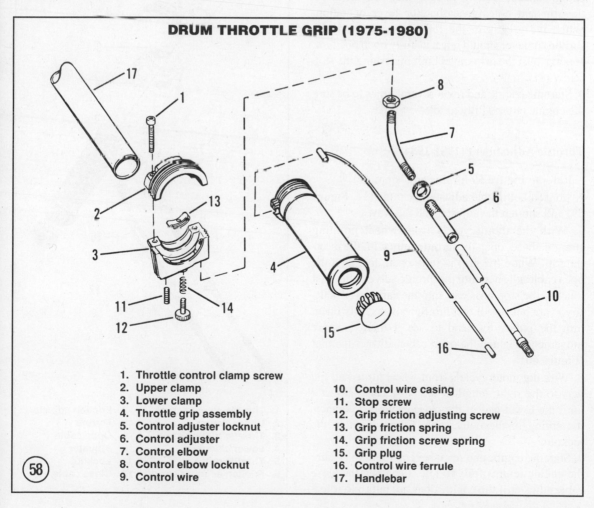

## DRUM THROTTLE GRIP (1975-1980)

1. Throttle control clamp screw
2. Upper clamp
3. Lower clamp
4. Throttle grip assembly
5. Control adjuster locknut
6. Control adjuster
7. Control elbow
8. Control elbow locknut
9. Control wire
10. Control wire casing
11. Stop screw
12. Grip friction adjusting screw
13. Grip friction spring
14. Grip friction screw spring
15. Grip plug
16. Control wire ferrule
17. Handlebar

(58)

3

grip is hard to turn, check the throttle cable from grip to carburetor. Make sure it is not kinked or chafed.

> *WARNING*
> *The adjusting screw (12, **Figure** 58) must not be so tight as to prevent the throttle grip from closing when it is released. This could cause a very dangerous riding condition.*

2. Support the bike so that the front wheel can be turned from side to side. Observe the throttle cable at the carburetor when turning the front wheel. The inner control wire (9) should not pull on the carburetor throttle. If it does, adjust cable free play by loosening the control adjuster locknut (5) and turning the control adjuster (6) to correct the overall cable length. Tighten the locknut and recheck the adjustment.

3. With the motorcycle's front wheel pointing straight ahead, open the throttle fully with the throttle grip and note the carburetor lever operation. When the grip is in the fully open position, the carburetor lever should reach its fully open position. If not, adjust the grip travel limit by turning the stop screw (11).

4. Start the engine and rev it several times to be sure the engine returns fully to idle.

### Throttle Adjustment (1981-1984)

Refer to **Figure 59** for this procedure.

1. Loosen both cable adjuster locknuts (11, **Figure 59**) and shorten the adjusters all the way.

2. With the motorcycle's front wheel pointing straight ahead, open the throttle fully with the throttle grip. While the grip is fully open, lengthen the open cable adjuster until the throttle valve pulley just touches the stop boss cast into the carburetor body. If you are not sure if the throttle valve is fully open, turn the pulley by hand to see if there is more movement. Tighten the open cable adjuster locknut (**Figure 60**).

3. With the motorcycle's front wheel turned all the way to the right, lengthen the close cable adjuster until the lower end of the close cable just contacts the spring in the outer cable fitting. Tighten the locknut.

4. Start the engine and rev it several times to be sure the engine returns fully to idle. Lengthen the close cable adjuster if the engine does not return to idle.

### Fuel Shutoff Valve/Fitter

Refer to Chapter Six for complete details on removal, cleaning and installation of the fuel shutoff valve.

### Fuel Line Inspection

Inspect the fuel lines from the fuel tank to the carburetor. If any are cracked or starting to deteriorate they must be replaced. Make sure the small hose clamps are in place and holding securely.

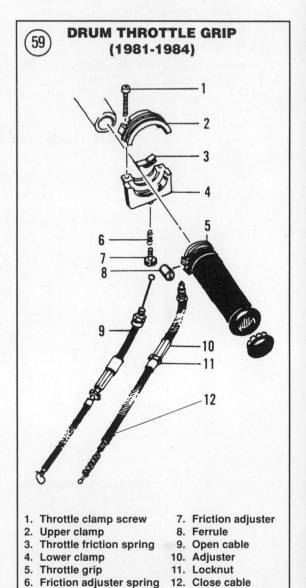

**(59) DRUM THROTTLE GRIP (1981-1984)**

| | |
|---|---|
| 1. **Throttle clamp screw** | 7. **Friction adjuster** |
| 2. **Upper clamp** | 8. **Ferrule** |
| 3. **Throttle friction spring** | 9. **Open cable** |
| 4. **Lower clamp** | 10. **Adjuster** |
| 5. **Throttle grip** | 11. **Locknut** |
| 6. **Friction adjuster spring** | 12. **Close cable** |

*WARNING*
*A damaged or deteriorated fuel line presents a very dangerous fire hazard to*

*both the rider and the bike if fuel should spill onto a hot engine or exhaust pipe.*

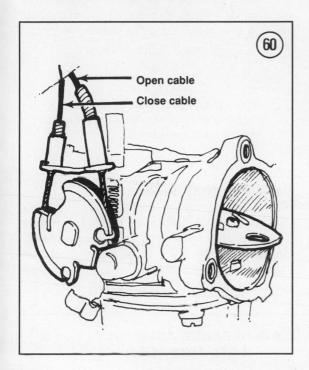

### Exhaust System

Check all fittings for exhaust leakage. Do not forget the crossover pipe connections. Tighten all bolts and nuts; replace any gaskets as necessary. Removal and installation procedures are described in Chapter Six.

### Air Cleaner Removal/Installation

A clogged air cleaner can decrease the efficiency and life of the engine. Never run the bike without the air cleaner installed; even minute particles of dust can cause severe internal engine wear.

The service intervals specified in **Tables 1-3** should be followed with general use. However, the air cleaner should be serviced more often if the bike is ridden in dusty areas.

The air filter on all models is installed on the right-hand side of the bike. See **Figure 61** or **Figure 62**. Remove the air filter cover and withdraw the

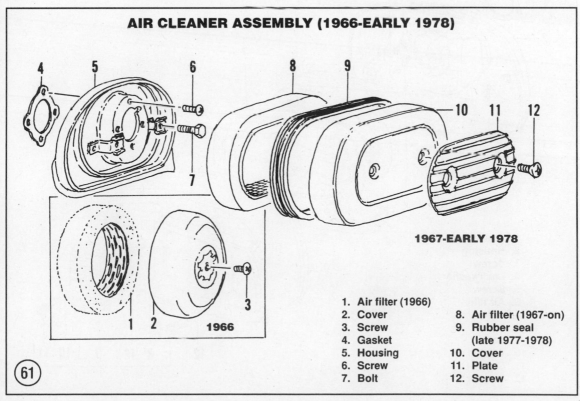

**AIR CLEANER ASSEMBLY (1966-EARLY 1978)**

1967-EARLY 1978

1966

1. Air filter (1966)
2. Cover
3. Screw
4. Gasket
5. Housing
6. Screw
7. Bolt
8. Air filter (1967-on)
9. Rubber seal (late 1977-1978)
10. Cover
11. Plate
12. Screw

filter. Clean the filter as described in this chapter. Reverse to install the air filter.

## Air Filter Cleaning

Refer to **Figure 61** or **Figure 62** for this procedure.

1A. *Paper filter*: Tap the filter lightly to remove dirt and dust on outside of filter. If element is oily or coated with dirt, replace it with a new filter.

1B. *Metal mesh or foam filter*: Remove wire mesh frame from inside filter. Wash filter in soap and water and allow to dry.

2. Inspect the element and make sure it is in good condition. Replace if necessary.

3A. *Paper filter*: Install paper filters dry; do not oil.

3B. *Metal mesh or foam filter*: Saturate the filter with engine oil. Work the oil over the entire filter surface with hands. Then squeeze the filter to remove all excess oil. Filter should be uniform in color to indicate that entire filter is oiled. Reinstall wire mesh frame into filter.

4. Clean out the inside of the air box with a shop rag and cleaning solvent. Remove any foreign matter that may have passed through a broken cleaner element.

*CAUTION*
*When cleaning the inside of air box, do not allow any dirt or other debris to run into the carburetor hoses.*

### Tappet Oil Screen Cleaning

At the intervals specified in **Tables 1-3**, remove the tappet oil screen located under the plug on the cam case near the rear cylinder tappet block. See **Figure 63**. Clean the screen in solvent. Replace the screen if damaged. Reverse to install.

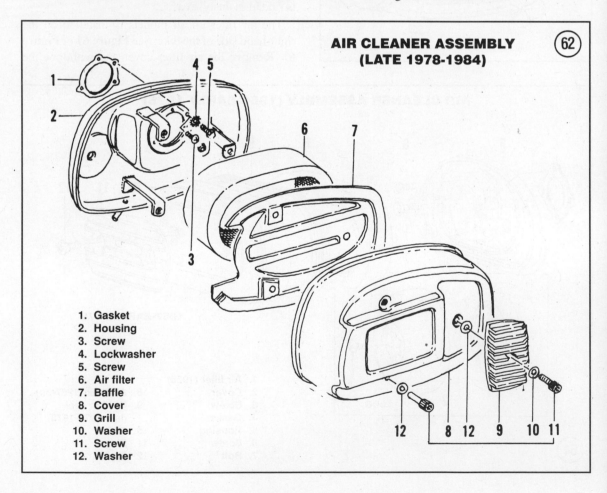

**AIR CLEANER ASSEMBLY (LATE 1978-1984)**  (62)

1. Gasket
2. Housing
3. Screw
4. Lockwasher
5. Screw
6. Air filter
7. Baffle
8. Cover
9. Grill
10. Washer
11. Screw
12. Washer

## Wheel Bearings

The wheel bearings should be cleaned and re-packed at the intervals specified in **Tables 1-3**.

Refer to Chapter Eight and Chapter Nine for complete service procedures.

## Steering Play

The steering head should be checked for looseness at the intervals specified in **Tables 1-3**.

1. Prop up the bike so that the front tire clears the ground.

2. Center the front wheel. Push lightly against the left handlebar grip to start the wheel turning to the right, then let go. The wheel should continue turning

under its own momentum until the forks hit their stop.

3. Center the wheel, then push lightly against the right handlebar grip. The wheel should turn to the left under its own momentum until the forks hit their stop.

4. If, with a light push in either direction, the front wheel will turn all the way to the stop, the steering adjustment is not too tight.

5. Center the front wheel and kneel in front of it. Grasp the bottoms of the 2 front fork slider legs. Try to pull the forks toward you, and then try to push them toward the engine. If no play is felt, the steering adjustment is not too loose.

6. If the steering adjustment is too tight or too loose, readjust it as described in Chapter Eight.

## Steering Head Bearings

The steering head bearings should be repacked at the intervals specified in **Tables 1-3**. Refer to Chapter Eight.

## Front Suspension Check

1. Apply the front brake and pump the fork up and down as vigorously as possible. Check for smooth operation and check for any oil leaks.

2. Make sure the upper and lower fork bridge bolts are tight

3. Check the tightness of the bolts securing the handlebar.

4. Check that the front axle pinch bolt is tight.

> *WARNING*
> *If any of the previously mentioned bolts and nuts are loose, refer to Chapter Eight for correct procedures and torque specifications.*

## Rear Suspension Check

1. Place the bike on the sidestand.

2. Push hard on the rear wheel sideways to check for side play in the rear swing arm bushings or bearings.

3. Check the tightness of the upper and lower shock absorber mounting nuts and bolts (**Figure 64**).

4. Make sure the rear axle nut is tight and the cotter pin is still in place.

5. Check the tightness of the rear brake torque arm bolts. Make sure the cotter pin is in place.

*WARNING*
*If any of the previously mentioned nuts or bolts are loose, refer to Chapter Nine for correct procedures and torque specifications.*

### Nuts, Bolts, and Other Fasteners

Constant vibration can loosen many fasteners on a motorcycle. Check the tightness of all fasteners, especially those on:
  a. Engine mounting hardware.
  b. Engine crankcase covers.
  c. Handlebar and front forks.
  d. Gearshift lever.
  e. Sprocket bolts and nuts.
  f. Brake pedal and lever.
  g. Exhaust system.
  h. Lighting equipment.

### SUSPENSION ADJUSTMENT

### Adjustable Fork (FL Models)

Adjustable front forks are installed on FL models with sidecar. Refer to **Figure 65** for this procedure.

#### *Damper adjustment*

The steering damper adjusting screw (1, **Figure 65**) is located on top of the steering head. To increase damping action, turn the steering damper clockwise; to decrease damping, turn the damper counterclockwise. When riding with sidecar, it is best to increase steering damping.

#### *Trail adjustment*

Trail can be decreased for use with a sidecar or increased for solo riding.
1. Support the motorcycle so that the front wheel clears the ground.
2. Remove the cotter pin and loosen the nut (29, **Figure 65**).
3. Disengage the bracket bolt washers (30) from their slots in the lower fork bracket (32).

4. Have an assistant steady the motorcycle. Then grasp the front wheel and either pull or push the forks as far forward or rearward as necessary for adjustment.
5. Engage tabs on bracket bolt washers (23, 25) in elongated holes so that the tabs face to the front of the holes. Tighten the nut and install a new cotter pin.

### Rear Shock Absorber Adjustment

On all models, the spring seat can be adjusted to suit rider preference. See **Figure 66**. Rotate the cam ring at the base of the spring to compress the spring (heavy loads) or extend the spring (light loads).

*NOTE*
*Use a spanner wrench or screwdriver to adjust the spring preload. Set both shocks to the same position.*

### TUNE-UP

A complete tune-up restores performance and power that is lost due to normal wear and deterioration of engine parts. Because engine wear occurs over a combined period of time and mileage, the engine tune-up should be performed at the intervals specified in **Tables 1-3**. More frequent tune-ups may be required if the bike is ridden primarily in stop-and-go traffic.

**Table 9** summarizes tune-up specifications.

Before starting a tune-up procedure, make sure to first have all the necessary new parts on hand.

Because different systems in an engine interact, the procedures should be done in the following order:
  a. Clean or replace the air filter element.
  b. Adjust pushrod clearance.
  c. Check engine compression.
  d. Check or replace the spark plugs.
  e. Check the ignition timing.
  f. Adjust carburetor idle speed.

To perform a tune-up on your Harley-Davidson, you will need the following tools:
  a. Spark plug wrench.
  b. Socket wrench and assorted sockets.
  c. Flat feeler gauge.
  d. Compression gauge.
  e. Spark plug wire feeler gauge and gapper tool.

## STEERING ASSEMBLY (FL ADJUSTABLE FORK)

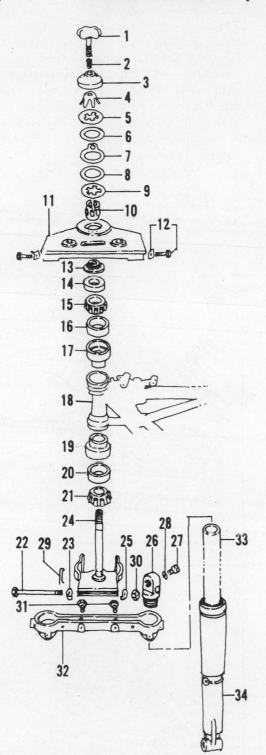

1. Steering damper adjusting screw
2. Spring
3. Spider spring cover
4. Spider spring
5. Pressure disc
6. Friction washer
7. Anchor plate
8. Friction washer
9. Pressure disc
10. Fork stem nut
11. Upper bracket
12. Bolt and washer
13. Head bearing nut
14. Dust shield
15. Bearing cone
16. Bearing race
17. Bearing cup
18. Frame
19. Bearing cup
20. Bearing race
21. Bearing cone
22. Bracket bolt
23. Bracket bolt washer
24. Steering stem and lower bracket
25. Bracket bolt washer
26. Fork cap
27. Filler screw
28. O-ring
29. Cotter pin
30. Nut
31. Bracket clamp bolt
32. Lower bracket
33. Fork tube
34. Slider

f. Ignition timing light.

## Air Cleaner

The air cleaner element should be cleaned or replaced prior to doing other tune-up procedures. Refer to *Air Filter Cleaning* as described in this chapter.

## Pushrod Adjustment

When properly assembled, the valve lifters maintain zero valve clearance while running, thus eliminating the need for periodic adjustment. Some valve clatter should be expected when starting the engine, until the lifters pump up. Refer to Chapter Four.

## Compression Test

At every tune-up, check cylinder compression. Record the results and compare them at the next check. A running record will show trends in deterioration so that corrective action can be taken before complete failure.

The results, when properly interpreted, can indicate general cylinder, piston ring and valve condition.

1. Warm the engine to normal operating temperature. Ensure that the choke valve and throttle valve are completely open.

2. Remove the spark plugs.

3. Connect the compression tester to one cylinder following manufacturer's instructions (**Figure 67**).

4. Have an assistant crank the engine over until there is no further rise in pressure.

5. Remove the tester and record the reading.

6. Repeat Steps 3-5 for the other cylinder.

When interpreting the results, actual readings are not as important as the difference between the readings. Standard compression pressure is shown in **Table 9**. Pressure should not vary from cylinder to cylinder by more than the amount specified in **Table 9**. Greater differences indicate worn or broken rings, leaky or sticky valves, blown head gasket or a combination of all.

If compression readings do not differ between cylinders by more than 10 psi (0.7 kg/cm$^2$), the rings and valves are in good condition.

If a low reading (10% or more) is obtained on one of the cylinders, it indicates valve or ring trouble. To determine which, pour about a teaspoon of engine oil through the spark plug hole onto the top of the piston. Turn the engine over once to distribute the excess oil, then take another compression test and record the reading. If the compression increases significantly, the valves are good, but the rings are defective on that cylinder. If compression does not increase, the valves require servicing. A valve could be hanging open but not burned or a piece of carbon could be on a valve seat.

*NOTE*
*If the compression is low, the engine cannot be tuned to maximum performance. The worn parts must be replaced and the engine rebuilt.*

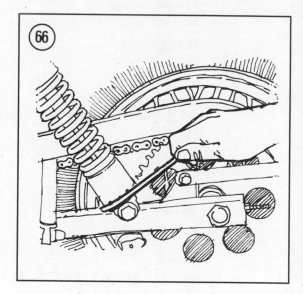

## Correct Spark Plug Heat Range

Spark plugs are available in various heat ranges that are hotter or colder than the spark plugs originally installed at the factory.

Select plugs in a heat range designed for the loads and temperature conditions under which the engine will operate. Using incorrect heat ranges can cause piston seizure, scored cylinder walls or damaged piston crowns.

In general, use a hotter plug for low speeds, low loads and low temperatures Use a colder plug for high speeds, high engine loads and high temperatures.

*NOTE*
*In areas where seasonal temperature variations are great, a "two-plug system"—a cold plug for hard summer*

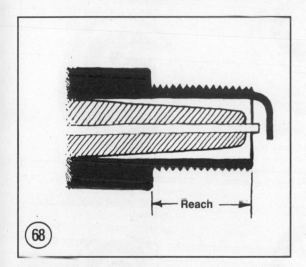

Reach

(68)

(69)

*riding and a hot plug for slower winter operation—may prevent spark plug and engine problems.*

The reach (length) of a plug is also important. A longer than normal plug could interfere with the valves and pistons, causing permanent and severe damage. Refer to **Figure 68**. The standard heat range spark plugs are listed in **Table 10**.

## Spark Plug Cleaning/Replacement

1. Grasp the spark plug leads as near to the plug as possible and pull them off the plugs.

2. Blow away any dirt that has accumulated in the spark plug wells (**Figure 69**).

*CAUTION*
*The dirt could fall into the cylinders when the plugs are removed, causing serious engine damage.*

3. Remove the spark plugs (**Figure 69**) with a spark plug wrench.

*NOTE*
*If plugs are difficult to remove, apply penetrating oil such as WD-40 or Liquid Wrench around base of plugs and let it soak in about 10-20 minutes.*

4. Inspect spark plug carefully. Look for plugs with broken center porcelain, excessively eroded electrodes and excessive carbon or oil fouling (**Figure 70**). Replace such plugs.

*NOTE*
*Spark plug cleaning with the use of a sand-blast type device is generally not recommended. While this type of cleaning is thorough, the plug must be perfectly free of all abrasive cleaning material when done. If not, it is possible for the cleaning material to fall into the engine during operation and cause damage.*

## Spark Plug Gapping and Installation

New plugs should be carefully gapped to ensure a reliable, consistent spark. You must use a special spark plug gapping tool.

⑦⓪ **SPARK PLUG CONDITION**

### NORMAL
- Identified by light tan or gray deposits on the firing tip.
- Can be cleaned.

### GAP BRIDGED
- Identified by deposit buildup closing gap between electrodes.
- Caused by oil or carbon fouling. If deposits are not excessive, the plug can be cleaned.

### OIL FOULED
- Identified by wet black deposits on the insulator shell bore and electrodes.
- Caused by excessive oil entering combustion chamber through worn rings and pistons, excessive clearance between valve guides and stems, or worn or loose bearings. Can be cleaned. If engine is not repaired, use a hotter plug.

### CARBON FOULED
- Identified by black, dry fluffy carbon deposits on insulator tips, exposed shell surfaces and electrodes.
- Caused by too cold a plug, weak ignition, dirty air cleaner, too rich a fuel mixture, or excessive idling. Can be cleaned.

### LEAD FOULED
- Identified by dark gray, black, yellow, or tan deposits or a fused glazed coating on the insulator tip.
- Caused by highly leaded gasoline. Can be cleaned.

### WORN
- Identified by severely eroded or worn electrodes.
- Caused by normal wear. Should be replaced.

### FUSED SPOT DEPOSIT
- Identified by melted or spotty deposits resembling bubbles or blisters.
- Caused by sudden acceleration. Can be cleaned.

### OVERHEATING
- Identified by a white or light gray insulator with small black or gray brown spots and with bluish-burnt appearance of electrodes.
- Caused by engine overheating, wrong type of fuel, loose spark plugs, too hot a plug, or incorrect ignition timing. Replace the plug.

### PREIGNITION
- Identified by melted electrodes and possibly blistered insulator. Metallic deposits on insulator indicate engine damage.
- Caused by wrong type of fuel, incorrect ignition timing or advance, too hot a plug, burned valves, or engine overheating. Replace the plug.

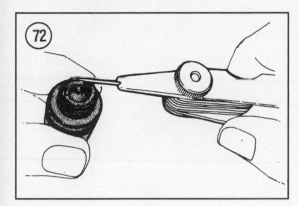

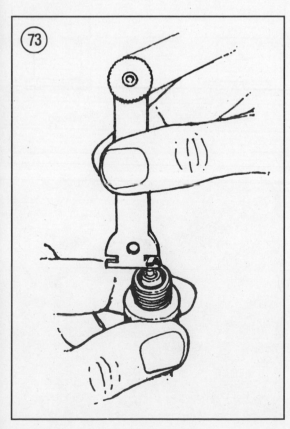

1. Remove the new plugs from the box. Screw in the small pieces that may be loose in each box (**Figure 71**).

2. Insert a wire gauge between the center and the side electrode of each plug (**Figure 72**). The correct gap is listed in **Table 10**. If the gap is correct, you will feel a slight drag as you pull the wire through. If there is no drag, or the gauge won't pass through, bend the side electrode *with the gapping tool* (**Figure 73**) to set the proper gap (**Table 10**).

3. Put a small drop of oil or anti-seize compound on the threads of each spark plug.

4. Screw each spark plug in by hand until it seats. Very little effort is required. If force is necessary, you have the plug cross-threaded or the spark plug threads in the cylinder head are damaged or contaminated with carbon or other debris; unscrew the plug and try again.

5. Tighten the spark plugs to 14 ft.-lb. (19.3 N•m). If you don't have a torque wrench, an additional 1/4 to 1/2 turn is sufficient after the gasket has made contact with the head. If you are reinstalling old, regapped plugs and are reusing the old gasket, tighten only an additional 1/4 turn.

*CAUTION*
*Do not overtighten. Besides making the plug difficult to remove, the excessive torque will squash the gasket and destroy its sealing ability.*

6. Install each spark plug wire. Make sure it goes to the correct spark plug.

**Reading Spark Plugs**

Much information about engine and spark plug performance can be determined by careful examination of the spark plugs. This information is more valid after performing the following steps.

1. Ride bike a short distance at full throttle in third or fourth gear.

2. Turn off kill switch before closing throttle and simultaneously pull in clutch. Coast and brake to a stop. *Do not* downshift transmission while stopping.

3. Remove spark plugs and examine them. Compare them to **Figure 70**.

If the insulator tip is white or burned, the plug is too hot and should be replaced with a colder one.

A too-cold plug will have sooty deposits ranging in color from dark brown to black. Replace with a hotter plug and check for too-rich carburetion or evidence of oil blow-by at the piston rings.

If any one plug is found unsatisfactory, discard both.

## BREAKER POINT IGNITION SERVICE

The expendable ignition parts (spark plugs, breaker points and condenser) should be replaced, and the ignition timing checked and adjusted if necessary, as part of a tune-up. If the bike is used primarily in stop-and-go city driving or is driven extensively at low engine speeds, these components should be cleaned, adjusted or replaced at more frequent intervals.

### Breaker Point Service

1. Check breaker point contact surfaces. Points with an even, overall gray color and only slight roughness or pitting need not be replaced. They can be dressed with a clean point file. Do not use sandpaper or emery cloth for dressing points and do not attempt to remove all irregularities—just remove scale or dirt.

2. Check the alignment of the points and correct as necessary. See **Figure 74**.

3. Check the point gap with a feeler gauge placed between the points (**Figure 75**) while the breaker arm rubbing block is on the high point of the cam lobe. The correct gap is listed in **Table 9**. The gap is correct when the gauge passes between the contact points with a slight drag.

*NOTE*
*Ignition timing should be checked whenever the circuit breaker points are adjusted. Changing the rear cylinder contact point gap will affect ignition timing.*

4A. *1966-1969*: If the points require adjustment, loosen the point plate lock screw (6, **Figure 76**). Then turn the eccentric adjust screw (7, **Figure 76**) with a screwdriver to open or close the point gap as required. When the point gap is correct, tighten the lock screw. Recheck gap to make sure it did not change when the screw was tightened.

4B. *1970-early 1978*: If the points require adjustment, loosen the point plate lock screw (A, **Figure**

77). Insert a screwdriver into the notch (B, **Figure 77**) beside the points and twist the screwdriver to open or close the point gap as required. When point gap is correct, tighten the lock screw. Recheck gap to make sure it did not change when the screw was tightened.

5. If points need replacing, proceed as follows:

 a. Loosen the wire stud screw at the points and disconnect the wires.

 b. Remove the breaker point lock screws and lift the point set from the breaker plate. Repeat for double contact points.

 c. Remove the condenser attaching screw and condenser.

 d. Wipe the breaker plate, cam and inside of cover with a lint-free cloth to remove all dirt and grease.

 e. Lightly lubricate the cam surfaces with breaker cam grease.

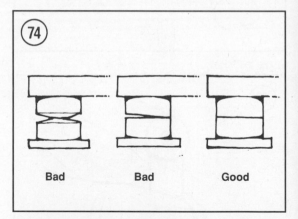

Bad          Bad          Good

f. Position the new breaker point set on the breaker plate and install the attaching screws. Position a new condenser on the breaker plate and install the attaching screw.

g. Attach the wire leads to the breaker point set. Tighten the nut or screw securely.

6. Align points and adjust gap as described in Steps 2-4.

### Ignition Timing Adjustment

Ignition timing should be checked and adjusted after point adjustment or replacement has been completed.

Refer to **Figure 76** or **Figure 77** for this procedure.

1. Remove the crankcase timing hole plug (**Figure 78**). Then install a clear plastic timing hole plug to prevent oil splash. These plugs can be purchased through a Harley-Davidson dealer.

2. Connect a timing light and tachometer according to manufacturer's instructions. Attach the timing light trigger lead to the front cylinder spark plug lead.

*CAUTION*
*Work carefully with the tachometer and timing light wire leads. The wires can be damaged if they contact the hot exhaust pipes.*

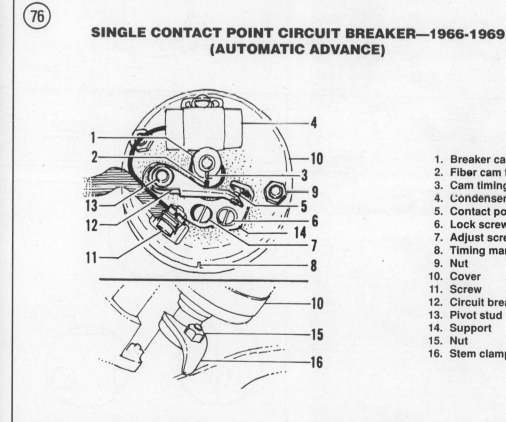

### SINGLE CONTACT POINT CIRCUIT BREAKER—1966-1969 (AUTOMATIC ADVANCE)

1. Breaker cam
2. Fiber cam follower
3. Cam timing mark
4. Condenser
5. Contact points
6. Lock screw
7. Adjust screw
8. Timing marks
9. Nut
10. Cover
11. Screw
12. Circuit breaker lever
13. Pivot stud
14. Support
15. Nut
16. Stem clamp

3. Start the engine and run at idle. Check the idle speed and compare to specifications in **Table 11**. If necessary, adjust engine speed to idle specification as described in this chapter.

4. Point the timing light at the timing light inspection hole (**Figure 79**). If the ignition is properly timed, the timing bar will be aligned as shown in **Figure 80** or **Figure 81**. If the marks are not correctly aligned, perform the following.

5A. *1966-1969*: Slightly loosen the circuit breaker clamp (**Figure 82**). Then rotate the circuit breaker housing as required until the ignition timing is correct. Tighten the clamp and recheck the timing.

5B. *1970-early 1978*: Remove the points cover from the right-hand side. Slightly loosen both timing plate screws (A, **Figure 83**). Then turn the timing plate (B, **Figure 83**) as required, using a screwdriver in the notch (C, **Figure 83**). Tighten the timing plate screws and recheck the timing. Reinstall the points cover.

6. Shut the engine off. Disconnect and remove all test equipment. Remove the clear plastic timing hole plug and reinstall the solid plug.

## BREAKERLESS IGNITION SERVICE (LATE 1978-1979)

### Sensor Air Gap

Inspect the sensor air gap according to the maintenance schedule.

1. Disconnect the spark plug leads at the spark plugs. Remove the spark plugs.

2. Remove the ignition timing cover.

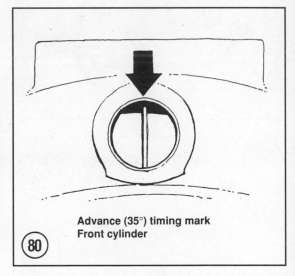

**Advance (35°) timing mark
Front cylinder**

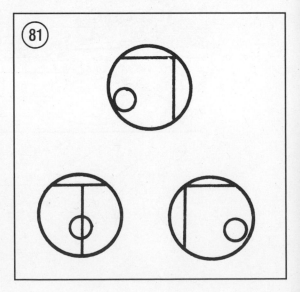

3. Turn the crankshaft and center the wide rotor lobe opposite the sensor (**Figure 84**).

4. Measure the gap between the sensor and the rotor with a flat non-magnetic feeler gauge. It should be 0.004-0.006 in. (0.10-0.15 mm). If the gap is incorrect, loosen the 2 screws that attach the sensor (**Figure 85**) and move the sensor until the gap is correct. Then, holding the sensor steady, tighten the screws.

> *NOTE*
> *If the engine doesn't run smoothly and spits back because of the standard lean fuel mixture, set the sensor air gap as close to 0.004 in. (0.10 mm) as possible. This will strengthen the ignition signal for better combustion.*

5. Rotate the crankshaft to center the small lobe opposite the sensor. Measure the gap as before and

correct it if necessary. Both gaps must be within the specified range.

6. Install the spark plugs and check the ignition timing as described in this chapter.

### Ignition Timing Inspection

> *NOTE*
> *Before starting this procedure, check all electrical connections related to the ignition system. Make sure all connections are tight and free of corrosion and that all ground connections are tight.*

1. Check the sensor air gap as described in this chapter.

2. Remove the plug from the timing hole on the left side of the engine (**Figure 78**). A clear plastic viewing plug is available from Harley-Davidson dealers to minimize oil spray. Make sure the plug doesn't contact the flywheel.

3. Connect a portable tachometer following the manufacturer's instructions. The bike's tach is not accurate enough in the low rpm range for this adjustment.

4. Connect a timing light to the front cylinder spark plug wire following the manufacturer's instructions.

> *CAUTION*
> *The use of a timing light with an inductive clamp-on pickup is recommended. If you must use another type, do not puncture the spark plug wire or cap, use an adaptor.*

5. Start the engine and allow it to idle at 2,000 rpm. If necessary, adjust idle as described in this chapter.

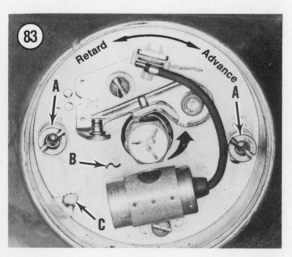

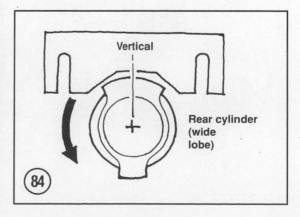

6. Aim the timing light at the timing inspection hole. At 2,000 rpm, the front cylinder's advance mark should appear in the center of the inspection window (**Figure 80**). If the mark is not in the correct position, loosen the timer plate screws (2, **Figure 85**) and turn the timer plate as required by prying the notch (8, **Figure 85**) with a screwdriver. Observe the front cylinder timing mark with the strobe light and tighten the timer plate screws when timing is correct.

7. Install the ignition timer cover and the inspection hole plug.

## BREAKERLESS IGNITION SERVICE (1980-ON)

### Ignition Timing Inspection and Adjustment

1. Remove the plug from the timing hole on the left side of the engine (**Figure 78**). A clear plastic viewing plug is available from Harley-Davidson dealers to minimize oil spray. Make sure the plug doesn't contact the flywheel.

2. Connect a portable tachometer following the manufacturer's instructions. The bike's tach is not accurate enough in the low rpm range for this adjustment.

3. Connect a timing light to the front cylinder spark plug wire following the manufacturer's instructions.

> *CAUTION*
> *The use of a timing light with an inductive clamp-on pickup is recommended.*

> *If you must use another type, do not puncture the spark plug wire or cap, use an adaptor.*

4. Start the engine and allow to idle at 2,000 rpm. If necessary, adjust idle as described in this chapter.

5. Aim the timing light at the timing inspection hole. At 2,000 rpm, the front cylinder's advance mark should appear in the center of the inspection window. **Figure 80** shows the advance mark for early 1980 engines up to engine number 1480-128-001. **Figure 86** shows the single drilled dot that shows full advance for the front cylinder on later engines. The double "figure 8" drilled marks represent full advance for the rear cylinder. If the proper mark does

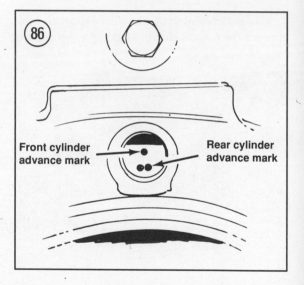

Front cylinder advance mark    Rear cylinder advance mark

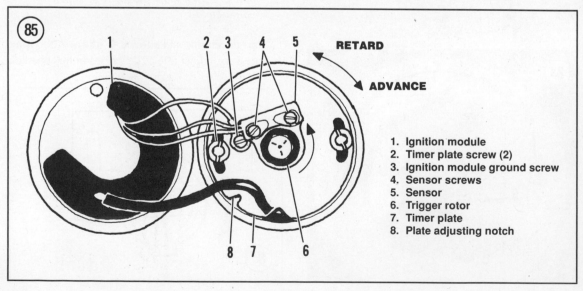

RETARD

ADVANCE

1. Ignition module
2. Timer plate screw (2)
3. Ignition module ground screw
4. Sensor screws
5. Sensor
6. Trigger rotor
7. Timer plate
8. Plate adjusting notch

not align, adjust the ignition timing as described in Steps 7-9.

6. If the ignition timing is correct, reinstall the timing hole plug.

7. Drill out the 2 pop rivets (12, **Figure 87**) with a 3/8 in. (9.5 mm) drill bit.

8. Referring to **Figure 87**, remove the outer cover (11), inner cover (10) and gasket (9).

9. Loosen the timing plate sensor plate screws (8, **Figure 87**) just enough to allow the plate to rotate. Start the engine and turn the plate as required so that the advanced mark is aligned as described in Step 5. Tighten the screws and recheck the timing.

10. Install the gasket and inner cover.

*NOTE*
*When installing pop rivets to secure the outer cover, make sure to use the headless type shown in Figure 88. The end of a normal pop rivet will break off on*

*installation and damage the timing mechanism.*

11. Install the outer cover with new rivets.

## CARBURETOR ADJUSTMENTS

### Model HD Carburetor Adjustment

Refer to **Figure 89** for this procedure.

1. Attach a tachometer to the engine following the manufacturer's instructions.

2. Check that the throttle lever (4) is fully closed when the throttle grip is closed.

3. Turn the low speed needle (1) clockwise until it seats lightly, then back it out 7/8 turn.

4. Turn the intermediate speed needle (2) in until it seats lightly, then back it out 7/8 turn.

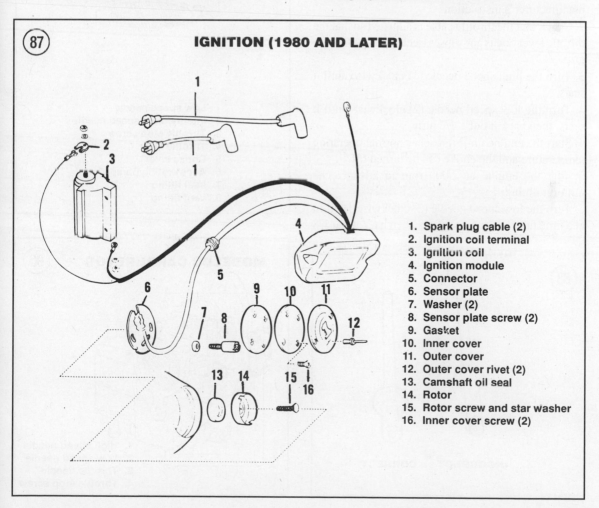

(87) **IGNITION (1980 AND LATER)**

1. Spark plug cable (2)
2. Ignition coil terminal
3. Ignition coil
4. Ignition module
5. Connector
6. Sensor plate
7. Washer (2)
8. Sensor plate screw (2)
9. Gasket
10. Inner cover
11. Outer cover
12. Outer cover rivet (2)
13. Camshaft oil seal
14. Rotor
15. Rotor screw and star washer
16. Inner cover screw (2)

5. Start the engine and run it until it reaches normal operating temperature and the choke can be turned off.

6. Slowly turn the intermediate speed needle (2) in the direction in which the highest engine speed is produced without missing or surging. Then turn the intermediate speed needle (2) counterclockwise 1/8 turn from that point.

7. Turn the throttle stop screw until the engine idles at 900-1,100 rpm.

8. Alternately adjust the low speed needle (1) and the throttle stop screw (3) to produced a smooth idle at 900-1,100 rpm.

9. Remove the tachometer.

### Model DC Carburetor (Adjustment)

Refer to **Figure 90** for this procedure.

1. Attach a tachometer to the engine following the manufacturer's instructions.

2. Check that the throttle cable is adjusted so that the throttle lever opens and closes freely with the throttle grip.

3. Turn the high speed needle (1) clockwise until it seats.

4. Turn the low speed needle (2) clockwise until it seats, then back it out 1 1/2 turns.

5. Start the engine until it reaches normal operating temperature and the choke can be turned off.

6. Idle the engine at 2,000 rpm to advance the ignition timing.

7. Turn the low speed needle (2) clockwise 1/8 turn at a time until the engine falters, then back it out 1/8

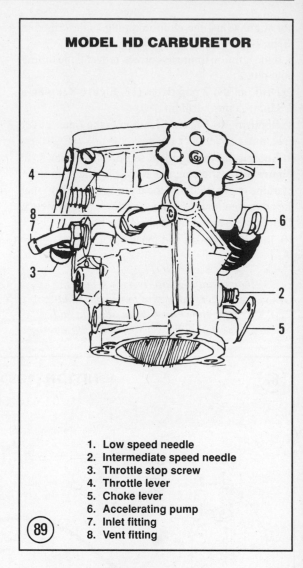

**MODEL HD CARBURETOR**

1. Low speed needle
2. Intermediate speed needle
3. Throttle stop screw
4. Throttle lever
5. Choke lever
6. Accelerating pump
7. Inlet fitting
8. Vent fitting

(89)

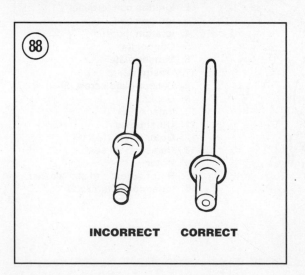

(88)

INCORRECT    CORRECT

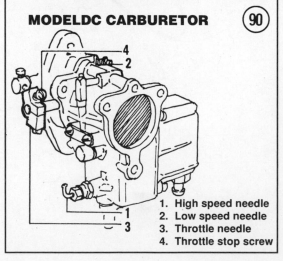

**MODELDC CARBURETOR** (90)

1. High speed needle
2. Low speed needle
3. Throttle needle
4. Throttle stop screw

turn or until the engine runs smoothly at idle speed with ignition fully advanced.

8. Adjust throttle stop screw (4) until engine idles at 900-1,100 rpm.

9. Repeat Step 7 and Step 8 as required.

10. Operate motorcycle at various speeds. Back out the high speed needle (1) and turn it in or out until best overall performance is obtained. This point should occur with needle set from 3/4 to 1 1/4 turns open.

### Bendix 16P12 Carburetor Adjustment

Refer to **Figure 91** for this procedure.

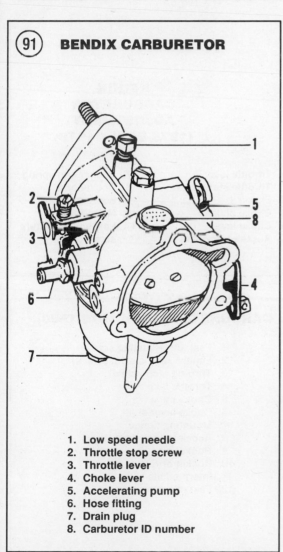

**91** **BENDIX CARBURETOR**

1. **Low speed needle**
2. **Throttle stop screw**
3. **Throttle lever**
4. **Choke lever**
5. **Accelerating pump**
6. **Hose fitting**
7. **Drain plug**
8. **Carburetor ID number**

1. Attach a tachometer to the engine following the manufacturer's instructions.

2. Turn low speed needle (1) clockwise until it seats lightly, then back it out 1 1/2 turns.

3. Run the engine until it reaches normal operating temperature and the choke can be turned off.

4. Adjust throttle stop screw (2) so that engine runs at 700-800 rpm with throttle fully closed.

5. Adjust low speed needle (1) as required to make engine accelerate and run smoothly at idle.

6. Adjust throttle stop screw (2), if necessary, to obtain idle of 700-900 rpm.

### Keihin Carburetor (1976-early 1978)

Refer to **Figure 92** for this procedure.

1. Attach a tachometer to the engine following the manufacturer's instructions.

2. Turn the low speed mixture screw (11) clockwise until it seats lightly. Then back it out 7/8 turn (1976-1977) or 1 1/2 turns (early 1978).

3. Start the engine and warm it to normal operating temperature.

4. Turn the throttle stop screw (12) and set the idle to 700-900 rpm.

5. Turn the idle mixture screw in or out as required, to get the highest and smoothest idle speed. The best performance will usually result if the mixture is slightly rich (screw turned in a little).

6. Rev the engine a couple of times and see if the idle speed is constant. Readjust the idle speed, if necessary, to 700-900 rpm by turning the throttle stop screw (12).

7. Adjust the accelerator pump volume, if necessary, with the rocker arm adjust screw (9). The standard setting is about 1/4 in. between the tip of the screw and its stop. Backing the screw out increases fuel volume. The pump stroke may also be adjusted by moving the rocker arm spring (7) to other than the standard center notch in the rocker arm (8).

### Keihin Carburetor (Late 1978-on)

Refer to **Figure 93** or **Figure 94** for this procedure.

1. Attach a tachometer to the engine following the manufacturer's instructions.

2. Start the engine and warm it to normal operating temperature. Check that the choke is off.

3. Set the idle speed to 900 rpm with the throttle stop screw.

4A. *Late 1978-1979*: The idle mixture screw has a limiter cap on it. Turn the limiter cap out (clockwise) to the leanest setting that gives a smooth idle.

*NOTE*

*The low-speed mixture screw limiter cap should not be removed unless the carburetor is being disassembled for cleaning. If the cap has been removed, turn the idle mixture screw in (clockwise) until it seats lightly, then back it out 1 1/8 turn on 74 cu. in. bikes or 3/4 turn on 80 cu. in. bikes. Install the limiter cap in its center position.*

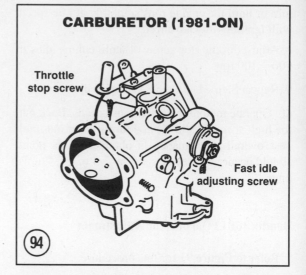

**CARBURETOR (1981-ON)**

Throttle stop screw

Fast idle adjusting screw

(94)

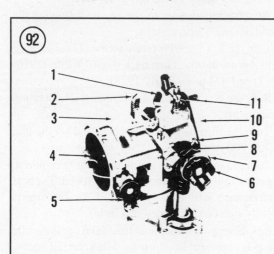

(92)

**KEIHIN CARBURETOR ADJUSTMENT (1976-EARLY 1978)**

1. Throttle lever
2. Throttle stop screw
3. Fitting
4. Choke plate
5. Choke lever
6. Accelerator pump lever
7. Rocker arm spring
8. Rocker arm
9. Screw
10. Mounting flange
11. Low speed mixture screw

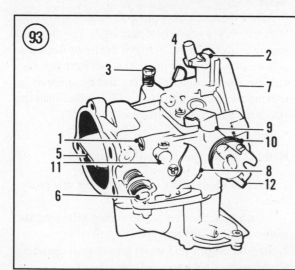

(93)

**CARBURETOR (LATE 1978-1980)**

1. Fast idle adjusting screw
2. Limiter cap (late 1978-1979)
3. Throttle stop screw
4. Throttle lever
5. Choke plate
6. Choke lever shaft
7. Mounting flange
8. Accelerating pump lever
9. Rocker arm
10. Rocker arm spring
11. Intermediate lever
12. Fast idle cam

*NOTE*
*Turning the idle mixture screw to the right (clockwise) leans the mixture. Turning it to the left (counterclockwise) enriches the mixture.*

4B. *1980-on*: The idle mixture is set and sealed at the factory. It is not intended to be adjustable.

a. Rev the engine a couple of times and see if the idle speed is constant. If necessary, readjust the idle speed to 900 rpm.

b. Pull the choke knob out to its first detent and turn the fast idle screw to set fast idle at 1,500 rpm. Push the choke knob all the way in and check that idle drops to 900 rpm.

## Table 1 MAINTENANCE SCHEDULE (1966-1969)*

| | |
|---|---|
| **Weekly** | Check tire pressure, wear.<br>Check battery electrolyte level. |
| **Every 300 miles** | Check engine oil level. |
| **Every 1,000 miles** | Adjust, lubricate drive chain (if chain oiler not used).<br>Clean air filter.<br>Check brake fluid level, refill if necessary. |
| **Every 2,000 miles** | Grease saddle post.<br>Grease saddle bar bearing.<br>Grease rear brake pedal bearing.<br>Grease foot shift lever bearing.<br>Grease hand clutch booster bearing.<br>Grease front and rear wheel hub thrust bearing. [1]<br>Grease foot clutch pedal bearing.<br>Oil hand levers.<br>Oil clutch control cable.<br>Oil front brake cable.<br>Oil throttle control cable.<br>Oil clutch booster lever rod clevis.[2]<br>Oil shifter control joints.<br>Oil saddle post roller and bolt.<br>Change engine oil and filter.<br>Clean tappet oil screen (if used).<br>Clean fuel strainer.<br>Check and adjust primary chain.[3]<br>Adjust front chain oiler.<br>Adjust rear chain oiler.<br>Adjust clutch.<br>Adjust brakes.<br>Clean, inspect and adjust breaker points. |
| **Every 5,000 miles or 1 year** | Grease throttle control spiral.<br>Grease front and rear wheel hub centers.[1]<br>Grease sidecar wheel hub center.[1]<br>Grease circuit breaker camshaft.<br>Grease speedometer and tachometer cables. |

(continued)

### Table 1 MAINTENANCE SCHEDULE (1966-1969)* (continued)

| | |
|---|---|
| **Every 5,000 miles or 1 year (continued)** | Replace spark plugs. |
| | Check ignition timing. |
| | Check generator brushes. |
| | Check shock bushings. |
| **Every 10,000 miles** | Grease generator bearings.[3] |
| **Every 50,000 miles** | Grease steering head bearings |

* This Harley-Davidson factory maintenance schedule should be considered as a guide to general maintenance and lubrication intervals. Harder than normal use and exposure to mud, water, high humidity, etc., will naturally dictate more frequent attention to most maintenance items.
1. 1966 models.
2. 1966-1967 models.
3. 1966-1969 models.

### Table 2 MAINTENANCE SCHEDULE (1970-EARLY 1978)

| | |
|---|---|
| **Weekly** | Check tire pressure, wear. |
| | Check battery electrolyte level. |
| **Every 300 miles** | Check engine oil level. |
| **Every 1,000 miles** | Check transmission oil level. |
| | Grease rear brake pedal bearing. |
| | Grease foot shift lever bearing. |
| | Grease speedometer drive. |
| | Grease foot clutch pedal bearing. |
| | Lubricate rear drive chain.[1] |
| | Lubricate clutch and brake hand levers. |
| | Oil clutch control cable. |
| | Oil front brake cable. |
| | Oil throttle control cable. |
| | Oil shift control joints. |
| | Oil seat post roller and bolt. |
| | Oil seat suspension bushings. |
| | Inspect, clean air filter. |
| | Inspect, adjust primary chain. |
| | Inspect, adjust drive chain. |
| | Check brake fluid level. |
| | Inspect, repair fuel and oil lines. |
| | Check tightness of nuts, bolts and fasteners. |
| | Check spoke tightness, tire pressure and tread wear of front and rear wheels. |
| | Check brake adjustment and lining wear. |
| | Check clutch adjustment. |
| **Every 2,000 miles** | Grease seat post. |
| | Grease seat bar bearing. |
| | Grease rear fork pivot bearing. |
| | Change engine oil and oil filter. |
| | Inspect, clean or replace fuel strainer. |
| | Clean tappet oil screen. |
| | Adjust primary chain. |
| | Check rear chain oiler. |
| | Inspect, clean circuit breaker points. |

(continued)

## Table 2 MAINTENANCE SCHEDULE (1970-EARLY 1978) (continued)

| | |
|---|---|
| **Every 2,000 miles (continued)** | Check ignition timing.<br>Check brake fluid level.<br>Check carburetor adjustment and idle speed.<br>Clean and regap spark plugs. |
| **Every 5,000 miles or yearly** | Grease throttle control.<br>Grease circuit breaker cam and advance unit.<br>Grease speedometer and tachometer cables.<br>Replace spark plugs.<br>Check, adjust ignition timing.<br>Check, adjust front and rear fork bearing adjustment.<br>Check rear shock bushings.<br>Change fork oil.<br>Check brake fluid level.<br>Clean fuel strainer screen and fuel tank. |
| **Every 10,000 miles** | Grease wheel bearings.<br>Grease rear fork bearings. |

1. If not equipped with automatic chain oiler.

## Table 3 MAINTENANCE SCHEDULE (LATE 1978-1984)

| | |
|---|---|
| **Every 500 miles or weekly** | Check engine oil level.<br>Lubricate drive chain, if required.<br>Check tire pressure, wear.<br>Check battery electrolyte level. |
| **Every 1,250 miles or monthly** | Check transmission oil level.<br>Inspect, clean air filter.<br>Inspect primary chain, adjust free play.<br>Inspect drive chain, adjust free play.<br>Check adjustment of drive chain oiler,<br>    if so equipped.<br>Inspect brakes, adjust rear pedal play;<br>    check pad and disc wear.<br>Inspect engine oil lines, brake fluid lines<br>    and fuel system for leaks.<br>Inspect wire wheels, adjust spoke tightness.<br>Check all nuts, bolts and fasteners, especially<br>    steering, suspension, axles and controls.<br>Check operation of all lights and switches. |
| **Every 2,500 miles or 6 months** | Change engine oil.<br>Change oil filter.<br>Check brake fluid level and condition.<br>Clean tappet oil screen.<br>Clean primary chain drain plug.<br>Replace spark plugs.<br>Check, adjust ignition timing.<br>Check, adjust sensor air gap (1978-1979).<br>Check carburetor, adjust idle and fast idle.<br>Clean fuel control valve screen in fuel tank.<br>Check clutch, adjust free play.<br>Check, adjust throttle and choke cables.<br>Grease speedometer drive, if so equipped. |

(continued)

### Table 3 MAINTENANCE SCHEDULE (LATE 1978-1984) (continued)

| | |
|---|---|
| **Every 2,500 miles or 6 months** | Lubricate control cables. |
| | Lubricate lever and pedal pivots. |
| | Lubricate seat post and/or linkage bushings. |
| | |
| **Every 5,000 miles or yearly** | Change transmission oil. |
| | Change fork oil. |
| | Grease throttle grip sleeve. |
| | Grease speedometer cables. |
| | Inspect/replace shock absorber rubber bushings. |
| | Lubricate ignition advance (1978-1979). |
| | Check/adjust steering bearing. |
| | Check/adjust swing arm bearing. |
| | |
| **Every 10,000 miles or 2 years** | Grease wheel bearings. |
| | Grease swing arm pivot. |
| | Inspect tension/adjust/align final drive belt, if so equipped. |
| | Inspect tension/adjust/replace primary drive belt, if so equipped. |
| | Lubricate compensating sprocket dampers. |
| | Grease clutch hub bearing. |

### Table 4 TIRE PRESSURE

| | Front | | Rear | | Sidecar | |
|---|---|---|---|---|---|---|
| | PSI | kg/cm$^2$ | PSI | kg/cm$^2$ | PSI | kg/cm$^2$ |
| **1966-1969** | | | | | | |
| Solo | | | | | | |
| 5.10 × 16 | 20 | 1.41 | 24 | 1.69 | | |
| 5.00 × 16 | 12 | .84 | 18 | 1.26 | | |
| With passenger | | | | | | |
| 5.10 × 16 | 20 | 1.41 | 26 | 1.83 | | |
| 5.00 × 16 | 12 | .84 | 20 | 1.41 | | |
| Rider with sidecar | | | | | | |
| passenger or 150 lb. load | | | | | | |
| 5.10 × 16 | 22 | 1.55 | 26 | 1.83 | 20 | 1.41 |
| 5.00 × 16 | 14 | .98 | 20 | 1.41 | 14 | .98 |
| **1970-early 1978** | | | | | | |
| FL series | | | | | | |
| Solo | 20 | 1.41 | 24 | 1.69 | | |
| With passenger | 20 | 1.41 | 26 | 1.83 | | |
| FX series | | | | | | |
| Solo | 24 | 1.69 | 26 | 1.83 | | |
| With passenger | 24 | 1.69 | 28 | 1.97 | | |
| FL series (rider with sidecar | | | | | | |
| passenger or 150 lb. load) | 22 | 1.55 | 26 | 1.83 | 20 | 1.41 |
| **Late 1978-on** | | | | | | |
| FL series | | | | | | |
| Solo | 20 | 1.41 | 24 | 1.69 | | |
| With passenter | 20 | 1.41 | 26 | 1.83 | | |
| FX series (except FXWG) | | | | | | |
| Solo | 24 | 1.69 | 26 | 1.83 | | |
| With passenger | 26 | 1.83 | 28 | 1.97 | | |
| FXWG | | | | | | |
| Solo | 30 | 2.11 | 26 | 1.83 | | |
| With passenger | 30 | 2.11 | 28 | 1.97 | | |
| | | (continued) | | | | |

## Table 4 TIRE PRESSURE (continued)

| | Front | | Rear | | Sidecar | |
|---|---|---|---|---|---|---|
| | PSI | kg/cm² | PSI | kg/cm² | PSI | kg/cm² |
| FL series (rider with sidecar passenger or 140 lb. load) | 22 | 1.55 | 26 | 1.83 | 20 | 1.41 |

CAUTION: Maximum pressure for all tires is 36 psi (2.53 kg/cm²).
NOTE: These pressures are based on 150 lb. (68.04 kg) rider weight. For each extra 50 lb. (22.68 kg), add 2 psi (0.14 kg/cm²) @ rear, 1 psi (0.07 kg/cm²) @ front and 1 psi (0.07 kg/cm² @ sidecar tire. Do not exceed maximum pressure of 36 psi (2.53 kg/cm²).

## Table 5 BATTERY STATE OF CHARGE

| Specifio gravity | State of charge |
|---|---|
| 1.110-1.130 | Discharged |
| 1.140-1.160 | Almost discharged |
| 1.170-1.190 | One-quarter charged |
| 1.200-1.220 | One-half charged |
| 1.230-1.250 | Three-quarters charged |
| 1.260-1.280 | Fully charged |

## Table 6 RECOMMENDED LUBRICANTS AND FLUIDS*

| | |
|---|---|
| Brake fluid | |
|   Prior to September, 1976 production | DOT 3 |
|   September, 1976 and later production | DOT 5 |
| Fork oil | |
|   1966-1969 | HD Hydra-Glide fork oil |
|   1970-on | HD type B |
| Battery top up | Distilled water |
| Transmission | HD transmission lubricant |
| Primary chaincase | HD primary chaincase |
| Engine oil | |
|   Below + 40° F | HD special light |
|   40-60° F | HD medium heavy |
|   + 60° F and up | HD regular heavy |
| Severe engine operating | |
|   conditions @ +80° F | HD extra heavy grade 60 |
| 10-90° F—normal and | |
|   severe operating conditions | HD Power Blend Super Premium |

* Lubricants recommended are Harley-Davidson brands.

## Table 7 ENGINE AND TRANSMISSION OIL CAPACITY

| | Quantity |
|---|---|
| Oil tank | 4 quarts (3.78 L, 3.3 imp. qt.) |
| Transmission | 1 1/2 pints (0.71 L, 0.62 imp. qt.) |

## Table 8 FRONT FORK OIL CAPACITY

|  | Quantity |
|---|---|
| **1966-1969** | |
| Wet | 6 1/2 oz. (192.2 cc, 6.76 imp. oz.) |
| Dry | 7 oz. (207 cc, 7.28 imp. oz.) |
| **1970-early 1978** | |
| FX Series | |
| 1970-1972 | |
| Wet | 5 1/2 oz. (162.6 cc, 5.72 imp. oz.) |
| Dry | 6 1/2 oz. (192.2 cc, 6.76 imp. oz.) |
| 1973-early 1978 | |
| Wet | 5 oz. (147.8 cc, 5.2 imp. oz.) |
| Dry | 6 oz. (177.4 cc, 6.24 imp. oz.) |
| FL Series | |
| 1970-early 1977 | |
| Wet | 6 1/2 oz. (192.2 cc, 6.76 imp. oz.) |
| Dry | 7 oz. (207 cc, 7.28 imp. oz.) |
| Early 1977-early 1978 | |
| Wet | 7 3/4 oz. (229.2 cc, 8.06 oz.) |
| Dry | 8 1/2 oz. (251.3 cc, 8.84 oz.) |
| **Late 1978-on** | |
| FL Series | |
| Wet | 7 3/4 oz. (229.2 cc, 8.06 imp. oz.) |
| Dry | 8 1/2 oz. (251.3 cc, 8.84 imp. oz.) |
| FX series (except FXWG) | |
| Wet | 5 oz. (147.8 cc, 5.2 imp. oz.) |
| Dry | 6 oz. 177.4 cc, 6.24 imp. oz.) |
| FXWG | |
| Wet | 9 1/4 oz. (273.5 cc, 9.62 imp. oz) |
| Dry | 10 oz. (295.7 cc, 10.4 imp. oz.) |

## TABLE 9 TUNE-UP SPECIFICATIONS

| | |
|---|---|
| **1966-1969** | |
| Breaker point gap | 0.020 in. (0.51 mm) |
| Dwell reading | 90° @ 2,000 rpm |
| Ignition timing | |
| Retarded | 5° BTDC |
| Automatic | 35° BTDC |
| **1970-early 1978** | |
| Breaker point gap | 0.020 in. (0.51 mm) |
| Ignition timing | |
| Retarded | 5° BTDC |
| Advance | 34-36° BTDC |
| **Late 1978-on** | |
| Ignition timer air gap | |
| Late 1978-1979 | 0.004-0.006 in. (0.102-0.152 mm) |
| 1980-on | No adjustment |
| Ignition timing | Electronic |
| Compression | |
| Standard | |
| 1966-early 1978 | 90 psi or over (6.33 kg/cm$^2$ or over) |
| Late 1978-1984 | 100 psi or over (7.03 kg/cm$^2$ or over) |
| Acceptable Variance between cylinders | |
| 1966-early 1978 | 29 psi or less (2.04 kg/cm$^2$ or over) |
| Late 1978-1984 | 10 psi or less (0.70 kg/cm$^2$ or over) |

### Table 10 SPARK PLUG TYPE AND GAP*

| | |
|---|---|
| Type | |
| 1966-1974 | HD No. 3-4 |
| 1975-1976 | HD No. 5-6 |
| 1977-1979 | HD No. 5A6 (standard) or HD No. 5R6 (resistor) |
| 1980-on | HD No. 5R6A or HD No. 5RL |
| Gap | |
| 1966-1969 | 0.025-0.030 in. (0.63-0.76 mm) |
| 1970-early 1978 | 0.028-0.033 in. (0.71-0.84 mm) |
| Late 1978-on | 0.038-0.043 in. (0.96-1.09 mm) |

* Spark plugs recommended are Harley-Davidson brands.

### Table 11 CARBURETOR IDLE SPEED

| Carburetor | Idle |
|---|---|
| HD | 900-1,100 rpm |
| Bendix | 700-900 rpm |
| Model DC | * |
| Keihin | 900 rpm |

* Adjust until engine idles and runs smoothly.

# CHAPTER FOUR

# ENGINE

The engine is an air-cooled 4-cycle, overhead-valve V-twin design. The engine has three major assemblies: top end, crankcase and gearcase. Viewed from the engine's right side, engine rotation is clockwise.

Both cylinders fire once in 720° of crankshaft rotation. The rear cylinder fires 315° after the front cylinder. The front cylinder fires again in another 405°. Note that one cylinder is always on its exhaust stroke when the other fires on its compression stroke.

This chapter provides complete service and overhaul procedures, including information for disassembly, removal, inspection, service and reassembly of the engine. **Tables 1-8** at the end of this chapter provide complete engine specifications.

Before starting any work, read the service hints in Chapter One. You will do a better job with this information fresh in your mind.

## ENGINE PRINCIPLES

**Figure 1** explains how the engine works. This will be helpful when troubleshooting or repairing the engine.

## SERVICING ENGINE IN FRAME

Many components can be serviced while the engine is mounted in the frame:

  a. Rocker arm cover.

  b. Cylinder head.

  c. Cylinder and pistons.

  d. Camshaft.

  e. Gearshift mechanism.

  f. Clutch.

  g. Transmission.

  h. Carburetor.

  i. Starter motor and gears.

  j. Generator or alternator and electrical systems.

## ENGINE

### Removal/Installation

1. Thoroughly clean the engine exterior of dirt, oil and foreign material, using one of the cleaners designed for the purpose.

> *NOTE*
> *Because Harley-Davidson models in this manual are not equipped with centerstands, the bike must be secured on blocks underneath the frame. Block the wheels to prevent the bike from rolling when removing the engine components.*

2. Disconnect the negative battery cable.

3. Drain the oil tank as described in Chapter Three.

4. Remove the seat.

5. Remove the instrument panel cover.

6. Remove the fuel tank(s) as described in Chapter Six.

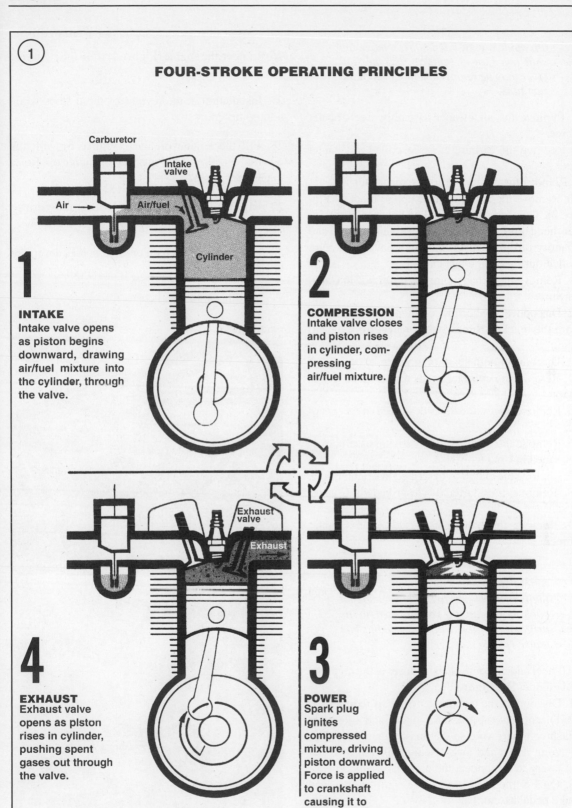

① **FOUR-STROKE OPERATING PRINCIPLES**

Carburetor

Intake valve

Air →   Air/fuel

Cylinder

**1 INTAKE**
Intake valve opens
as piston begins
downward, drawing
air/fuel mixture into
the cylinder, through
the valve.

**2 COMPRESSION**
Intake valve closes
and piston rises
in cylinder, com-
pressing
air/fuel mixture.

Exhaust valve

Exhaust

**4 EXHAUST**
Exhaust valve
opens as piston
rises in cylinder,
pushing spent
gases out through
the valve.

**3 POWER**
Spark plug
ignites
compressed
mixture, driving
piston downward.
Force is applied
to crankshaft
causing it to
rotate.

4

*NOTE*
*On models with hand shift, remove the shift lever lower bolts so that the shift lever may be removed with the left hand fuel tank.*

7. Remove the air cleaner assembly. See Chapter Three.

8. Remove the carburetor as described in Chapter Six.

9. Remove the carburetor intake manifold.

10. Remove the upper engine support bracket (**Figure 2**). There are shims installed between the cylinder head and the frame lug. Note location and quantity of these shims; they will have to be reinstalled during engine installation.

11. Remove the exhaust pipes as described in Chapter Six.

12. Disconnect the horn wires, then remove the horn from the bracket. Remove the bracket from the engine.

13. Disconnect both spark plug wires.

14. Remove the cylinder heads as described in this chapter.

15. Remove the cylinders as described in this chapter.

16. Remove the generator or alternator assembly as described in Chapter Seven.

17. Remove the clutch as described in Chapter Five.

18. Remove the starter as described in Chapter Seven.

19. Remove the battery as described in Chapter Three.

*CAUTION*
*Before disconnecting the oil hoses in Step 20, make sure to label hoses so that each hose can be reconnected to its original position. Failure to do this can cause engine damage.*

20. Label and disconnect all oil hoses at the oil pump and engine. See **Figure 3**.

21. Disconnect the tachometer cable, if required.

22. Disconnect the master cylinder and side plate attachment hardware so that master cylinder is free to swing down and away from crankcase. It is not necessary to disconnect the brake hydraulic line.

23. On FX models, remove the right footpeg and brake pedal assembly.

24. Take a final look all over the engine to make sure everything has been disconnected.

25. Loosen all engine mount bolts and remove them.

26. Remove the engine (**Figure 4**) from the right side of the frame.

27. Installation is the reverse of these steps while noting the following.

28. Fill the engine oil tank (**Figure 5**) with the recommended type and quantity of engine oil. Refer to Chapter Three.

29. Adjust the clutch and throttle cables as described in Chapter Three.

30. Adjust the drive chain or drive belt as described in Chapter Three.

31. Start the engine and check for leaks.

## ROCKER ARM COVER/ CYLINDER HEAD

Refer to **Figure 6** and **Figure 7** for these procedures.

### Removal

This procedure describes removal of the rocker arm cover and cylinder head. The rocker arm cover and cylinder head must be removed together. The unit can be removed with the engine in the frame.

1. Perform Steps 1-12 under *Engine Removal.*

2. Remove the spark plugs.

3. Disconnect the oil feed lines at the cylinder heads. See **Figure 8** and **Figure 9**. Label the lines for reconnection.

4. Using a screwdriver between the cylinder fins and the pushrod cover spring cap retainer (**Figure 10**),

pry the retainer downward and remove it. Repeat for each pushrod.

5. Lift the lower pushrod covers (A, **Figure 11**) upward and secure each with a bent coat hanger as shown in **Figure 11**. Rotate the engine until both valves are closed in the cylinder head to be removed. Valve position can be determined by observing the tappet (B, **Figure 11**) position. If the tappets are lower in the tappet guide, the valves for that cylinder head are closed. Remove the coat hangers.

6. Loosen the cylinder head bolts in a criss-cross pattern (**Figure 12**) and remove them.

*NOTE*
*Before performing the following steps, mark the individual parts during removal so that they can be reinstalled into their original positions.*

7. Tap the cylinder head with a rubber mallet to free it.

8. Lift the cylinder head enough to remove the pushrods and pushrod covers.

9. Remove the cylinder head (**Figure 13**) and cylinder head gasket.

10. Repeat Steps 1-10 and remove the opposite cylinder head.

11. Disassemble and inspect the rocker arm/cylinder head assembly as described in this chapter.

### Installation

1. Clean the cylinder head (**Figure 14**) and cylinder (**Figure 15**) mating surfaces of any gasket residue.

2. Install a new cylinder head gasket (**Figure 16**), making sure the oil return hole in the gasket lines up with the oil hole in the cylinder head.

3. Install the cylinder head/rocker arm cover assembly (**Figure 13**).

4. Coat cylinder head bolt threads with oil and install them finger-tight only.

*CAUTION*
*Failure to follow the torque pattern and sequence in Step 5 may cause cylinder head distortion and gasket leakage.*

5. Tighten the cylinder head bolts in a criss-cross pattern to the specifications in **Tables 5-8**.

6. Install and adjust pushrods as described in this chapter.

7. Reverse Steps 1-3 under *Cylinder Head Removal.*

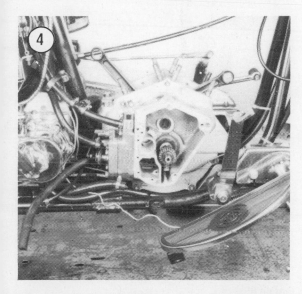

**ROCKER ARM AND
PUSHROD ASSEMBLY**

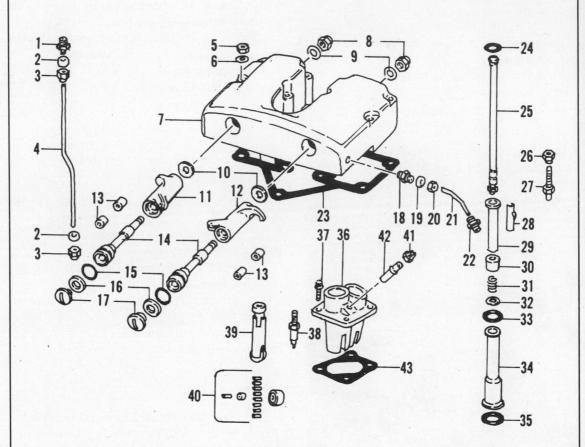

1. Oil fitting
2. Oil line sleeve
3. Fitting
4. Oil line
5. Nut
6. Washer
7. Rocker arm cover
8. Nut
9. Washer
10. Spacer
11. Rocker arm
12. Rocker arm
13. Bushing
14. Rocker arm shaft
15. O-ring

16. Washer
17. Rocker arm shaft screw
18. Oil line fitting
19. Sleeve
20. Nut
21. Oil line
22. Oil line fitting
23. Gasket
24. O-ring
25. Pushrod
26. Nut
27. Tappet adjusting screw
28. Spring cap retainer
29. Upper pushrod cover

30. Cap
31. Spring
32. Washer
33. O-ring
34. Lower pushrod cover
35. O-ring
36. Tappet guide
37. Screw
38. Pushrod hydraulic
39. Tappet and roller
40. Tappet roller kit
41. Clamp
42. Oil fitting (1981 models)
43. Gasket

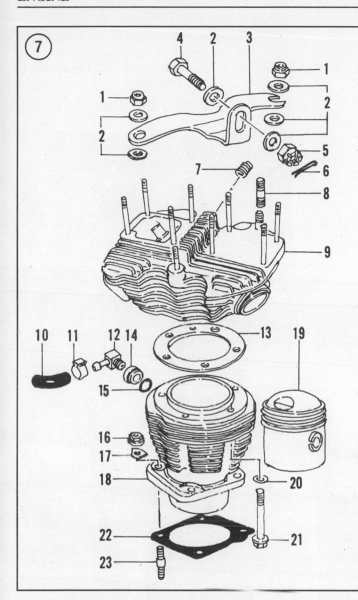

**CYLINDER HEAD AND
CYLINDER ASSEMBLY**

**4**

1. Nut
2. Washer
3. Engine brace
4. Bolt
5. Nut
6. Cotter pin
7. Helicoil insert
8. Stud
9. Cylinder head
10. Hose (1981-1982)
11. Clamp (1981-1982)
12. Fitting (1981-1982)
13. Cylinder head gasket
14. Spacer (1981-1982)
15. O-ring (1981-1982)
16. Nut
17. Spacer
18. Cylinder
19. Piston
20. Washer
21. Bolt
22. Gasket
23. Stud

## Removal/Installation

Refer to **Figure 6** for this procedure.

1. Remove the rocker arm cover nuts and washers (**Figure 6**).

2. Tap the rocker arm cover with a plastic mallet and remove it from the cylinder head.

3. Remove the rocker arm shaft screw and O-ring (**Figure 17**). On 1981 and later models, remove the spacer washer also. Discard the O-ring.

4. Turn the rocker arm cover over so that the rocker arm shaft and rocker arms are exposed.

5. Tap the rocker arm shaft (**Figure 18**) out of the cover.

6. Lift the rocker arm (**Figure 19**) from the rocker arm cover.

7. Remove the spacer (**Figure 20**) from the rocker arm cover. See **Figure 21**.

8. Remove any rocker arm shims (C, **Figure 22**), if used.

9. Repeat Steps 3-8 for the opposite rocker arm.

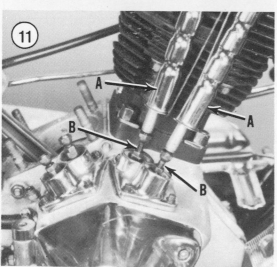

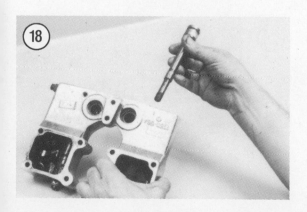

10. Inspect the rocker arms as described in this chapter.

11. Installation is the reverse of these steps while noting the following.

12. Coat all bearing surfaces with engine oil.

13. Install the spacer (**Figure 21**) and any shims if used.

14. Position the rocker arm (**Figure 23**) in the cover. Make sure the spacer and shims (if used) are not displaced.

4

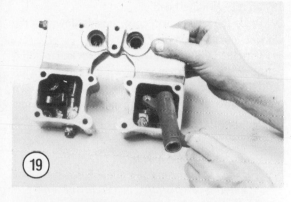

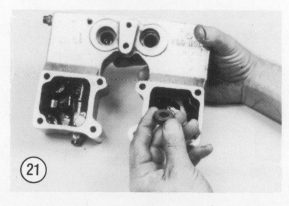

15. Lightly tap the rocker arm shaft (**Figure 18**) through the rocker arm cover and into the rocker arm and spacer.

16. Install the rocker arm shaft washer and locknut (**Figure 24**). Tighten the locknut to specification in **Tables 7**.

17. Check that the rocker arm rotates freely on the shaft. Remove the rocker arm shaft and reinstall if any binding is detected.

18. Install a new O-ring, spacer washer (1981-on models) and cap screw (**Figure 17**). Tighten the cap screw to specification in **Tables 7**.

19. Install a new rocker arm cover gasket (A, **Figure 25**) on the cylinder head.

20. Place the rocker arm cover upside down on the workbench as shown in **Figure 26**.

21. Turn the cylinder head upside down and install it on the rocker arm cover as shown in **Figure 27**. Assembling the cylinder head onto the rocker arm cover in this manner will prevent the rocker arms from falling down and jamming against the valve stems.

> *CAUTION*
> *If the rocker arm and cylinder head assemblies are installed in a right-side up manner, make sure the rocker arms do not jam against the valve stems when the rocker arm cover is installed.*

22. Install the rocker arm cover nuts (**Figure 6**) and washers and tighten to specifications in **Tables 5-8**.

**Rocker Arm Inspection**

1. Clean all parts in solvent. Blow compressed air through all oil passages to make sure they are clear.

2. Examine the rocker arm pads and ball sockets (**Figure 28**) for pitting and excessive wear; replace the rocker arms if necessary.

3. Measure the rocker arm shaft outside diameter (**Figure 29**) where it rides in the rocker arm. Then

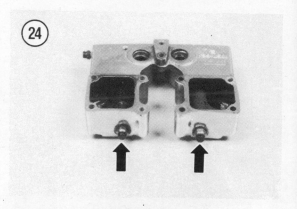

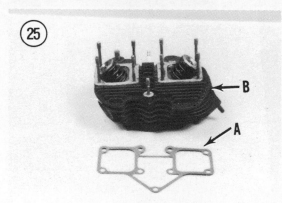

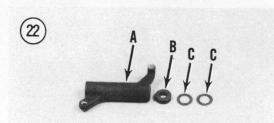

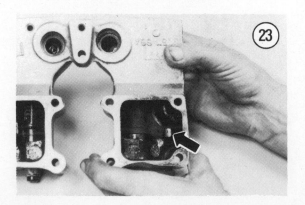

measure the rocker arm shaft bore inside diameter (**Figure 6**). Replace the bushings if the clearance exceeds the specifications in **Table 2** or **Table 4**. Rocker arm bushing replacement is described in this chapter.

### Rocker Arm Bushing Replacement

Replace worn or damaged rocker arm bushings as follows.

1. Press or drive the old rocker arm bushings from the rocker arm. If the bushings prove difficult to remove, perform the following:

    a. Thread a 9/16-18 in. tap into the bushing (15, **Figure 6**).

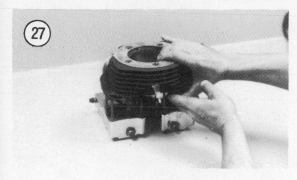

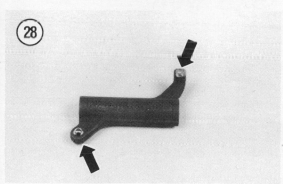

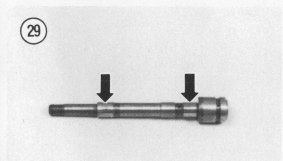

b. Remove the tap.

c. Thread a 9/16-18 in. bolt into the bushing. Then carefully tap on the bolt and drive the bushing out.

> *NOTE*
> *If there are not enough threads in the bushing to engage the bolt, it will be necessary to reinstall the tap to drive the bushing out.*

2. Using rifle cleaning brushes, clean the rocker arm bushing bore thoroughly of all foreign material. Wash the rocker arm thoroughly and dry with compressed air.

3. Press the new bushings into the rocker arm. Make sure the oil holes in the bushings are correctly aligned with those in the rocker arm and that the split portion of the bushing faces toward the top of the arm. Press the bushing flush with the edge of the arm.

4. The new bushings must be reamed with Harley-Davidson tool HD-94804-57 (**Figure 30**). Have a Harley-Davidson dealer do this for you if you do not have access to the reamer.

### Cylinder Head Inspection

1. Without removing valves, remove all carbon deposits from the combustion chambers (**Figure 31**) with a wire brush. A blunt screwdriver or chisel may be used if care is taken not to damage the head, valves and spark plug threads.

> *CAUTION*
> *If the combustion chambers are cleaned while the valves are removed, make sure to keep the scraper or wire brush away from the valve seats to prevent damag-*

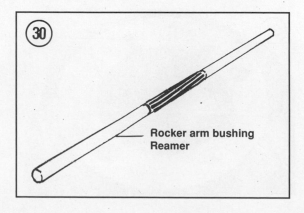

Rocker arm bushing Reamer

*ing the seat surfaces. A damaged or even slightly scratched valve seat will cause poor valve seating.*

2. Examine the spark plug threads in the cylinder head for damage. If damage is minor or if the threads are dirty or clogged with carbon, use a spark plug thread tap (**Figure 32**) to clean the threads following the manufacturer's instructions. If thread damage is severe, refer further service to a Harley-Davidson dealer or competent machine shop.

3. After all carbon is removed from combustion chambers and valve ports and the spark plug thread holes are repaired, clean the entire head in solvent.

*NOTE*
***Figure 33*** *and* ***Figure 34*** *show the cylinder removed from the engine. Cylinder removal is not necessary for this procedure.*

4. Clean away all carbon on the piston crowns (**Figure 33**). Do not remove the carbon ridge at the top of the cylinder bore (**Figure 34**).

5. Check for cracks in the combustion chamber and exhaust ports. A cracked head must be replaced.

6. After the head has been thoroughly cleaned, place a straightedge across the gasket surface at several points. Measure warp by inserting a feeler gauge between the straightedge and cylinder head at each location. There should be no warpage; if a small amount is present, refer service to a Harley-Davidson dealer.

7. Check the rocker arm cover mating surface using the procedure in Step 6. There should be no warpage.

8. Check the valves and valve guides as described under *Valves and Valve Components* in this chapter.

9. Check the pushrods as described in this chapter.

## PUSHRODS

### Removal (All Models)

Refer to **Figure 7**.

1. Using a screwdriver between the cylinder fins and the pushrod cover spring cap retainer (**Figure 35**), pry the retainer downward and remove it.

2. Lift the lower pushrod cover (A, **Figure 36**) upward and secure with a bent coat hanger as shown in **Figure 36**.

3. Rotate the engine until the tappet (B, **Figure 36**) for the pushrod to be removed just starts upward.

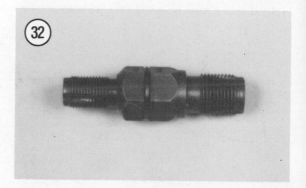

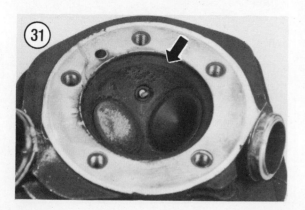

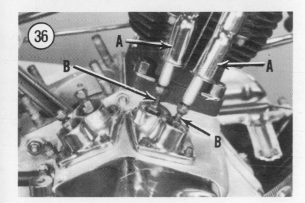

4. Loosen the pushrod locknut (**Figure 37**) and turn the adjusting screw (**Figure 37**) to obtain slack in the pushrod.

5. Lift the pushrod assembly (**Figure 38**) out of the engine.

6. Remove and discard the upper (**Figure 39**) and lower (**Figure 40**) O-rings.

*NOTE*
*Figure 41 shows the pushrod O-ring position with the rocker arm cover removed for clarity. It is not necessary to remove the rocker arm cover for this procedure.*

7. Repeat Steps 1-6 to remove remaining pushrods and O-rings.

**Disassembly/Inspection/Assembly**

1. Remove the pushrod from the pushrod cover assembly.

2. Disassemble the pushrod cover as follows:

   a. Remove the lower pushrod cover (**Figure 42**).

   b. Remove the middle O-ring (**Figure 43**).

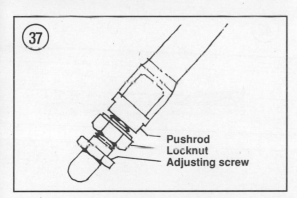

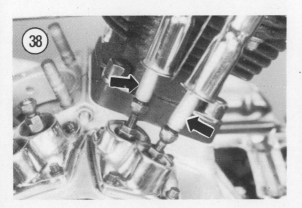

Pushrod
Locknut
Adjusting screw

c. Remove the washer (**Figure 44**).

d. Remove the spring (**Figure 45**).

e. Remove the cover spring cap (**Figure 46**).

3. Check the pushrod ends for wear.

4. Unscrew the tappet adjusting screw from the pushrod. Check the adjusting screw-to-tappet end for wear or damage. Check the tappet adjusting screw threads in the end of the pushrod. Replace worn or damaged parts.

5. Roll the pushrods on a flat surface, such as a piece of glass, and check for bending. Replace the pushrod if bent.

6. Reverse Step 4 and Step 2 to reassemble the pushrod. Install a new middle O-ring (**Figure 43**) during reassembly. Push the lower pushrod cover into the cover spring cap to seat the O-ring. See **Figure 47**.

**Pushrod Installation and Adjustment**

The pushrods must be adjusted whenever they are removed or after assembling the engine. Before adjusting the pushrod, its valve must be fully closed. Valve position can be determined by observing the corresponding pushrod on the other cylinder. For example, when the rear cylinder intake valve lifter is raised fully, the front cylinder intake valve is fully closed, and the front intake pushrod can be adjusted.

1. Before installing or adjusting the pushrods, the hydraulic lifters (**Figure 48**) must be clean and free of all oil. See *Tappets and Tappet Guides* in this chapter.

2. Install new upper (**Figure 39**) and lower (**Figure 40**) pushrod O-rings.

3. Rotate the engine until the front cylinder exhaust tappet just starts upward.

4. Install the rear cylinder exhaust pushrod and pushrod cover.

5. Raise the lower pushrod cover and secure it with a coat hanger as shown in **Figure 36**. This allows easy access to the pushrod adjustment screw.

6. Rotate the engine until the front cylinder intake pushrod tappet just starts upward.

7. Install the rear cylinder intake pushrod and pushrod cover.

8. Repeat for front cylinder pushrods and covers.

9. Adjust the pushrods as follows.

*NOTE*
*The engine must be dead cold when adjusting the pushrods.*

*NOTE*
*Hydraulic lifters are standard on all models covered in this manual.*

10A. *Hydraulic lifters:*

a. Each lifter must be adjusted when its pushrod is at its lowest position. See paragraph at the

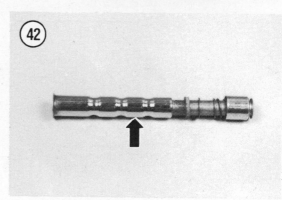

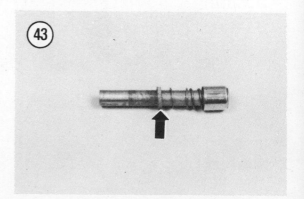

beginning of this procedure. Then turn the pushrod adjusting screw clockwise (downward) until the pushrod's ball end is seated in the hydraulic lifter with no appreciable play. See **Figure 37**.

b. Hold the pushrod across its flats with a wrench. Then turn the adjusting screw clockwise (downward) slowly until the hydraulic lifter is completely compressed.

c. Turn the adjusting screw counterclockwise (upward) 1 3/4 turns (1966-early 1978) or 1

1/2 turns (late 1978-1984) to complete the adjustment. Then turn the locknut against the pushrod. Tighten the locknut to 6-11 ft.-lb. (8.3-15.2 N•m).

d. Reposition engine and repeat for remaining pushrods.

*NOTE*
*Solid lifters are not factory installed on 1966-1984 models covered in this manual. However, they were used on other early model Harley-Davidsons. Because solid lifters are occasionally used to replace stock hydraulic lifters, solid lifter adjustment is described here for the convenience of the manual user.*

10B. *Solid lifters:* Position lifters as described in Step 10A. Then turn the adjuster (**Figure 37**) until the pushrod can be turned freely by hand with no up or down play. Tighten the locknut and recheck the adjustment.

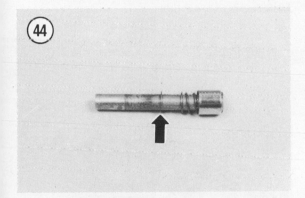

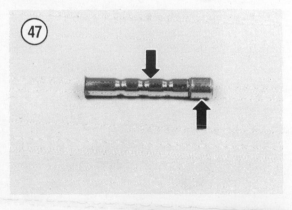

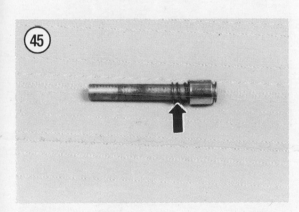

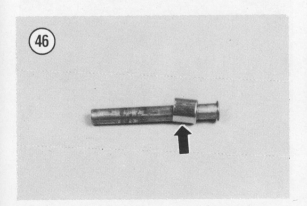

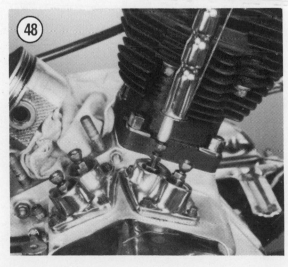

11. Using a screwdriver as shown in **Figure 49**, lift the upper pushrod cover up (**Figure 50**) and seat it against the upper O-ring. See **Figure 51**.

12. Install the pushrod retainer as shown in **Figure 52**.

## TAPPETS AND TAPPET GUIDES

All models are factory equipped with roller tappets and hydraulic lifters. Hydraulic lifters consist of a piston, cylinder and check valve. During operation, hydraulic lifters pump full of engine oil, thus taking up all play in the valve train. When the engine is off, the hydraulic lifters will "leak down" as oil escapes from the hydraulic unit. When the engine is started, it is normal for the tappets to click until they refill with oil. If the tappets stop clicking after the engine is run for a few minutes, they are working properly.

### Removal

During removal, store tappets in proper sequence for installation in their original positions. Refer to **Figure 53** for this procedure.

1. Remove the pushrods as described in this chapter.

2. On late 1981 and early 1982 models, disconnect the oil hose (**Figure 53**) from the tappet guide.

3. Lift the hydraulic lifter (**Figure 54**) from the tappet and roller assembly.

4. Remove the 4 tappet guide screws (**Figure 55**). Loosen the tappet guide by striking it lightly with a plastic-tipped hammer.

5. Use your fingers and push the tappet and roller assembly against the side of the tappet guide to hold it in position. Then lift the tappet guide away from the gearcase while still holding the tappets in position. This will prevent the tappets from falling into the gearcase.

6. Pull the tappet and roller assembly out from the bottom of the guide (**Figure 56**). **Figure 57** shows solid lifters.

7. Remove the lower pushrod O-rings or cork washers (**Figure 58**) from the tappet guide.

### Disassembly/Inspection

*CAUTION*
*Hydraulic lifters are selectively fitted, and their parts must not be inter-*

52

55

4

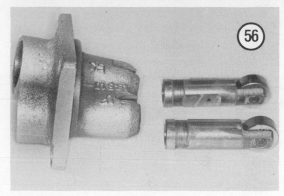

56

### HYDRAULIC TAPPET ASSEMBLY

53

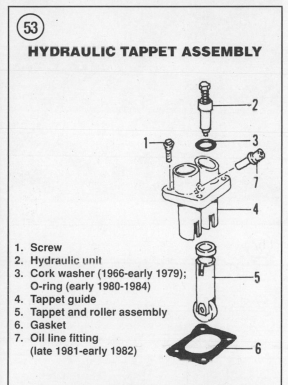

2
1
3
7
4
5

1. Screw
2. Hydraulic unit
3. Cork washer (1966-early 1979);
   O-ring (early 1980-1984)
4. Tappet guide
5. Tappet and roller assembly
6. Gasket
7. Oil line fitting
   (late 1981-early 1982)

6

57

54

58

*changed with other units. Store each unit in a separate container.*

1. Twist the hydraulic lifter and pull the piston and spring from the cylinder.

2. Soak the hydraulic lifter components in clean engine oil. Then remove and place on a clean towel and allow oil to drain from components.

3. Blow out all tappet guide and hydraulic oil passages with compressed air.

4. Clean the tappet guide oil channel openings with a piece of wire.

5. Check the tappet rollers (**Figure 59**) for pitting, scoring, galling or excessive wear. If the rollers are worn excessively, check the cam lobes (**Figure 60**) for the same wear conditions. The cam lobes can be observed through the tappet guide hole in the crankcase (**Figure 61**). Replace the cam, if necessary, as described in this chapter.

> *NOTE*
> *Figure 60 shows the cam removed for clarity. Cam removal is not necessary for this procedure.*

6. Check the roller end clearance by grasping the tappet assembly in one hand and attempting to move the roller back and forth. **Table 2** and **Table 4** list wear specifications. Replace the tappet assembly if necessary.

**Reassembly/Installation**

1. Slide the tappets through the bottom of the tappet guide so that the tappet oil holes face to the center of the guide (**Figure 62**).

> *NOTE*
> *If the tappets are not installed correctly, the exhaust tappets cannot fill with engine oil.*

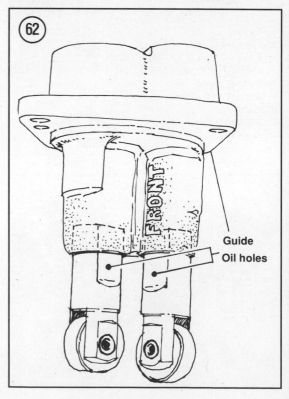

Guide

Oil holes

2. Install a new tappet guide gasket (**Figure 63**). Do not use any type of sealer on the gasket.

3. Hold the tappets with your fingers as during removal and install the tappet guide onto the gearcase. Do not allow the tappets to drop into the gearcase.

4. Install the hydraulic units and tappet guide retaining screws.

5. Install new lower pushrod O-rings or cork washers. See 3, **Figure 53**.

6. Tighten the tappet guide retaining screws securely.

7. On late 1981-early 1982 models, reconnect the oil line at the tappet fitting (7, **Figure 53**).

8. Install and adjust the pushrods as described under *Pushrods* in this chapter.

**Tappet Oil Screen**

The tappet oil screen should be cleaned at specified service intervals. See Chapter Three for complete information.

> *CAUTION*
> *Failure to clean the tappet oil screen as described in Chapter Three will cause premature tappet wear.*

## VALVES AND VALVE COMPONENTS

General practice among those who do their own service is to remove the cylinder heads and take them to a Harley-Davidson dealer or machine shop for inspection and service. Since the cost is low relative to the required effort and equipment, this is the best approach, even for experienced mechanics.

Refer to **Figure 64** for this procedure.

> *CAUTION*
> *All component parts of each valve assembly must be kept together. Do not mix with like components from other valves or excessive wear may result.*

1. Remove the cylinder head(s) as described in this chapter.

2. Remove the exhaust valve stem pads on early models (if so equipped). See 17, **Figure 5**.

3. Install a valve spring compressor squarely over the upper valve collar with other end of tool placed against valve head (**Figure 65**).

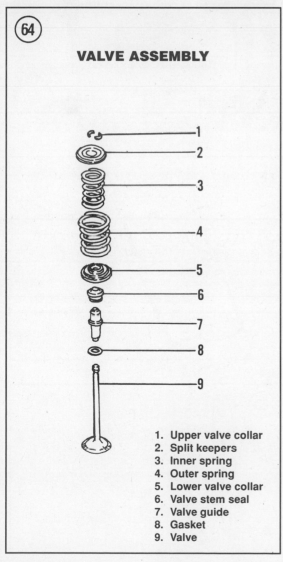

**VALVE ASSEMBLY**

1. Upper valve collar
2. Split keepers
3. Inner spring
4. Outer spring
5. Lower valve collar
6. Valve stem seal
7. Valve guide
8. Gasket
9. Valve

4. Tighten valve spring compressor until split valve keepers separate (**Figure 66**). Lift out split keepers with needlenose pliers.

5. Gradually loosen valve spring compressor and remove from head. Lift off the upper valve collar.

> *CAUTION*
> *Remove any burrs from the valve stem grooves before removing the valve (**Figure 67**). Otherwise the valve guides will be damaged.*

6. Remove inner and outer springs.

7. Remove the valve stem seals, if so equipped.

8. Remove the valve.

9. Remove the lower spring collar.

10. Repeat Steps 2-9 and remove remaining valves.

### Inspection

1. Clean valves with a wire brush and solvent.

2. Inspect the contact surface (**Figure 68**) of each valve for burning. Minor roughness and pitting can be removed by lapping the valve as described in this chapter. Excessive unevenness of the contact surface is an indication that the valve is not serviceable. The contact surface of the valve may be ground on a valve grinding machine, but it is best to replace a burned or damaged valve with a new one.

3. Inspect the valve stems for wear and roughness.

4. Remove all carbon and varnish from the valve guides with a stiff spiral wire brush.

> *NOTE*
> *Step 5 and Step 6 require special measuring equipment. If you do not have the required measuring devices, proceed to Step 8.*

5. Measure valve stem outside diameter where it rides in the valve guide with a micrometer (**Figure 69**).

6. Measure each valve guide at top, center and bottom with a small hole gauge.

7. Subtract the measurement made in Step 5 from the measurement made in Step 6 above. The difference is the valve guide-to-valve stem clearance. See specifications in **Table 2** or **Table 4** for correct clearance. Replace any guide or valve that is not within tolerance.

8. Insert each valve in its guide. Hold the valve just slightly off its seat and rock it sideways. If it rocks more than slightly, the guide is probably worn and

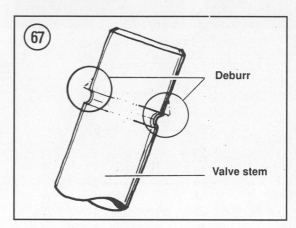

Deburr

Valve stem

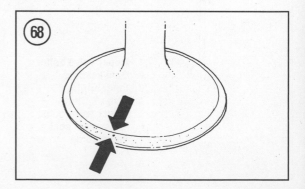

should be replaced. As a final check, take the head to a dealer and have the valves and valve guides measured.

9. Measure the valve spring heights with a vernier caliper (**Figure 70**). All should be of length specified in **Table 2** or **Table 4** with no bends or other distortion. Replace defective springs.

10. Check the valve spring retainer and valve keepers. If they are in good condition, they may be reused.

11. Inspect valve seats. If worn or burned, they must be reconditioned. This should be performed by your dealer or local machine shop. Seats and valves in near-perfect condition can be reconditioned by lapping with fine carborundum paste.

### Installation

1. Coat the valve stems with oil and insert into cylinder head.

2. Install the bottom spring collar.

3. Install a new valve stem seal (if so equipped).

4. Install valve springs. Then install the upper valve spring collar.

5. Push down on upper valve spring collar with the valve spring compressor and install valve keepers. After releasing tension from compressor, examine valve keepers and make sure they are seated correctly.

### Valve Guide Replacement

When guides are worn so that there is excessive stem-to-guide clearance or valve tipping, they must be replaced. Replace all, even if only one is worn. This job should be done only by a Harley-Davidson dealer or qualified specialist as special tools are required.

### Valve Seat Reconditioning

This job is best left to your dealer or local machine shop. They have the special equipment and knowledge for this exacting job. You can still save considerable money by removing the cylinder heads and taking just the heads to the shop.

### Valve Lapping

Valve lapping is a simple operation which can restore the valve seal without machining if the amount of wear or distortion is not too great.

1. Smear a light coating of fine grade valve lapping compound on seating surface of valve (**Figure 71**).

2. Insert the valve into the head.

3. Wet the suction cup of the lapping stick (**Figure 72**) and stick it onto the head of the valve. Lap the valve to the seat by spinning tool between hands while lifting and moving valve around seat 1/4 turn at a time.

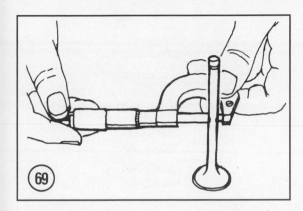

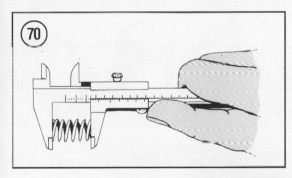

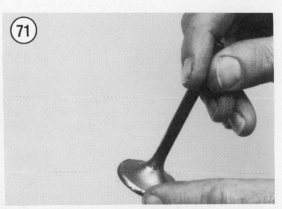

4. Wipe off valve and seat frequently to check progress of lapping. Lap only enough to achieve a precise seating ring around valve head (**Figure 73**).

5. Closely examine valve seat in cylinder head. It should be smooth and even with a smooth, polished seating "ring."

6. Thoroughly clean the valves and cylinder head in solvent to remove all grinding compound. Any com-

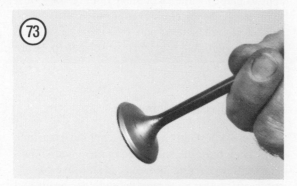

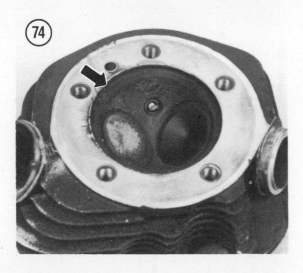

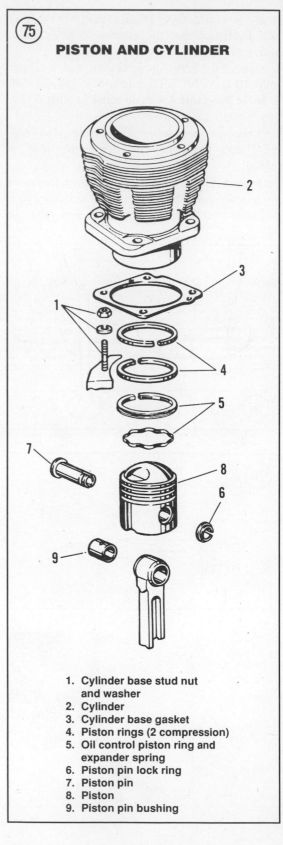

**PISTON AND CYLINDER**

1. Cylinder base stud nut and washer
2. Cylinder
3. Cylinder base gasket
4. Piston rings (2 compression)
5. Oil control piston ring and expander spring
6. Piston pin lock ring
7. Piston pin
8. Piston
9. Piston pin bushing

pound left on the valves or the cylinder head will end up in the engine and will cause damage.

7. After the lapping has been completed and the valve assemblies have been reinstalled into the head the valve seal should be tested. Check the seal of each valve by pouring solvent into each of the intake and exhaust ports. There should be no leakage past the seat. If leakage occurs, combustion chamber (**Figure 74**) will appear wet. If fluid leaks past any of the seats, disassembly that valve assembly and repeat the lapping procedure until there is no leakage.

## CYLINDER

Refer to **Figure 75**, typical for this procedure.

### Removal

1. Remove the cylinder head as described in this chapter.
2. Remove all dirt and foreign material from the cylinder base.
3. Turn the engine over until the piston is at bottom dead center (BDC).
4. Remove the cylinder base nuts (A, **Figure 76**) and washers.
5. On late 1981 to early 1982 models, disconnect the oil hose fitting at the cylinder.
6. Loosen the cylinder (B, **Figure 76**) by tapping around the perimeter with a rubber or plastic mallet.
7. Pull the cylinder straight up and remove the remaining cylinder head bolts (C, **Figure 76**) and the cylinder.
8. Stuff clean shop rags into the crankcase opening (**Figure 77**) to prevent objects from falling into the crankcase.
9. Repeat Steps 1-8 for the other cylinder.

### Inspection

The following procedure requires the use of highly specialized and expensive measuring instruments. If such instruments are not readily available, have the measurements performed by a dealer or qualified machine shop.

1. Thoroughly clean the cylinder with solvent and dry with compressed air.
2. Using a gasket scraper, remove all gasket residue from the cylinder surfaces.
3. Inspect the top (**Figure 78**) and bottom (**Figure 79**) cylinder machined surfaces for nicks, scratches or burrs. Remove minor damage with a fine-cut file.
4. Measure the cylinder bores with a cylinder gauge (**Figure 80**) or inside micrometer at the points shown in **Figure 81**.
5. Measure in 2 axes—in line with the piston pin and at 90° to the pin. If the taper or out-of-round is greater than specifications (**Table 2** or **Table 4**), the cylinders must be rebored to the next oversize and new pistons and rings installed. Rebore both cylinders even though only one may be worn.

*NOTE*
*A feeler gauge and a piston compression ring can be used to determine taper if you are not equipped with a bore gauge. To check, install the compression ring in the cylinder at the upper limit of ring travel. Measure the ring gap. Then push the ring down to the lower limit of ring travel and remeasure the ring gap. The top gap minus the bottom gap measurement, divided by 3, will give an approximation of the cylinder's taper.*

6. Hold a light in one end of the cylinder and check the cylinder wall from the opposite end. Check for scratches and scoring; if evident, the cylinders should be rebored.

7. Lightly oil the cylinder bore to prevent rust after performing Steps 1-6.

**Installation**

1. **Figure 82** identifies the front and rear cylinders.

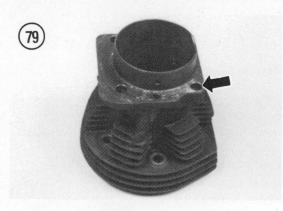

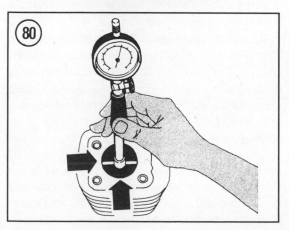

2. If the base gasket is stuck to the bottom of the cylinder it should be removed and the cylinder surface (**Figure 79**) cleaned thoroughly.

3. Check that the top cylinder surface (**Figure 78**) is clean of all old gasket material.

4. Install a new cylinder base gasket in the crankcase. Make sure all holes align.

5. Turn the engine over until the piston is at bottom dead center (BDC). Guide the pistons so they don't hang up on other parts.

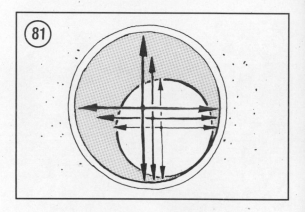

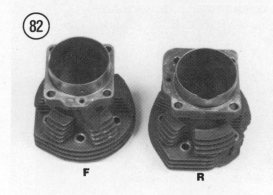

F          R

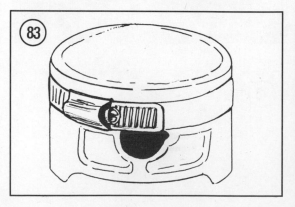

6. Lubricate the cylinder bores and pistons liberally with engine oil.

7. Compress the rings with aircraft type hose clamps of appropriate diameter (**Figure 83**). Tighten the hose clamps just enough to compress the rings.

*CAUTION*
*Don't tighten the clamps any more than necessary to compress the rings. If the rings can't slip through easily, the clamps may gouge the rings and piston.*

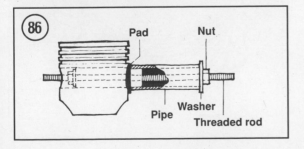
Pad  Nut
Pipe  Washer
Threaded rod

8. Carefully align the cylinder over the piston and slide it down. Once the rings are positioned in the cylinder, remove the hose clamps installed in Step 7.

9. Rotate the cylinder as necessary and slide it over the crankcase studs.

10. Install the cylinder spacers (**Figure 84**) (if so equipped) and nuts. If cylinder spacers are used, install them with the stamped side facing up. Tighten the base nuts to specifications in **Tables 7**. On late 1981 and early 1982 models, attach the oil hoses to the cylinder fittings. If the fittings were removed, reconnect using a new O-ring. Make sure the fittings are aligned correctly.

11. Install the cylinder head as described in this chapter.

## PISTONS AND PISTON RINGS

### Piston Removal/Installation

1. Remove the cylinder head and cylinder as described in this chapter.

2. Stuff the crankcase with clean shop rags to prevent objects from falling into the crankcase.

3. Lightly mark the pistons with a F (front) or R (rear) so they will be installed into the correct cylinders. Also mark the piston crown with an arrow pointing to the front of the bike. Because the piston pins are offset in the pistons, the pistons must not be installed backwards.

4. Remove the piston rings as described under *Piston Ring Replacement* in this chapter.

*WARNING*
*Because the piston pin retaining rings are highly compressed in the piston pin ring groove, safety glasses must be worn during their removal and installation.*

5. Various types of piston pin circlips (**Figure 85**) have been used. On some models, the circlips can be removed with a pair of circlip pliers. On other models, use an awl and pry the piston pin retaining rings out of the piston. Place your thumb over the hole to help prevent the rings from flying out during their removal.

6. Support the piston and push out the piston pin. If the piston is seized, the pin may be difficult to remove. Use a homemade tool as shown in **Figure 86**.

7. Inspect the piston as described in this chapter.

8. Installation is the reverse of these steps. Note the following.

9. Coat the connecting rod bushing, piston pin and piston with assembly oil.

10. Place the piston over the connecting rod. If you are installing old parts, make sure the piston is installed on the correct rod as marked during removal. On 1966-early 1978 models, note that there is a web on one piston pin boss inside the piston (**Figure 87**). On these models, install both pistons so that these webs are on the right-hand side of the engine.

11. Insert the piston pin until it starts into the connecting rod bushing. Hold the rod so that the lower end does not take any shock. If the pin does not slide easily, use the homemade tool (**Figure 86**) but eliminate the piece of pipe. Drive the pin in until it is centered in the piston.

*NOTE*
*An easy way of heating the piston to ease piston pin installation is to wrap the piston in shop rags that have been soaked in hot water.*

12. Install *new* piston pin retaining rings. Make sure each ring seats fully in the piston groove. Turn the retaining ring so that its gap is away from the slot at the bottom.

13. Install rings as described under *Piston Ring Replacement* in this chapter.

**Piston Inspection**

1. Carefully clean the carbon from the piston crown with a soft scraper (**Figure 88**). Do not remove or damage the carbon ridge around the circumference of the piston above the top ring.

*CAUTION*
*Do not wire brush piston skirts.*

2. Examine each ring groove for burrs, dented edges and wide wear. Pay particular attention to the top compression ring groove. Because the top ring groove is closer to the combustion chamber and thus subjected to higher temperatures and pressures, it tends to wear more than the lower ring grooves.

3. Using a broken piston ring, remove all carbon from the ring grooves (**Figure 89**).

4. Measure piston-to-cylinder clearance as described under *Piston Clearance* in this chapter.

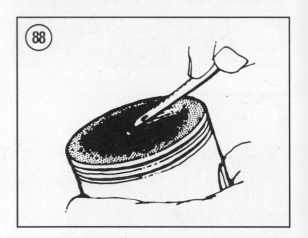

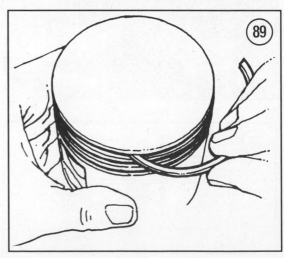

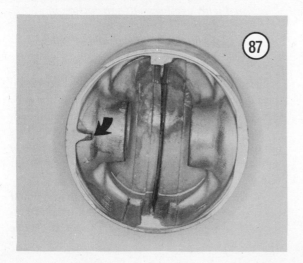

5. If damage or wear indicate piston replacement, select a new piston as described under *Piston Clearance* in this chapter.

## Piston Clearance

1. Make sure the piston and cylinder walls are clean and dry.

2. Measure the inside diameter of the cylinder bore at a point 1/2 in. (12.7 mm) from the upper edge with a bore gauge (**Figure 80**).

3. Measure the outside diameter of the piston across the skirt (**Figure 90**) at right angles to the piston pin.

*NOTE*
*Late 1983 and later models are equipped with barrel shape pistons and*

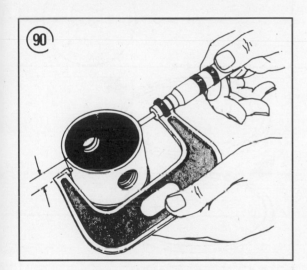

they may be difficult to measure as described in Step 3.

4. Piston clearance is the difference between the maximum piston diameter and the minimum cylinder diameter. Subtract the dimension of the piston from the cylinder dimension. If the clearance exceeds the dimension listed in **Table 2** or **Table 4**, the cylinder should be rebored to the next oversize and a new piston and rings installed.

*NOTE*
*The new pistons should be obtained before the cylinders are rebored so the pistons can be measured individually; slight manufacturing tolerances must be taken into account to determine the actual bore size.*

## Piston Ring Replacement

*NOTE*
*The second piston ring's lower inner edge is tapered on 80 cu. in. (340 cc) engines from case number 1479-345-165 on and on 74 cu. in. (1200 cc) engines from case number 179-023-001 on. See **Figure 91**. This type of ring has a dot or dash marked on the side that must face up. Both rings may have one side stamped "TOP."*

1. Remove the old rings with a ring expander tool or by spreading the ring ends with your thumbs and lifting the rings up evenly (**Figure 92**).

2. Check end gap of each ring. To check ring, insert the ring into the bottom of the cylinder bore and square it with the cylinder wall by tapping it with the piston. The ring should be pushed in about 1/2 in. (12.7 mm). Insert a feeler gauge as shown in **Figure 93**. Compare gap with **Table 2** or **Table 4**. Replace ring if gap is too large.

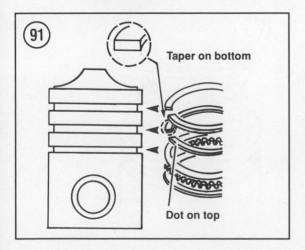

**Taper on bottom**

**Dot on top**

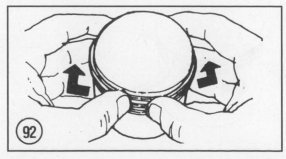

*NOTE*
*The oil control ring expander spacer is
unmeasurable. If the oil control ring
rails show wear, both rails and the
spacer should be replaced as a set.*

3. Roll each ring around its piston groove as shown
in **Figure 94** to check for binding. Minor binding
may be cleaned up with a fine-cut file.

*NOTE*
*Install all rings with their markings fac-
ing up.*

4. Install oil ring in oil ring groove with a ring
expander tool or by spreading the ends with your
thumbs.

5. Install the 2 compression rings carefully with a
ring expander tool or by spreading the ends with
your thumbs. Oversize compression rings are
stamped with a number to indicate their size. If
installing oversize compression rings, check the
number to make sure the correct rings are being
installed. The ring numbers should be the same as
the piston oversize number.

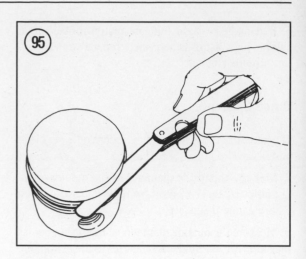

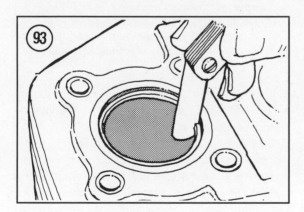

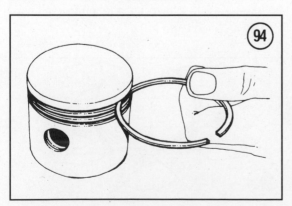

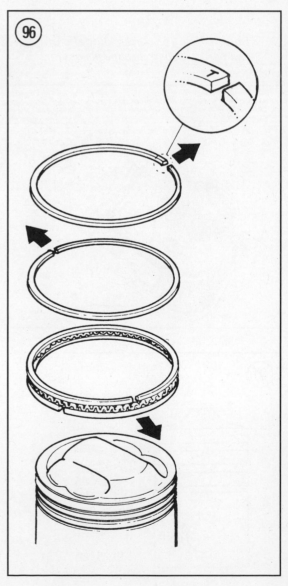

6. Measure the piston ring side clearance as shown in **Figure 95**. Compare to specifications in **Table 2** or **Table 4**.

7. Stagger ring gaps around piston every 120° as shown in **Figure 96**.

## OIL PUMP

The oil pump is installed at the rear of the gearcase. The pump consists of 2 sections: a feed pump which supplies oil under pressure to the engine components, and a scavenger pump which returns oil to the oil tank from the engine.

The oil pump may be removed as a unit only with the engine removed. However, the oil pump can be disassembled with the engine mounted in the frame.

### Removal/Disassembly

Refer to **Figure 97** (1966-1967) or **Figure 98** (1968-on) for this procedure.

*NOTE*
*Label all gears and Woodruff keys during removal as they must be installed in their original positions.*

**With engine in frame**

1. Drain the engine oil tank as described in Chapter Three.

*CAUTION*
*When disconnecting the oil lines in Step 2, make sure to tag each line so that it may be returned to its original position.*

2. Disconnect the oil lines from the pump. Plug the ends of the lines to prevent oil leakage and contamination.

3. Remove the bolts securing the oil pump cover and remove the cover. See 3, **Figure 97** or 3, **Figure 98**.

4. Remove the circlip (5), gear (6), Woodruff key (7) and idler gear (8).

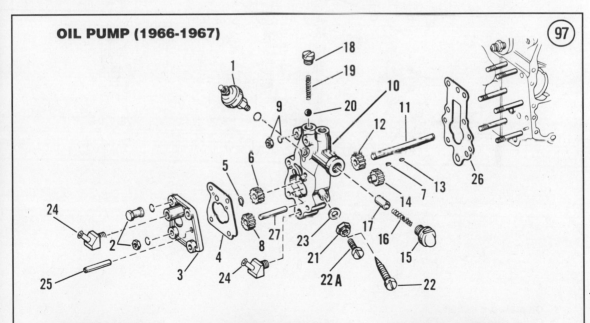

**OIL PUMP (1966-1967)** ⑨⑦

| | | |
|---|---|---|
| 1. Oil pressure switch | 11. Oil pump drive shaft | 19. Check valve spring |
| 2. Cover stud nut | 12. Drive gear | 20. Check valve ball |
| 3. Oil pump cover | 13. Woodruff key | 21. Locknut |
| 4. Gasket | 14. Idler gear | 22. Chain oiler adjusting screw |
| 5. Circlip | 15. Bypass valve plug | 22A. Chain oiler screw (1966-1967) |
| 6. Drive gear | 16. Bypass valve spring | 23. Washer |
| 7. Woodruff key | 17. Bypass valve spring | 24. Oil line nipple (1966-1967) |
| 8. Idler gear | plunger | 25. Chain oiler pipe |
| 9. Nuts and washers | 18. Check valve spring | 26. Body gasket |
| 10. Oil pump body | cover screw | 27. Idler gear shaft |

*CAUTION*
*When performing the following steps, make sure the drive gear shaft is not pushed into the gearcase, as the Woodruff key on the end of the shaft could fall into the gearcase.*

5. Remove the oil pump body mounting nuts (9) or bolts. Then slide the oil pump body off the drive gear shaft.

6. Remove the drive gear (12), Woodruff key (13) and the idler gear (14).

7. Remove the check valve spring cover screw (18), check valve spring (19) and the check valve ball (20).

8. Remove the bypass valve plug (15), bypass valve spring (16) and the bypass valve spring plunger (17).

*Engine removed*

This procedure is performed on 1968 and later model oil pumps. Refer to **Figure 97** for 1966-1967 models.

1. Remove the gearcase as described in this chapter.

2. Remove the check valve ball as follows:

   a. Remove the check valve spring cover screw (**Figure 99**).

   b. Remove the check valve spring (**Figure 100**).

   c. Remove the check valve ball (**Figure 101**).

3. Remove the bypass valve plunger as follows:

   a. Remove the bypass valve plug (**Figure 102**).

   b. Remove the bypass valve spring (**Figure 103**).

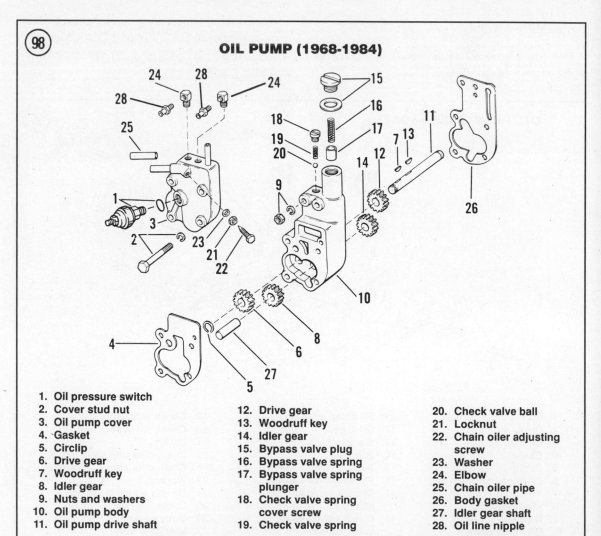

**98**

**OIL PUMP (1968-1984)**

1. Oil pressure switch
2. Cover stud nut
3. Oil pump cover
4. Gasket
5. Circlip
6. Drive gear
7. Woodruff key
8. Idler gear
9. Nuts and washers
10. Oil pump body
11. Oil pump drive shaft
12. Drive gear
13. Woodruff key
14. Idler gear
15. Bypass valve plug
16. Bypass valve spring
17. Bypass valve spring plunger
18. Check valve spring cover screw
19. Check valve spring
20. Check valve ball
21. Locknut
22. Chain oiler adjusting screw
23. Washer
24. Elbow
25. Chain oiler pipe
26. Body gasket
27. Idler gear shaft
28. Oil line nipple

c. Remove the bypass valve plunger (**Figure 104**).

4. Remove the 4 nuts or bolts and remove the oil pump cover (**Figure 105**).

5. Remove the oil pump drive gear snap ring (**Figure 106**).

6. Pull the oil pump partway out. Then take thc oil pump drive gear (**Figure 107**) and its Woodruff key off the end of the oil pump drive shaft.

7. Remove the oil pump body with the drive shaft as shown in **Figure 108**.

8. If necessary, disassemble the oil pump as follows:

a. Remove the oil pump cover.

b. Referring to **Figure 98**, remove the circlip (5), gear (6), Woodruff key (7) and gear (8). See **Figure 109**.

c. Remove the idler gear (**Figure 110**).

d. Remove the drive gear (**Figure 111**).

e. Remove the Woodruff key (**Figure 112**) from the drive shaft.

f. Remove the drive shaft (**Figure 113**).

**4**

## Inspection

1. Clean all parts thoroughly in solvent.

2. Check the check valve ball and spring (**Figure 114**) for wear or damage. Replace the check valve spring cover screw O-ring if damaged. Replace parts as necessary.

3. Check the bypass valve spring plunger and spring (**Figure 115**) for wear and damage.

4. Check the drive shaft (A, **Figure 116**) for cracks, scoring or wear. Also check the drive shaft keyways (B, **Figure 116**) for cracking. Replace the drive shaft if necessary.

5. Check the oil pump gears (**Figure 117**) and the drive gear (**Figure 118**) for cracks, scoring or excessive wear.

6. Replace the oil seal (A, **Figure 119**) if it appears worn or damaged. The seal lip must face toward the feed gears.

7. Check the oil pump idler gear shaft (B, **Figure 119**). If it is loose, replace the oil pump body as slippage will allow metal in the oil pump. A new shaft can be pressed in, but in most cases, the oil pin

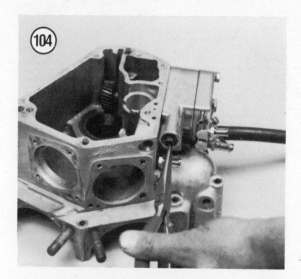

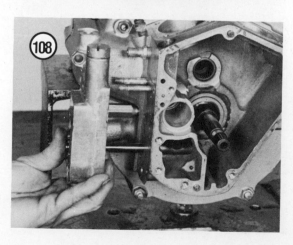

has enlarged and distorted the shaft bore in the housing.

8. Check the oil pump body machined surface (**Figure 120**) for nicks or gouging.

9. Assemble the pump and pump idler gears. Lay a straightedge across the gears and measure the height of the gear in relation to the gasket surface. The gear faces should extend above the pump body 0.003-0.004 in. (0.07-0.10 mm). Perform this check for both the feed and scavenger gears. If this clearance is incorrect, the oil pump must be replaced.

### Assembly/Installation

1. Assembly is the reverse of disassembly. Note the following.

*CAUTION*
*Never use "homemade" gaskets to check and reassemble the oil pump. Factory gaskets are made to a specified thickness with holes placed accurately to pass oil through the oil pump. Gaskets of the incorrect thickness can cause loss of oil pressure and severe engine damage.*

2. Coat all parts with fresh engine oil prior to installation.

3. Install new snap rings.

*NOTE*
*The bolts used to secure the oil pump use a 1/4-24 in. thread (**Figure 121**). If the bolts are stripped or if the threads in the gearcase are stripped, make sure to use the correct size bolts or tap for replacement or repair. The 1/4-24 thread is a non-standard size.*

*CAUTION*
*Do not overtighten the bolts in Step 4 as this eliminates the pump gear side*

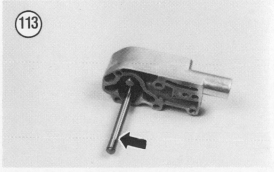

*clearance which could cause the pump to seize, resulting in engine seizure. Note that the torque setting is in.-lb., not ft.-lb.*

4. Tighten the oil pump body and oil pump cover nuts or bolts to 45-50 in.lb. (5.2-5.7 N•m).

5. Refill the engine oil tank as described in Chapter Three.

6. Connect the oil lines to the oil pump using new clamps.

## GEARCASE COVER AND TIMING GEARS

### Removal/Disassembly (1966-1969)

Refer to **Figure 122** for this procedure.

1. Remove the pushrods, tappet guide and tappets as described in this chapter.

2. Remove the oil screen cap (1, **Figure 122**) and gasket from the top of the gearcase. Then remove the spring.

3. Rotate the oil screen (4) until the notch in screen aligns with the housing key. Then remove the screen (4) and seals (1966).

4. Remove the cover screws (6).

5. Remove the generator as described in Chapter Seven.

6. Remove the circuit breaker as described in Chapter Seven.

7. Remove the cover (8) and gasket. If necessary, tap the cover with a plastic-tipped hammer to loosen it.

*NOTE*
*When removing the idler gear spacer (10) and the circuit breaker gear spacer (11) in Step 9, make sure to mark each spacer so it can be returned to its origi-*

*nal position. The spacers may appear to be the same, but one may be thicker than the other.*

8. Remove the idler gear (10) and the circuit breaker gear (11) spacers.

9. Remove the breather valve spacing washer (12) from the breather valve and gear.

10. Remove the cam gear (13), cam gear spacing washer (14) and the cam gear thrust washer (15).

11. Remove the breather valve and gear (16).

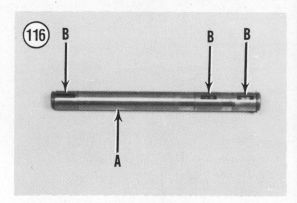

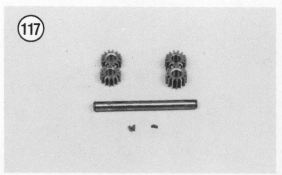

12. Remove the idler gear (18) and the circuit breaker gear (17).

13. Remove the pinion gear shaft nut (19). This nut has a left-hand thread.

14. Remove the pinion gear (20) with a puller (**Figure 123**).

> *NOTE*
> *Harley-Davidson makes a special pinion gear puller (HD 96830-51).*

15. Remove the pinion gear key (21).

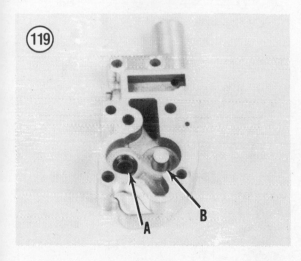

16. Remove the pinion gear spring (22), gear shaft pinion spacer (23), oil pump pinion shaft gear (24) and key (25).

17. Remove the breather screen (26) and breather separator (27) from the gearcase.

18. Remove the oiler drive gear shaft spring ring (28).

19. Remove the oiler drive gear (29) and key (30).

### Inspection

1. Clean all parts in solvent. If necessary, clean the gearcase with solvent, making sure not to allow solvent to enter the engine cases.

2. Clean the oil screen (4) of all debris. Small particles or fibers may partially or totally plug it, yet it may appear clean. Replace the screen if there is any doubt about its condition.

3. Clean the breather screen (26) of all debris. Replace the screen if necessary.

4. Measure clearance between camshaft bushing (37) and cover pinion gear bushing (38). Have a Harley-Davidson dealer replace bushings if clearance exceeds 0.0025 in. (0.063 mm). Bushings must be reamed after installation.

5. Replace roller bearings if clearance between bearings and shaft exceeds 0.004 in. (0.10 mm).

### Assembly

1. Assembly is the reverse of disassembly. Note the following.

2. Adjust the breather gear end play as follows:

   a. Assemble the breather valve and gear (16) with a new gearcase cover gasket (9). Install the gear in the gearcase.

   b. Install the breather valve spacer washer (**Figure 124**).

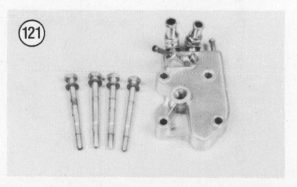

## GEARCASE COMPONENTS (1966-1969)

122

1. Oil screen cap
2. Cap gasket
3. Spring
4. Oil screen
5. Oil screen seal (1966)
6. Gear cover screw
7. Generator fastening screw
8. Gear cover
9. Gear cover gasket
10. Idler gear spacer
11. Circuit breaker gear spacer
12. Breather valve spacer washer
13. Cam gear
14. Can gear spacing washer
15. Cam gear thrust washer
16. Breather valve and gear
17. Circuit breaker gear
1C. Idler gear
19. Gear shaft nut
20. Pinion gear
21. Pinion gear key
22. Pinion gear spring
23. Gear shaft pinion spacer
24. Oil pump pinion shaft gear
25. Oil pump pinion shaft gear key
26. Breather screen
27. Breather separator
28. Oiler drive gear shaft spring ring
29. Oiler drive gear
30. Oiler drive gear key
31. Needle roller cam shaft bearing
32. Circuit breaker gear stud
33. Idler gear stud
34. Idler gear bushing
35. Circuit breaker gear bushing
36. Circuit breaker gear bushing
37. Gearcase cover cam shaft bushing
38. Gearcase cover pinion gear bushing

c. Place a straightedge across the gearcase. Then measure clearance between the spacer washer and straightedge, using a feeler gauge.

d. Subtract 0.006 in. (0.15 mm) from the clearance obtained in "c." This distance is the amount that the new gasket will compress.

e. The correct end play should be between 0.001-0.005 in. (0.02-0.13 mm). Spacer washers are available in thicknesses of 0.110, 0.115, 0.120 and 0.125 in. (2.79, 2.92, 3.05 and 3.17 mm).

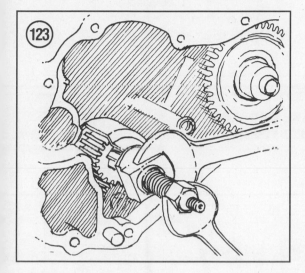

If the end play exceeds 0.005 in. (0.13 mm) install a thicker spacer washer.

3. Check and adjust cam gear end play as follows:

a. Install the cam gear (13) with its spacer washer (14) and thrust washer (15).

b. Install the gearcase cover and screws. Tighten the screws in a criss-cross pattern to compress the gasket evenly.

c. Measure camshaft end play through the tappet guide hole, using a feeler gauge. End play should be 0.001-0.005 in. (0.02-0.13 mm).

d. If the end play exceeds 0.005 in. (0.13 mm), install a thicker spacer washer. Spacer washers are available in 0.005 in. (0.13 mm) increments from 0.050 to 0.095 in. (1.27-2.41 mm).

4. Remove the gearcase cover installed in Step 3.

5. After determining proper end play measurements, the breather, cam, pinion and circuit breaker gear timing marks must be aligned as shown in **Figure 124**.

6. After assembling and timing gears, turn the gear cluster by hand. The gears should rotate freely without any binding or tight spots. If the cluster is tight, make sure the gear spacer (10) and the circuit breaker gear spacer (11) are not switched.

7. During final assembly, install the gasket using a non-hardening gasket sealer.

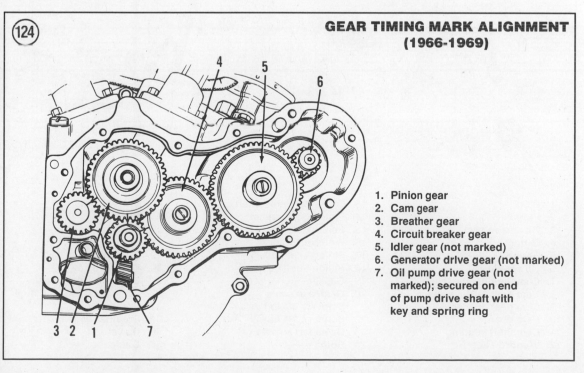

**GEAR TIMING MARK ALIGNMENT (1966-1969)**

1. Pinion gear
2. Cam gear
3. Breather gear
4. Circuit breaker gear
5. Idler gear (not marked)
6. Generator drive gear (not marked)
7. Oil pump drive gear (not marked); secured on end of pump drive shaft with key and spring ring

8. Pour approximately 1/4 pint (0.12 L, 0.10 imp. qt.) of engine oil through the tappet guide hole to provide initial gear train lubrication.

**Removal (1970-on)**

Refer to **Figure 125** for this procedure.

1. Remove the pushrods, valve tappets and guides as described in this chapter.

2. Remove the tappet oil screen cap (3, **Figure 125**) and O-ring. Then remove the spring and screen (6).

3. Remove the circuit breaker or electronic ignition timer and sensor plate as described in Chapter Seven.

4. Place an oil drain pan underneath the gearcase cover.

*NOTE*
*The gearcase cover screws are different lengths. Mark each screw during removal.*

5. Remove the gearcase cover screws and remove the gearcase cover (**Figure 126**). Discard the gasket.

*CAUTION*
*The gearcase cover is located by snug dowel pins and must be worked off carefully. Do not pry the gearcase cover with*

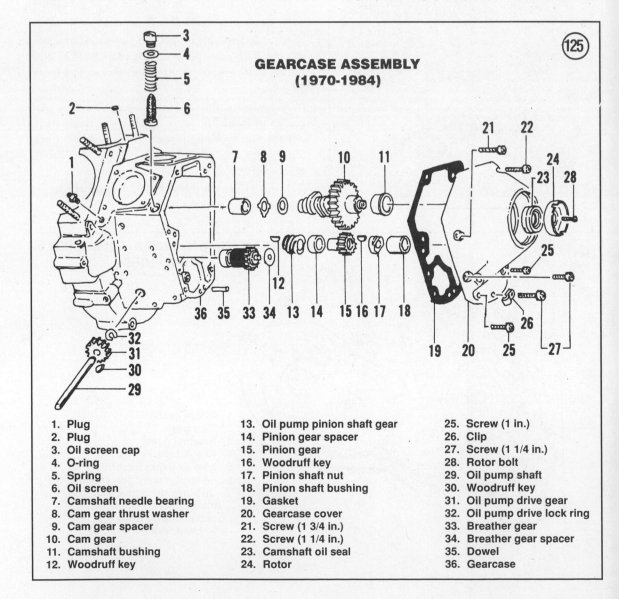

**GEARCASE ASSEMBLY
(1970-1984)**

125

| | | |
|---|---|---|
| 1. Plug | 13. Oil pump pinion shaft gear | 25. Screw (1 in.) |
| 2. Plug | 14. Pinion gear spacer | 26. Clip |
| 3. Oil screen cap | 15. Pinion gear | 27. Screw (1 1/4 in.) |
| 4. O-ring | 16. Woodruff key | 28. Rotor bolt |
| 5. Spring | 17. Pinion shaft nut | 29. Oil pump shaft |
| 6. Oil screen | 18. Pinion shaft bushing | 30. Woodruff key |
| 7. Camshaft needle bearing | 19. Gasket | 31. Oil pump drive gear |
| 8. Cam gear thrust washer | 20. Gearcase cover | 32. Oil pump drive lock ring |
| 9. Cam gear spacer | 21. Screw (1 3/4 in.) | 33. Breather gear |
| 10. Cam gear | 22. Screw (1 1/4 in.) | 34. Breather gear spacer |
| 11. Camshaft bushing | 23. Camshaft oil seal | 35. Dowel |
| 12. Woodruff key | 24. Rotor | 36. Gearcase |

*any metal tool. If necessary, tap the cover lightly with a soft-faced hammer at the point where the cover projects beyond the crankcase.*

6. Remove the cam gear (**Figure 127**).

7. Remove the cam gear spacer and cam gear thrust washer (**Figure 128**).

8. Remove the breather gear washer (**Figure 129**) and the breather gear (**Figure 130**).

> *NOTE*
> *The pinion gear shaft nut has left-hand threads.*

9. Remove the pinion gear shaft nut (**Figure 131**) using the socket (HD-94555-55A), if necessary. Then remove the pinion gear (**Figure 132**) using the pinion gear puller (HD-96830-51A).

10. Remove the following parts in order:
 a. Woodruff key (**Figure 133**).
 b. Gear shaft pinion spacer (**Figure 134**).
 c. Oil pump pinion shaft gear (**Figure 135**).
 d. Woodruff key (**Figure 136**).

11. If required, remove the oil pump drive gear and the oil pump as described in this chapter.

**Inspection**

1. Thoroughly clean gearcase compartment, cover and components with solvent. Blow out all oil passages with compressed air. Make sure that all traces of gasket compound are removed from the gasket mating surfaces.

2. Check the oil screen (6, **Figure 125**) to make sure it is not blocked or damaged. Test the screen by holding it upside down and filling it with engine oil. Watch the screen to see that the oil flows evenly through the screen. If not, replace the screen.

3. Check the pinion gear and cam gear bushings in the gearcase cover for grooving, pitting or other wear. If the bushings appear visibly worn, have them replaced by a Harley-Davidson dealer. If the bushings appear okay, perform Step 4.

4. Insert the pinion and cam gears in their respective bushings. Then check the running clearance with a dial indicator. Refer to **Table 2** or **Table 4** for the correct clearances. If the clearance is excessive, have the bushings replaced by a Harley-Davidson dealer.

5. Inspect the camshaft (A, **Figure 137**) small ends at the bearing surface and near the camshaft lobes. If there is evidence of scoring or excessive wear re-

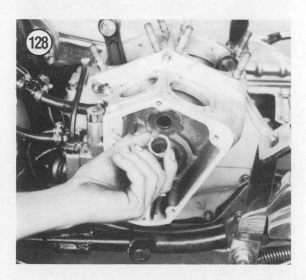

place the camshaft and its needle bearing. Have the needle bearing replaced by a Harley-Davidson dealer.

6. Inspect the camshaft lobes (B, **Figure 137**).If there is evidence of scoring or excessive wear, replace the camshaft.

7. Check the cam gear oil seal (**Figure 138**) in the gearcase cover. If worn, carefully pry it out of the case. Install a new seal by driving it into the gearcase cover using a suitable size drift or socket placed on the outer edge of the seal.

8. Inspect the breather gear (**Figure 139**) teeth for damage. Also check the screen for debris or damage and clean with solvent if necessary. Replace the breather gear if necessary.

9. Check all gears for signs of wear or damage; replace if necessary. If the cam gears appear okay, check the gear mesh as follows:

a. Assemble the cam gear (10, **Figure 125**) and pinion gears (13 and 15, **Figure 125**) in the gearcase. Do not install the cam gear spacer washer (9, **Figure 125**).

b. Install the gearcase cover (**Figure 126**). Install a minimum of 3 gearcase cover screws. Tighten the screws securely.

c. Check the gear mesh through the tappet guide hole (**Figure 140**) by hand. Mesh is correct when there is no play between the gears and

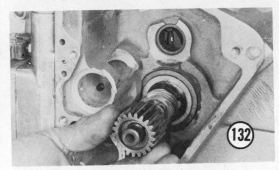

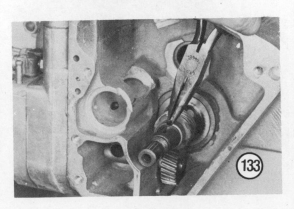

the cam gear can be moved back and forth with slight drag.

d. If the gear mesh is incorrect, replace the cam and pinion gears.

## Assembly

1. Assembly is the reverse of disassembly. Note the following.

2. Before final assembly of gearcase components, check the breather gear end play as follows:

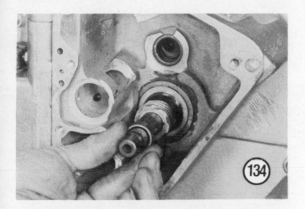

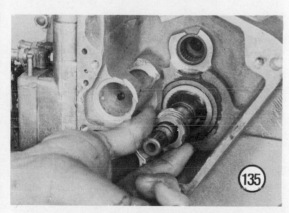

a. Install the breather gear (**Figure 130**) and a new cover gasket to the gearcase.

b. Install the spacer (**Figure 129**) on the breather gear.

c. Lay a straightedge across the gearcase at the breather gear spacer. Then using a feeler gauge, measure the clearance between the straightedge and the spacer. See **Figure 141**.

d. Subtract 0.006 in. (0.15 mm) from the clearance determined in sub-step "c." This is the amount the gearcase gasket will compress.

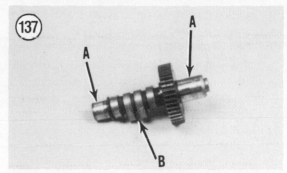

e. An end play clearance of 0.001-0.016 in. (0.03-0.41 mm) is correct. If the clearance exceeds this amount, install a thicker spacer (**Figure 142**). Spacers are available in thicknesses of 0.110, 0.115, 0.120 and 0.125 in. (2.80, 2.92, 3.05 and 3.17 mm).

3. Before final assembly of gearcase components, check the cam gear end play as follows:

a. Install the can gear thrust washer, spacer (**Figure 128**) and the cam gear (**Figure 127**).

b. Install the gearcase cover (**Figure 126**). Install a minimum of 4 gearcase cover screws. Tighten the screws securely.

c. Measure the camshaft end play between the gear shaft and the thrust washer with a feeler gauge inserted through the gearcase tappet hole (**Figure 143**).

d. An end play clearance of 0.001-0.016 in. (0.03-0.41 mm) is correct. If the clearance exceeds this amount, install a suitable size spacer (**Figure 144**) to bring the clearance within specifications. Spacers are available in 0.005 in. (0.13 mm) increments from 0.050 to 0.095 in. (1.27-2.41 mm).

4. Slide the cam gear spacer and thrust washer onto the end of the cam gear.

5. Install the oil pump pinion shaft gear Woodruff key (**Figure 136**).

6. Slide the oil pump pinion shaft gear (**Figure 135**) onto the shaft, making sure to align the keyway in the gear with the Woodruff key.

7. Install the pinion gear spacer (**Figure 134**) onto the end of the pinion gear.

8. Install the oil pump drive gear Woodruff key. Then install the oil pump drive gear and secure it with a *new* circlip.

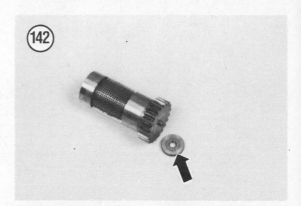

9. Install the breather gear washer onto the end of the breather gear (**Figure 129**).

10. Install the cam, breather and pinion gears so that the timing marks on the gears align as shown in **Figure 145**.

*NOTE*
*The pinion shaft has left-hand threads.*

11. Install the pinion shaft nut and tighten it to 35-45 ft.-lb. (48.3-62.1 N•m). Use the pinion shaft nut socket (HD-94555-55A) to tighten the pinion shaft nut.

12. After tightening the pinion shaft nut, check that the pinion shaft spacer has noticeable end play. If not, remove the pinion shaft nut and the pinion shaft

(145)

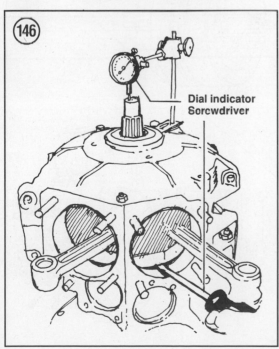

(146)

Dial indicator
Screwdriver

gear and reinstall. Make sure the timing marks in Step 10 are correct.

13. Turn the gear train to make sure all gears rotate freely. Any binding should be corrected before completing engine reassembly.

*NOTE*
*On late 1977-1984 models, the pinion and cam gears are color coded to match the gear pitch. If the pinion or cam gear requires replacement, replace it with a gear having the same pitch color code.*

14. Coat a new gearcase gasket with a non-hardening gasket sealer and install it.

15. Install the gearcase and screws. Tighten the screws to 80-110 in.-lb. (9.2-12.6 N•m).

16. Pour approximately 1/4 pint (0.12 L, 0.10 imp. qt.) of engine oil through the tappet guide hole to provide initial gear train lubrication.

## CRANKCASE

Crankcases must be disassembled to service the crankshaft, connecting rod bearings, pinion shaft bearings and sprocket shaft bearings. This section describes basic checks and procedures that can be performed in the home shop. Bearing and crankshaft service should be referred to a Harley-Davidson shop equipped to handle such repairs.

### Crankshaft End Play Check

It is recommended that crankshaft end play be measured before completely disassembling the crankcase. Crankshaft end play is a measure of sprocket shaft bearing wear.

*NOTE*
*Do not remove the engine sprocket or compensating sprocket when performing Step 1. These parts must be installed to preload the bearing races.*

1. Remove the engine from the frame as described in this chapter.

2. Position the crankcase so that the sprocket shaft is vertical (**Figure 146**).

3. Attach a dial indicator so that the probe touches against the end of the crankshaft (**Figure 146**).

4. Using a large screwdriver, pry the flywheel upward and note the end play registering on the dial

indicator. If end play exceeds limit in **Table 2** or **Table 4**, the inner bearing spacer (4, **Figure 147** or 4, **Figure 148**) must be replaced as follows:

a. 1966-1968: On these models, only 1 factory spacer is available. To correct end play, shims may be used. The factory spacer thickness is 0.003 in. (0.08 mm).

b. 1969-on: On these models, adjust end play by selecting a different spacer from the chart in **Table 8**.

## Disassembly

Refer to **Figure 147** (1966-1968) or **Figure 148** (1969-1984) for this procedure.

1. Remove the engine from the frame as described in this chapter.

2. Disassemble and remove the gearcase assembly as described in this chapter.

3. Check the flywheel end play as described in this chapter.

*NOTE*
*When removing the crankcase bolts and studs in Step 4, note that the top center stud (4, **Figure 149**) and the right bottom studs (5, **Figure 149**) are matched and fitted to the crankcase holes for correct crankcase alignment. During their removal, mark them so they can be reinstalled in their original positions.*

4. See **Figure 149**. Remove the crankcase bolts and studs.

5. Lay the crankcase assembly on wood blocks so that right-hand side faces up.

6. Tap the crankcase with a plastic mallet and remove the right-hand crankcase half.

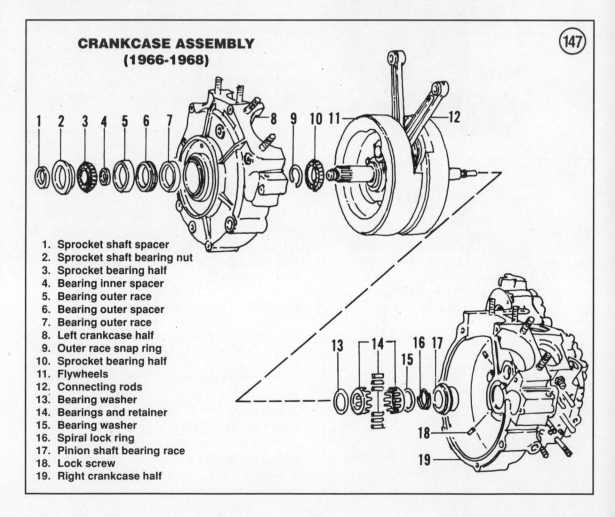

**CRANKCASE ASSEMBLY
(1966-1968)**

1. Sprocket shaft spacer
2. Sprocket shaft bearing nut
3. Sprocket bearing half
4. Bearing inner spacer
5. Bearing outer race
6. Bearing outer spacer
7. Bearing outer race
8. Left crankcase half
9. Outer race snap ring
10. Sprocket bearing half
11. Flywheels
12. Connecting rods
13. Bearing washer
14. Bearings and retainer
15. Bearing washer
16. Spiral lock ring
17. Pinion shaft bearing race
18. Lock screw
19. Right crankcase half

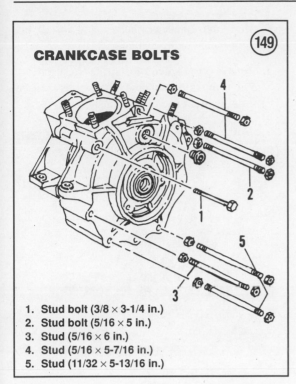

**CRANKCASE BOLTS** (149)

1. Stud bolt (3/8 × 3-1/4 in.)
2. Stud bolt (5/16 × 5 in.)
3. Stud (5/16 × 6 in.)
4. Stud (5/16 × 5-7/16 in.)
5. Stud (11/32 × 5-13/16 in.)

7. Remove the pinion shaft spiral lock ring (**Figure 150**).

8. Grasp the 2 bearing washers and remove the washers, bearings and retainers as an assembly (**Fig-**

(150)

4

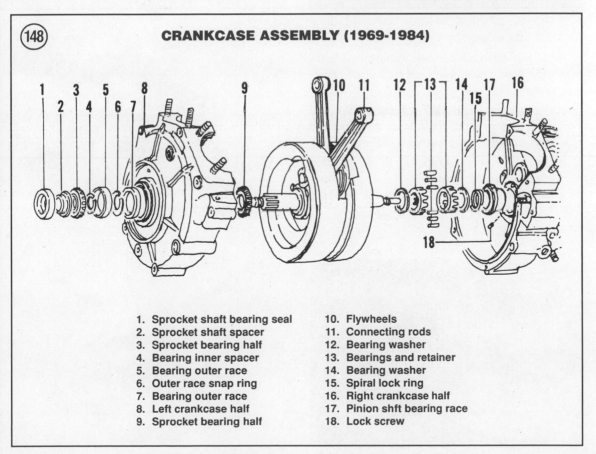

(148) **CRANKCASE ASSEMBLY (1969-1984)**

1. Sprocket shaft bearing seal
2. Sprocket shaft spacer
3. Sprocket bearing half
4. Bearing inner spacer
5. Bearing outer race
6. Outer race snap ring
7. Bearing outer race
8. Left crankcase half
9. Sprocket bearing half
10. Flywheels
11. Connecting rods
12. Bearing washer
13. Bearings and retainer
14. Bearing washer
15. Spiral lock ring
16. Right crankcase half
17. Pinion shft bearing race
18. Lock screw

ure **151**). Store the complete assembly in a plastic bag.

*NOTE*
*Further disassembly of the crank-case/flywheel assembly is not recom-mended. A hydraulic press is required to separate the crankshaft from the left-hand crankcase half. Furthermore, ad-ditional equipment is required to install and line ream new bearings properly.*

9. Check connecting rod side play with a feeler gauge as shown in **Figure 152**. If the side play is not within the specifications in **Table 2** or **Table 4**, refer service to a Harley-Davidson dealer.

10. Refer crankcase to a Harley-Davidson dealer for inspection and repair.

11. Installation is the reverse of these steps. Note the following:

   a. If the crankshaft was removed, have it pressed into the left-hand crankcase half by a Harley-Davidson dealer.

   b. Make sure rods are aligned as shown in **Figure 153** before assembling the right-hand crank-case.

   c. Referring to **Figure 147** or **Figure 148**, install the bearing washer, bearings and retainer and the second bearing washer. See **Figure 151**.

   d. Install a new spiral lock ring (**Figure 150**) in the pinion shaft groove.

   e. Coat the crankcase mating surfaces with Har-ley-Davidson Crankcase Sealant or RTV sili-cone sealant.

   f. Align the crankcase halves and install the right-hand crankcase half.

   g. Referring to **Figure 149**, tap the No. 4 and No. 5 bolts into the crankcase. These bolts are used for alignment and must be installed first.

h. Install the remaining studs and bolts (**Figure 149**).

i. Install the stud nuts and tighten in a criss-cross pattern until all are snug. Then tighten the bolts to 15-19 ft.-lb. (20.7-26.2 N•m) and tighten the nuts to 15-19 ft.-lb. (20.7-26.2 N•m).

j. On 1966-1968 models, tighten the sprocket shaft bearing nut (2, **Figure 147**). Then install the sprocket shaft spacer (1, **Figure 147**).

k. On 1969-1984 models, install the sprocket shaft spacer (2, **Figure 148**). Then install a new sprocket shaft bearing seal (**Figure 154**).

## ENGINE BREAK-IN

Following cylinder servicing (new pistons, new rings, etc.) and major lower end work, the engine should be broken in just as though it were new. The performance and service life of the engine depends greatly on a careful and sensible break-in.

For the first 500 miles (800 km), no more than one-third throttle should be used and the speed should be varied as much as possible within the one-third throttle limit. Prolonged, steady running at one speed, no matter how moderate, is to be avoided as is hard acceleration.

Following the 500-mile (800 km) service, increasingly more throttle can be used but full throttle should not be used until the motorcycle has covered at least 1,000 miles (1,600 km) and then it should be limited to short bursts until 1,500 miles (2,400 km) have been logged.

During engine break-in, oil consumption will be higher than normal. It is important to frequently check and correct the oil level in the tank, making sure to maintain a 1 in. (25.4 mm) air gap in the top of the tank.

### 500 Mile (800 km) Service

It is essential that the oil tank be drained, flushed and refilled and the oil filter serviced after the first 500 mile (800 km) break-in period. In addition, it is a good idea to repeat this service at the completion of break-in (about 1,500 miles [2,400 km]) to ensure that all of the particles produced during break-in are removed from the lubrication system. The small added expense may be considered a smart investment that will pay off in increased engine life.

**Table 1 GENERAL ENGINE SPECIFICATIONS (1966-EARLY 1978)**

|  | Specification |
| --- | --- |
| Type | Air cooled, 4-stroke, 45° V-twin |
| Number of cylinders | 2 |
| Bore and stroke | |
|   1200 cc | 3.438 × 3.968 in. (87.32 × 100.79 mm) |
| Displacement | |
|   1200 cc | 73.66 cu. in. (1200 cc) |
| Compression ratio | |
|   FLH | 8:1 |
|   FL | 7.25:1 |
| Lubrication | Force-feed oiling system |

**Table 2 ENGINE SPECIFICATIONS (1966-EARLY 1978)**

| | Specification | |
| --- | --- | --- |
| | in. | mm. |
| Valves | | |
|   Fit in guide | | |
|     Exhaust | | |
|       1966-1969 | 0.004-0.006 | 0.10-0.15 |
|       1970-early 1978 | 0.0035-0.0055 | 0.089-0.140 |
| | (continued) | |

**Table 2 ENGINE SPECIFICATIONS (1966-EARLY 1978) (continued)**

| | Specification | |
| --- | --- | --- |
| | in. | mm |
| Intake | | |
| 1966-1969 | 0.002-0.004 | 0.05-0.10 |
| 1970-early 1978 | 0.0018-0.0038 | 0.046-0.096 |
| Valve spring free length | | |
| Outer | | |
| 1966-1969 | | |
| FL | 1 13/16 | 46.04 |
| FLH | 1 31/32 | 49.99 |
| 1970-early 1978 | 1 31/32 | 49.99 |
| Inner | | |
| 1966-1969 | | |
| FL | 1 15/32 | 37.29 |
| FLH | 1 25/64 | 35.31 |
| 1970-early 1978 | 1 23/64 | 34.52 |
| Rocker arm | | |
| Fit in bushing-bearing | | |
| clearance | 0.0005-0.002 | 0.013-0.051 |
| End clearance | 0.004-0.025 | 0.10-0.63 |
| Cylinder | | |
| Taper | 0.002 | 0.05 |
| Out-of-round | 0.003 | 0.08 |
| Piston | | |
| Clearance | 0.001-0.002 | 0.03-0.05 |
| Piston pin fit | Light hand press @ 70° | |
| Piston rings | | |
| End gap | 0.010-0.020 | 0.25-0.51 |
| Side clearance | | |
| Compression | 0.004-0.005 | 0.10-0.13 |
| Oil control ring | 0.003-0.005 | 0.08-0.13 |
| Connecting rod | | |
| Piston pin fit | 0.0008-0.0012 | 0.020-0.030 |
| Side play | | |
| 1966-1969 | 0.006-0.010 | 0.15-0.25 |
| 1970-early 1978 | 0.005-0.025 | 0.13-0.63 |
| Crankpin fit | | |
| 1966-1969 | 0.0006-0.001 | 0.02-0.03 |
| 1970-early 1978 | 0.001-0.0015 | 0.03-0.04 |
| Tappets | | |
| Crankcase guide fit | | |
| 1966-1969 | 0.002 | 0.05 |
| 1970-early 1978 | 0.0025 | 0.063 |
| Fit in guide | | |
| 1966-1969 | 0.001-0.002 | 0.03-0.05 |
| 1970-early 1978 | 0.001-0.003 | 0.03-0.08 |
| Roller fit | 0.0005-0.001 | 0.013-0.03 |
| Roller end clearance | 0.008-0.010 | 0.20-0.25 |
| Gearcase | | |
| Timer gear end play | 0.003-0.007 | 0.08-0.18 |
| Idler gear end play | 0.003-0.020 | 0.08-0.51 |
| Breather gear end play | 0.001-0.005 | 0.03-0.13 |
| Cam gear shaft | | |
| Fit in bushing | | |
| 1966-1969 | 0.001-0.0015 | 0.03-0.04 |
| 1970-early 1978 | 0.0008-0.0018 | 0.020-0.046 |
| Fit in bearing | 0.0005-0.003 | 0.013-0.08 |
| Cam gear end play | 0.001-0.005 | 0.03-0.13 |

(continued)

### Table 2 ENGINE SPECIFICATIONS (1966-EARLY 1978) (continued)

| | Specification | |
| --- | --- | --- |
| | in. | mm |
| Oil pump drive shaft crankcase bushing clearance | 0.0008-0.0012 | 0.020-0.030 |
| Circuit breaker and idler gear play on shafts | 0.001-0.0015 | 0.03-0.04 |
| Flywheels | | |
| Runout (at rim) | 0.003 | 0.08 |
| Runout (at mainshaft) | 0.001 | 0.03 |
| Sprocket shaft bearing | | |
| Cut fit in crankcase | | |
| 1966-1969 | 0.0015-0.0035 | 0.04-0.09 |
| 1970-early 1978 | 0.0012-0.0032 | 0.030-0.081 |
| Cone fit on shaft | 0.0002-0.0015 | 0.005-0.038 |
| End play | | |
| 1966-1969 | 0.0005-0.006 | 0.013-0.152 |
| 1970-early 1978 | 0.001-0.006 | 0.03-0.15 |
| Pinion shaft bearings | | |
| Roller bearing fit | 0.0004-0.0008 | 0.010-0.020 |
| Cover bushing fit | 0.0005-0.0012 | 0.013-0.030 |

4

### Table 3 GENERAL ENGINE SPECIFICATIONS (LATE 1978-1984)

| Specification | |
| --- | --- |
| Type | Air cooled, 4-stroke, 45°, V-twin. |
| Number of cylinders | 2 |
| Bore and stroke | |
| 1200 cc | 3.438 × 3.968 in. (87.32 × 100.79 mm) |
| 1340 cc | 3.498 × 4.250 in. (88.85 × 107.95 mm) |
| Displacement | |
| 1200 cc | 73.66 cu. in. (1200 cc) |
| 1340 cc | 81.6 cu. in. (1340 cc) |
| Compression ratio | |
| 1200 cc | 8:1 |
| 1340 cc | |
| Late 1978-1980 | 8:1 |
| 1981-1984 | 7.4:1 |
| Lubrication | Force-feed oiling system |

### Table 4 ENGINE SPECIFICATIONS (LATE 1978-1984)

| | Specification | | Wear limit | |
| --- | --- | --- | --- | --- |
| | in. | mm | in. | mm |
| Cylinder head | | | | |
| Valve guide fit | 0.004-0.002 | 0.10-0.05 | — | — |
| Valve seat in head | 0.006-0.004 | 0.15-0.10 | — | — |
| Valves | | | | |
| Fit in guide | | | | |
| Exhaust | | | | |
| Late 1978-early 1981 | | | | |
| without seals | 0.0035-0.0055 | 0.089-0.140 | 0.0075 | 0.190 |
| Late 1981-on with seals | 0.0014-0.0031 | 0.035-0.079 | 0.004 | 0.102 |
| Intake | | | | |
| Late 1978-1979 | 0.0018-0.0038 | 0.046-0.096 | 0.0038 | 0.096 |
| 1980-early 1981 | | | | |
| without seals | 0.0021-0.0040 | 0.053-0.102 | 0.0060 | 0.152 |
| Late 1981-on with seals | 0.0009-0.0026 | 0.023-0.066 | 0.0035 | 0.089 |
| Seat width | 0.050-0.090 | 1.27-2.29 | — | — |
| | (continued) | | | |

## Table 4 ENGINE SPECIFICATIONS (LATE 1978-1984) (continued)

| | Specification | | Wear limit | |
|---|---|---|---|---|
| | in. | mm | in. | mm |
| Valve spring free length | | | | |
| Outer | | | | |
| Late 1978-early 1982 | 1.900-1.960 | 48.3-49.78 | — | — |
| Late 1982-on | 1.720-1.780 | 43.69-45.21 | — | — |
| Inner | | | | |
| Late 1978-early 1982 | 1.36 | 34.52 | — | — |
| Late 1982-on | 1.50-1.56 | 38.1-39.6 | — | — |
| Seat width | 0.050-0.090 | 1.27-2.29 | — | — |
| Valve stem extension from cylinder head boss | 1.600-1.645 | 40.64-41.78 | — | — |
| Rocker arm | | | | |
| Fit in bushing | 0.0005-0.002 | 0.013-0.051 | 0.0035 | 0.089 |
| End clearance | 0.004-0.025 | 0.10-0.63 | 0.025 | 0.63 |
| Cylinder | | | | |
| Taper | 0.002 | 0.05 | — | — |
| Out-of-round | 0.003 | 0.08 | — | — |
| Piston | | | | |
| Clearance | 0.0020-0.0025 | 0.051-0.063 | 0.005 | 0.13 |
| Piston pin fit | | | | |
| Late 1978-early 1983 | 0.0008-0.001 | 0.020-0.025 | — | — |
| Late 1983-on | 0.00015-0.00065 | 0.0038-0.0165 | — | — |
| Piston rings | | | | |
| End gap | | | | |
| Compression rings | | | | |
| Late 1978-early 1983 | 0.010-0.020 | 0.25-0.51 | 0.031 | 0.79 |
| Late 1983-on | 0.008-0.015 | 0.20-0.38 | 0.015 | 0.38 |
| Oil control ring | 0.010-0.045 | 0.25-1.14 | 0.045 | 1.14 |
| Side clearance | | | | |
| Compression | 0.004-0.005 | 0.10-0.13 | 0.006 | 0.15 |
| Oil control ring | 0.003-0.005 | 0.08-0.13 | 0.006 | 0.15 |
| Connecting rod | | | | |
| Piston pin fit | | | | |
| Late 1978-early 1983 | — | — | 0.0008-0.0010 | 0.020-0.025 |
| Late 1983-on | — | — | 0.00015-0.00065 | 0.0038-0.0165 |
| Side play | 0.005-0.025 | 0.13-0.63 | 0.030 | 0.76 |
| Crankpin fit | 0.001-0.0015 | 0.03-0.04 | 0.0017 | 0.04 |
| Tappets | | | | |
| Crankcase guide fit | 0.0025 | 0.063 | 0.0025 | 0.063 |
| Fit in guide | 0.001-0.002 | 0.03-0.05 | 0.003 | 0.08 |
| Roller fit | 0.0005-0.001 | 0.013-0.030 | 0.0012 | 0.030 |
| Roller end clearance | 0.008-0.010 | 0.20-0.25 | 0.015 | 0.38 |
| Gearcase | | | | |
| Breather end play | 0.001-0.016 | 0.03-0.41 | 0.016 | 0.41 |
| Cam gear shaft | | | | |
| Fit in bushing | 0.0008-0.0018 | 0.020-0.048 | 0.003 | 0.08 |
| Fit in bearing | 0.0005-0.003 | 0.01-0.08 | 0.003 | 0.08 |
| Cam gear end play | 0.001 | 0.03 | 0.016 | 0.41 |
| Oil pump drive shaft crankcase bushing | 0.0008 | 0.020 | 0.0025 | 0.063 |
| Flywheels | | | | |
| Runout (at rim) | 0.000-0.006 | 0.00-0.15 | 0.006 | 0.15 |
| Runout (at flywheel) | 0.000-0.002 | 0.00-0.05 | 0.002 | 0.05 |
| Sprocket shaft bearing | | | | |
| Cup fit in crankcase | 0.0012-0.0032 | 0.030-0.081 | — | — |
| Cone fit on shaft | 0.0002-0.0015 | 0.005-0.038 | — | — |
| Pinion shaft bearings | | | | |
| Roller bearing fit | 0.0004-0.0008 | 0.010-0.020 | 0.0008 | 0.020 |
| Cover bushing fit | 0.0005-0.0012 | 0.013-0.030 | 0.0025 | 0.063 |

## Table 5 TIGHTENING TORQUES (1966-1969)

| Item | ft.-lb. | N·m |
|---|---|---|
| Rocker arm cover | 15 | 20.7 |
| Cylinder head | 65 | 89.7 |
| Oil pump | 50 in.-lb. | 5.7 |
| Flywheel assembly | | |
|   Gear shaft nut | 100 | 138 |
|   Sprocket shaft nut | 100 | 138 |
|   Crank pin nut | 175 | 241.5 |

## Table 6 TIGHTENING TORQUES (1970-EARLY 1978)

| Item | ft.-lb. | N·m |
|---|---|---|
| Rocker arm cover | 15 | 20.7 |
| Cylinder head | 65 | 89.7 |
| Oil pump | 45-50 in.-lb. | 5.2-5.7 |
| Cylinders | 32-36 | 44.2-49.7 |
| Flywheel assembly | | |
|   Gear shaft nut | 170 | 234.6 |
| Flywheel assembly (continued) | | |
|   Sprocket shaft nut | | |
|     1970-1971 | 170 | 234.6 |
|     1972-early 1978 | 400 | 552 |
|   Crank pin nut | 175 | 241.5 |
| Engine mounting bolts | 35-40 | 48.3-55.2 |
| Chain housing to transmission studs | 30-35 | 41.4-48.3 |
| Chain case to engine mount bolts | 18-22 | 24.8-30.4 |
| Upper engine mounting bracket | 35-40 | 48.3-55.2 |
| Tappet guide screws | 10 | 13.8 |
| Crankshaft lock plates | 20-24 in.-lb. | 2.3-2.7 |
| Crankcase | | |
|   Stud nuts | 12-15 | 16.6-20.7 |
|   Bolts | 22-26 | 29.9-35.9 |

## Table 7 TIGHTENING TORQUES (LATE 1978-1984)

| Item | ft.-lb. | N·m |
|---|---|---|
| Sprocket shaft nut | | |
|   1980-early 1981 | 300-400 | 414-552 |
|   Late 1981-on | 290-320 | 400.2-441.6 |
| Crank pin nut | | |
|   1980-early 1981 | 150-250 | 207-345 |
|   Late 1981-on | 180-210 | 248.4-289.8 |
| Pinion shaft nut | | |
|   1980-early 1981 | 120-160 | 165.6-220.8 |
|   Late 1981-on | 140-170 | 193.2-234.6 |
| Pinion gear nut | 35-45 | 48.3-62.1 |
| Oil pump cover nut or bolt | | |
|   With plastic gasket | 45-50 in.-lb. | 5.2-5.7 |
|   With black paper gasket | 90-120 in.-lb. | 10.3-13.8 |
|   With white paper gasket | 50-60 in.-lb. | 5.7-6.9 |
| Tappet guide bolts | 65-105 in.-lb. | 7.5-12.1 |
| Rocker cover nuts | 12-15 | 16.6-20.7 |
| Cylinder head bolts | 55-75 | 75.9-103.5 |
| Cylinder base nuts | 32-40 | 44.2-55.2 |
| Crankcase stud nut | 12-15 | 16.6-20.7 |
| Crankcase bolt | 22-26 | 30.4-35.9 |
| Tappet adjusting locknut | 6-11 | 8.3-15.2 |
| Gear case cover screws | 90-120 | 124.2-165.6 |

(continued)

4

### Table 7 TIGHTENING TORQUES (LATE 1978-1984) (continued)

| Item | ft.-lb. | N·m |
|---|---|---|
| Rocker arm shaft screw | | |
|   Late 1978-1981 | 8-12 | 11-16.6 |
| Rocker arm shaft locknut | 12-18 | 16.6-24.8 |
| Timer screws | | |
|   Inner cover | 15-30 in.-lb. | 1.7-3.4 |
|   Sensor plate | 15-30 in.-lb. | 1.7-3.4 |
| Trigger rotor screw | 43-48 in.-lb. | 4.9-5.5 |
| Tappet screen plug | 90-160 in.-lb. | 10.3-18.4 |
| Spark plug | 18-28 | 24.8-38.6 |
| Upper engine mounting bracket nut | 35-40 | 48.3-55.2 |

### Table 8 INNER BEARING SPACER SIZE

| Part No. | Spacer size (in.)* |
|---|---|
| 9120 | 0.0925-0.0915 |
| 9121 | 0.0945-0.0935 |
| 9122 | 0.0965-0.0955 |
| 9123 | 0.0985-0.0975 |
| 9124 | 0.1005-0.0995 |
| 9125 | 0.1025-0.1015 |
| 9126 | 0.1045-0.1035 |
| 9127 | 0.1065-0.1055 |
| 9128 | 0.1085-0.1075 |
| 9129 | 0.1101-0.1095 |
| 9130 | 0.1125-0.1115 |
| 9131 | 0.1145-0.1135 |
| 9132 | 0.1165-0.1155 |
| 9133 | 0.1185-0.1175 |
| 9134 | 0.1205-0.1195 |

* Multiply specification by 25.4 to find millimeter (mm) equivalent.

# CLUTCH, PRIMARY DRIVE, TRANSMISSION AND KICKSTARTER

This chapter describes service procedures for the clutch, primary drive, transmission and kickstarter.

Clutch, primary drive and transmission specifications are listed in **Tables 1-6** at the end of the chapter.

## CLUTCH

All clutches covered in this manual are of multiple disc construction with alternate plates splined to the clutch hub and clutch shell. When the clutch is engaged, springs force all plates together, causing them to turn as a unit, thereby transmitting engine power to the transmission and rear wheel.

Refer to **Figure 1** (1966-1969) or **Figure 2** (1970-1984) when performing procedures in this section.

### Removal

1. Disconnect the battery negative cable to prevent accidental starter operation, if so equipped.

2. Remove the left footboard mounting bolts and move the footboard out of the way, if necessary.

3. Remove the shift lever and the left footpeg bracket, if necessary.

4. Remove the primary chain case cover screws. Then tap the cover (**Figure 3**) with a plastic-faced mallet to break the gasket seal. Pull the cover off of the engine and transmission housings.

5. *Belt drive models:* Pry the compensating sprocket lockwasher away from the sprocket nut (**Figure 4**).

6. The engine must be locked to prevent the crankshaft from turning when loosening the motor sprocket nut (**Figure 5**) or the compensating sprocket nut (1, **Figure 6**). Lock the engine by shifting the transmission into gear. Then hold the motor sprocket with Harley-Davidson tool 94557-55 (**Figure 7**) or compensating sprocket cover with a chain wrench. Wrap the compensating sprocket cover with shim stock to prevent damage from the chain wrench. If these tools are not available, an air gun and socket will be necessary.

7. After locking the engine, use a large socket and breaker bar or air gun to loosen the motor sprocket nut (**Figure 5**) or the compensating sprocket nut (1, **Figure 6**).

8. Remove the nut. On models with compensating sprocket, remove the cover, cup (if used) and sliding cam. See **Figure 6**.

9. Remove the pushrod adjusting screw locknut. See **Figure 8** and **Figure 9**.

10. Place a flat washer (1/8 in. [3.17 mm] thick, 1 3/4 in. [44.45 mm] in diameter and 3/8 in. [9.52 mm] inside diameter) over the pushrod adjusting screw (**Figure 10**). Then reinstall the adjusting screw locknut removed in Step 9.

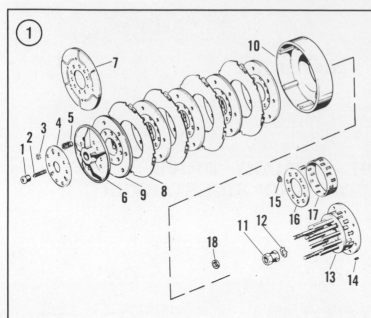

**①**

## CLUTCH (1966-1969)

1. Pushrod adjusting screw locknut
2. Adjusting screw
3. Spring tension adjusting nut
4. Spring collar
5. Springs
6. Pressure plate
7. Spring disc (1966-1967)
8. Steel disc
9. Friction disc
10. Cluch shell
11. Clutch hub nut
12. Hub nut lockwasher
13. Clutch hub
14. Clutch hub key
15. Bearing plate spring
16. Bearing plate
17. Bearing retainer
18. Hub nut seal

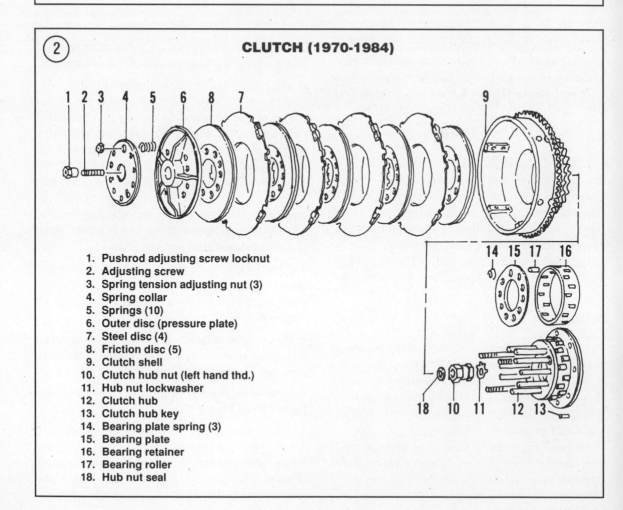

**②**

## CLUTCH (1970-1984)

1. Pushrod adjusting screw locknut
2. Adjusting screw
3. Spring tension adjusting nut (3)
4. Spring collar
5. Springs (10)
6. Outer disc (pressure plate)
7. Steel disc (4)
8. Friction disc (5)
9. Clutch shell
10. Clutch hub nut (left hand thd.)
11. Hub nut lockwasher
12. Clutch hub
13. Clutch hub key
14. Bearing plate spring (3)
15. Bearing plate
16. Bearing retainer
17. Bearing roller
18. Hub nut seal

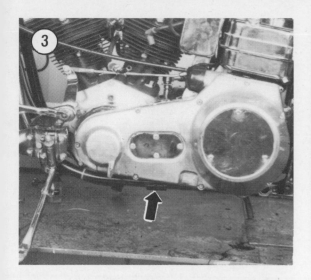

5

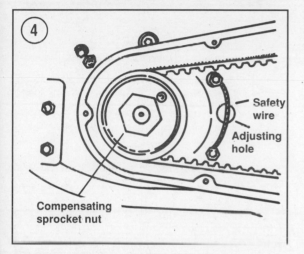

Safety wire

Adjusting hole

Compensating sprocket nut

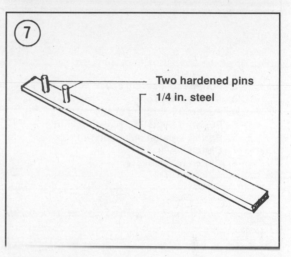

Two hardened pins

1/4 in. steel

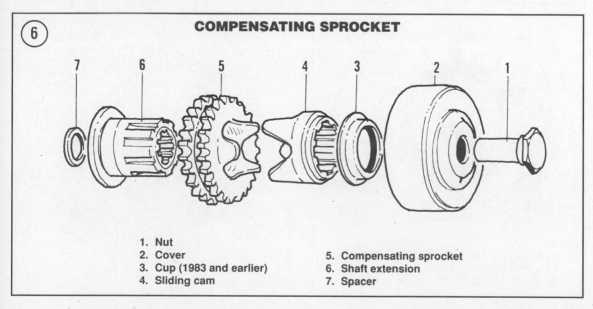

## COMPENSATING SPROCKET

7  6  5  4  3  2  1

1. Nut
2. Cover
3. Cup (1983 and earlier)
4. Sliding cam
5. Compensating sprocket
6. Shaft extension
7. Spacer

11. Refer to **Figure 10**. Tighten the pushrod adjusting screw locknut until the clutch spring adjusting nuts are loose. Then remove the clutch spring nuts.

*NOTE*
*Do not disassemble the parts in Step 12 unless replacement is required. Disassembly is described under **Clutch Inspection** in this chapter.*

12. Remove the pressure plate, clutch springs and the releasing disc as an assembly. See **Figure 1** or **Figure 2**.

13. *1966-1967:* Remove the spring disc (7, **Figure 1**).

14. Remove the friction (**Figure 11**) and steel (**Figure 12**) clutch plates in order.

*NOTE*
*Step 15 and Step 16 apply to models with a primary chain.*

15. Referring to **Figure 13**, remove the primary chain adjuster bolt and remove the chain adjuster assembly. See **Figure 14**.

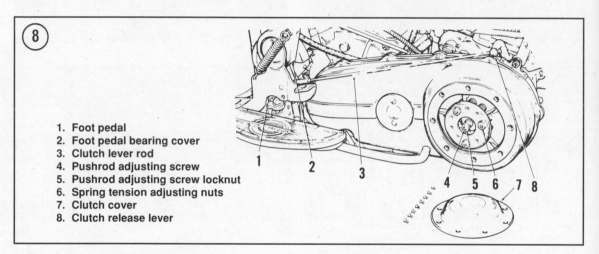

1. Foot pedal
2. Foot pedal bearing cover
3. Clutch lever rod
4. Pushrod adjusting screw
5. Pushrod adjusting screw locknut
6. Spring tension adjusting nuts
7. Clutch cover
8. Clutch release lever

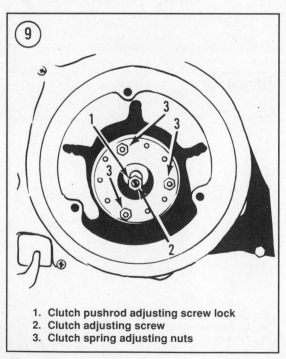

1. Clutch pushrod adjusting screw lock
2. Clutch adjusting screw
3. Clutch spring adjusting nuts

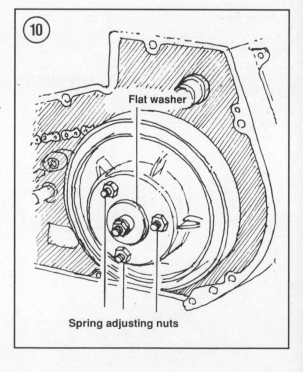

Flat washer

Spring adjusting nuts

16. Remove the oil hose from the primary chain adjuster fitting, if so equipped.

17. Remove the clutch shell, compensating sprocket and primary chain or belt as one unit (**Figure 15**).

*NOTE*
*The clutch hub nut to be removed in Step 18 uses left-hand threads.*

18. Pry back the clutch hub lockwasher tab. Then remove the clutch hub nut (A, **Figure 16**) and lock-washer.

19. Attach the clutch hub puller (HD-95960-41A) to the clutch hub (**Figure 17**). Then turn the puller's center bolt clockwise and remove the clutch hub (B, **Figure 16**).

20. Remove the clutch hub Woodruff key (**Figure 18**) from groove in main shaft.

**Inspection**

1A. *Primary chain models:* Clean all clutch parts in a non-oil based solvent and thoroughly dry with compressed air.

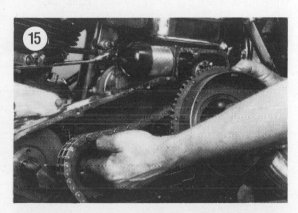

1B. *Primary belt models*: Wipe clutch parts clean with a shop cloth.

2. If disassembly of the pressure plate assembly is required, perform the following.

a. Install 3 bolts through the original pressure plate-to-plate hub bolt holes. The bolts should be long enough to allow removal of the parts while under compression.

b. Secure each bolt with a flat washer and nut. Tighten all nuts in a criss-cross pattern until the clutch springs compress slightly.

c. Remove the adjuster locknut and remove the adjuster screw.

d. Loosen the 3 nuts in a criss-cross pattern. Loose the nuts 1/2 to 1 turn at a time to release spring tension.

e. After loosening the nuts, remove the washers and 3 bolts. The separate the pressure plate and remove the clutch springs.

3. Measure the free length of each clutch spring as show in **Figure 19**. Replace any springs that are too short (**Table 1**).

4. Inspect steel clutch plates (A, **Figure 20**) for warping or wear grooves. Replace if necessary.

5. *1966-early 1981*: Steel plates on these models are equipped with a disc buffer on the edge of each plate slot. The disc buffer consists of a housing with a spring loaded ball. Depress the balls to check their condition. If they don't spring bak, replace them as follows:

a. The disc buffer is secured with 2 rivets. Using a hand grinder, grind the rivet head off.

b. Tap the rivets out with a suitable size drift punch.

c. Position the new disc buffer on the clutch plate.

d. Secure with 2 new rivets.

6. Inspect friction plates (**Figure 21**) for uneven wear, a shiny appearance or signs of oil soaking. Also check the plates for worn or grooved lining surfaces. Replace excessively worn or burned pates. Refer to specification in **Table 1**.

7. Check for loose clutch plate rivets; replace if necessary.

8. Check the clutch shell inner bearing race (A, **Figure 22**) for grooves, wear or damage. Also check the clutch shell plate tabs (B, **Figure 22**) for looseness or damage. If found, replace the clutch shell.

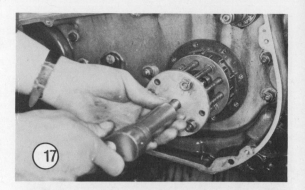

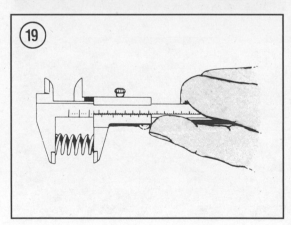

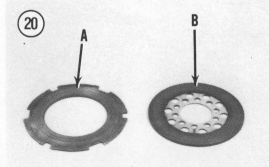

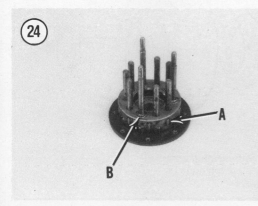

9. Check the clutch shell gear teeth (**Figure 23**) for wear or damage. Replace the clutch shell if necessary.

10. Spin the clutch hub roller assembly (A, **Figure 24**) by hand. If bearing assembly appears rough, disassemble it by removing the 3 bearing plate springs (B, **Figure 24**). Then slide the bearing plate off the hub pins and remove the bearing retainer. Check all parts for wear or damage; replace parts as required.

11. If so equipped, pry the pushrod seal out of the hub nut. Install a new seal by tapping it into place.

12. Inspect the pushrod for wear or bending. Replace if necessary.

13. Inspect the primary chain or belt as described in this chapter.

14. Inspect the compensating sprocket as described in this chapter.

### Installation

1. If any component was replaced, check the primary chain or primary belt alignment as described in this chapter.

2. *Belt drive:* If the compensating sprocket was disassembled, assemble it as described under *Compensating Sprocket* in this chapter.

3. Install the Woodruff key (**Figure 18**) in the main shaft keyway.

4. Install the pushrod. Install a new clutch pushrod oil seal on early models.

5. Slide the clutch hub assembly (**Figure 16**) onto the main shaft.

> *NOTE*
> *The clutch hub nut to be tightened in Step 6 uses left-hand threads.*

6. Install clutch hub lockwasher and nut. Tighten the nut to 50-60 ft.-lb. (69-82.8 N•m). Bend the lockwasher tab over the nut to lock it.

7. Install the clutch shell, primary chain or belt and the motor sprocket or compensating sprocket as an assembly. See **Figure 15**.

8. Install the compensating sprocket nut and tighten it to 80-100 ft.-lb. (110-138 N•m). On models with motor sprocket, tighten the sprocket nut securely.

9. Install the friction and steel clutch plates in the order shown in **Figure 1** or **Figure 2**. Note the following:

    a. On models which use the disc buffers on the steel plates, stagger the plates around the clutch shell.

    b. Install the steel clutch plates with the side stamped *OUT* facing outward.

10. If the pressure plate unit was disassembled, assemble as follows:

    a. Place the clutch hub on the workbench so that the bolts face up (**Figure 24**).

    b. Install the retaining disc on the hub.

    c. Install the clutch springs on the hub pins and studs.

    d. Place the pressure plate over the clutch spring. Because of stud hole arrangement, plate collar will fit only one way.

    e. Screw the pushrod adjusting locknut onto the adjusting screw until the screw is flush with the top of the nut. Install a 1 3/4 in. (44.45 mm) washer under the nut and thread the adjusting screw into the releasing disc.

    f. Tighten the nut to compress the clutch springs.

    g. Install the 3 clutch spring adjusting nuts.

    h. Remove the adjusting screw locknut and remove the 1 3/4 in. (44.45 mm) washer. Then reinstall the locknut.

    i. Tighten the 3 adjusting nuts in a criss-cross pattern until the distance from the releasing disc to the pressure plate is 31/32 in. (24.6 mm) (1966-1967) or 1 1/32 in. (26.2 mm)

(1968-on). Tighten the adjusting locknut to maintain this distance.

11. Install the primary chain adjuster (**Figure 13**).

12. Adjust the primary drive chain or belt as described in Chapter Three.

13. Install the primary chain case cover using a new gasket.

14. Install all parts previously removed.

## PRIMARY CHAIN

### Removal/Installation

    Remove the primary chain as described under *Clutch Removal*.

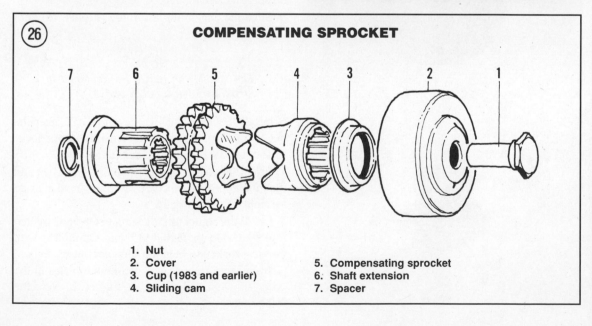

**COMPENSATING SPROCKET**

1. Nut
2. Cover
3. Cup (1983 and earlier)
4. Sliding cam
5. Compensating sprocket
6. Shaft extension
7. Spacer

## Inspection

Refer to the procedures in Chapter Three to lubricate and adjust the chain. If the chain cannot be adjusted within the specifications in Chapter Three, it must be replaced.

Always replace the primary drive chain (A, **Figure 25**) if it is worn or damaged. Attempting to repair a worn chain can cause expensive engine damage.

## Adjustment Shoe Replacement

If the primary chain cannot be adjusted properly and the adjustment shoe appears worn, replace it as follows.

1. Remove the primary chain case cover as described under *Clutch Removal* in this chapter.

2. See B, **Figure 25**. Remove the top shoe bracket bolt and remove the bracket.

3. Pry back the locking tabs and remove the adjusting shoe mounting bolts. Remove the old adjusting shoe and install a new one. Lock the new adjusting shoe in place by bending the lockwasher tabs over the mounting bolts.

4. Adjust the primary chain as described in Chapter Three.

5. Install the primary chain case cover as described under *Clutch Installation* in this chapter.

## Alignment

The clutch sprocket is aligned with the motor or compensating sprocket by a number of spacers (7, **Figure 26**) placed between the alternator rotor hub and the compensating sprocket hub. The same spacers should be reinstalled any time the compensating sprocket is removed. However, if the primary chain is wearing on one side or if new clutch components were installed that could affect alignment, perform the following.

1. Disconnect the negative battery cable.

2. Remove the primary chain, compensating sprocket and clutch shell as described under *Clutch Removal* in this chapter.

3A. *1966-1969:* Determine spacer thickness as follows:

   a. Measure the distance from the chain cover surface to the clutch disc friction surface (A, **Figure 27**).

   b. Add 0.200 in. (5.08 mm) to measurement A.

   c. Measure the distance from the chain cover surface to the Timken bearing inner race (1966-1968) or to the shield washer (1969). See B, **Figure 27** for your model.

   d. Subtract measurement A from B to obtain spacer thickness.

   e. Spacers are available in thicknesses of 0.516, 0.546, 0.576, 0.606, 0.636 and 0.666 in. (13.11, 13.87, 14.63, 15.39, 16.15 and 16.92 mm).

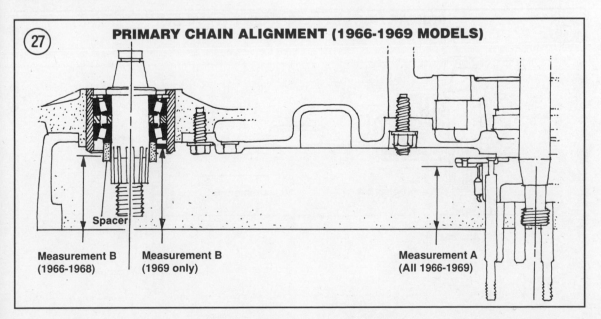

**PRIMARY CHAIN ALIGNMENT (1966-1969 MODELS)**

Measurement B (1966-1968)

Measurement B (1969 only)

Measurement A (All 1966-1969)

Spacer

3B. *1970-on:* Determine spacer thickness as follows:

  a. Install the clutch hub.

  b. Measure the distance from the alternator rotor hub to the primary drive housing gasket surface (A, **Figure 28**).

  c. Measure the distance from the clutch disc friction surface to the primary drive housing gasket surface (B, **Figure 28**).

  d. Subtract measurement B from A to obtain spacer thickness C.

  e. Select the proper spacer thickness from **Table 3**.

4. Reinstall all parts as described under *Clutch Installation* in this chapter.

## PRIMARY BELT

Toothed, rubber primary belts were introduced on the 1980 FXB (**Figure 29**). The belts should not be lubricated in any manner, but belt tension must be inspected according to the maintenance schedule in Chapter Three.

On belt primary drive models, the primary chain oiler is eliminated. This keeps chain oil from leaking onto the clutch and allows the "dry" clutch to run drier. The new clutch hub bearings on these models require greasing according to the maintenance schedule in Chapter Three. Also, the rubber dampers at the crankshaft at the crankshaft compensating sprocket require periodic lubrication according to the maintenance schedule (Chapter Three).

### Inspection

The drive belts have a built-in polyethylene lubricant coating that burnishes off during break-in. It is normal to find this material collecting in the primary housing; it is not a sign of belt wear or damage.

Referring to **Figure 30**, check the belt for cracked or damaged teeth, wear or cracks on the belt face. If the belt is worn, or damaged on one side only, check the belt alignment as described in this chapter. Replace the belt if damaged. Do not apply any lubricants. Inspect the belt tension as described in Chapter Three.

### Alignment

The clutch sprocket is aligned with the compensating sprocket by a number of spacers placed between the alternator rotor hub and the compensating sprocket hub (20, **Figure 29**). The same spacers should be reinstalled any time the compensating sprocket is removed. However, if the primary belt is wearing on one side or if new clutch components were installed that could affect alignment, perform the following.

1. Disconnect the negative battery cable.

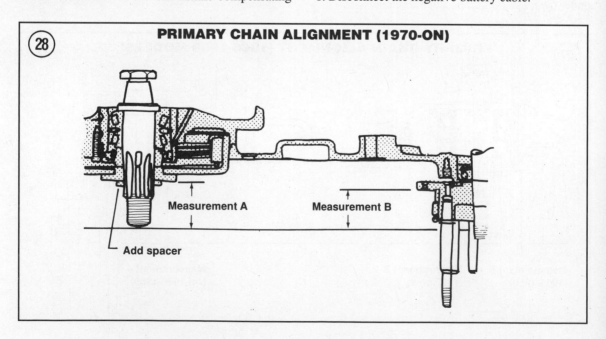

**(28)** **PRIMARY CHAIN ALIGNMENT (1970-ON)**

Measurement A

Measurement B

Add spacer

2. Remove the primary belt, compensating sprocket and clutch shell as described under *Clutch Removal* in this chapter.

3. Measure the distance from the alternator rotor hub to the primary drive housing gasket surface (A, **Figure 28**).

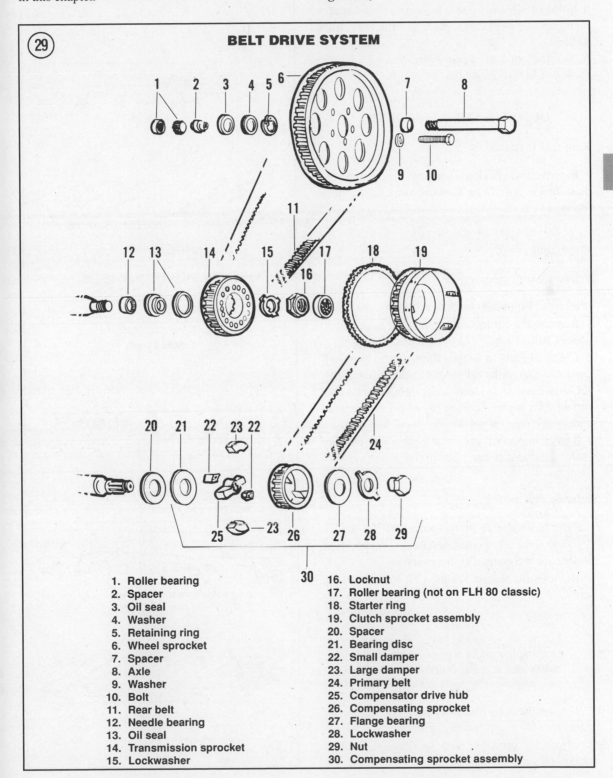

**BELT DRIVE SYSTEM**

1. Roller bearing
2. Spacer
3. Oil seal
4. Washer
5. Retaining ring
6. Wheel sprocket
7. Spacer
8. Axle
9. Washer
10. Bolt
11. Rear belt
12. Needle bearing
13. Oil seal
14. Transmission sprocket
15. Lockwasher
16. Locknut
17. Roller bearing (not on FLH 80 classic)
18. Starter ring
19. Clutch sprocket assembly
20. Spacer
21. Bearing disc
22. Small damper
23. Large damper
24. Primary belt
25. Compensator drive hub
26. Compensating sprocket
27. Flange bearing
28. Lockwasher
29. Nut
30. Compensating sprocket assembly

5

4. Measure the distance from the clutch hub friction surface to the primary drive housing gasket surface (B, **Figure 28**).

5. Subtract measurement B from A to obtain spacer thickness C. Select the proper spacer thickness from **Table 4**.

6. Reinstall all parts as described under *Clutch Installation* in this chapter.

## COMPENSATING SPROCKET

### Removal/Installation

Remove and install the compensating sprocket as described under *Clutch Removal* and *Clutch Installation*.

### Inspection

#### Primary chain models

Refer to **Figure 26** for this procedure.

1. Remove the compensating sprocket as described under *Clutch Removal* in this chapter.

2. Clean all parts in lacquer thinner, then blow dry. Do not use an oil-based solvent to clean these parts.

3. Check the sprocket teeth, shaft splines and the cam surfaces for wear. Replace any worn or damaged part.

4. If any component was replaced, check the primary chain alignment as described in this chapter.

#### Primary belt models

Refer to **Figure 29** for this procedure.

1. Remove the compensating sprocket as described under *Clutch Removal* in this chapter.

2. Remove the flange bearing (27, **Figure 29**) set screw.

> *NOTE*
> *The set screw is used to align the hole in the outer plate with the hole in the hub in case a puller is needed to remove the compensating sprocket assembly.*

3. Remove the outer plate, rubber dampers and inner plate from the compensating sprocket.

4. Check all parts for wear or damage. Replace individual parts as necessary.

**PRIMARY BELT INSPECTION**

Broken belt

Missing teeth

Cracked teeth

Severe wear or cracks on belt face

Belt wear or damage on one side only

Tooth wear

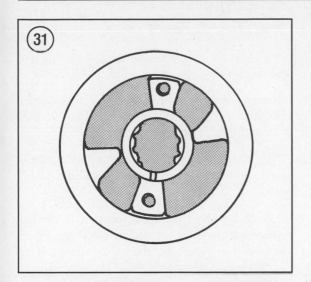

5. The rubber dampers (22 and 23, **Figure 29**) should be lubricated every 10,000 miles with Harley-Davidson "Poly-Oil."

6. Assemble the compensating sprocket in the order shown in **Figure 29**. Install the rubber dampers as shown in **Figure 31**.

7. Align the hole in the outer plate with the hole in the hub and install the set screw.

8. If any component was replaced, check the primary belt alignment as described in this chapter.

## PRIMARY CHAIN/BELT CASE

### Removal/Installation

1. Remove the primary chain case cover, clutch, primary chain and compensating sprocket as described under *Clutch Removal* in this chapter.

2. Remove the solenoid and plunger as described in Chapter Seven.

3. Disconnect the starter drive housing as described under starter motor removal in Chapter Seven.

4. Cut any safety wire (**Figure 32**) used to lock engine case bolts.

*NOTE*
*Label all hoses before removal so that they can be reinstalled in their original positions.*

5. On primary chain models, disconnect the chain oiler hose at the oil pump.

6. Remove the primary case-to-engine case bolts (**Figure 33**).

7. Remove the primary case-to-transmission case mounting bolts (**Figure 34**).

8. Remove the bolt at the rear of the primary case.

9. Label any oil or vent hoses at the primary case. Then disconnect the hoses.

10. Lightly tap the primary case with a plastic mallet and remove it. The 2 washers that fall behind the primary case when it is removed belong on the 2 front transmission case-to-primary case studs. Reinstall the washers at this time.

11. Check the gasket surface for cracks or damage. Inspect the primary case bearing for wear or damage. Have a Harley-Davidson dealer replace the bearing if necessary.

12. Installation is the reverse of these steps. Note the following.

13. On primary chain models, replace the alternator O-ring (**Figure 35**) if worn or damaged.

14. Install the primary case with a new gasket.

15. Install the primary case-to-transmission case bolts (**Figure 34**) finger tight at this time.

16. Install the 4 primary case-to-engine case bolts. Install the 2 bolts with drilled heads into the rear mounting holes (**Figure 33**). Tighten these bolts to 18-22 ft.-lb. (24.8-30.4 N•m). Safety wire the 2 rear bolts as shown in **Figure 36** and **Figure 37**.

> *CAUTION*
> *Always install safety wire so that it tightens the bolt. **Figure 38** shows the correct way to safety wire two bolts by the double twist method. Always use a stainless steel wire approved for safety wiring.*

17. Align the transmission case so that the inner primary case does not bind on the main shaft or any mounting bolts.

18. Tighten the primary case-to-transmission case bolts (**Figure 34**) to 18-22 ft.-lb. (24.8-30.4 N•m).

19. Tighten the rear primary case bolt securely.

20. Install the starter drive housing and the solenoid and plunger as described in Chapter Seven.

21. Install the clutch, primary chain and compensating sprocket as described under *Clutch Installation* in this chapter.

22. Perform the primary housing vacuum check in the following procedure.

**Primary Housing Vacuum Check**

The primary housing must be checked for air tightness after reassembly.

1. Remove one of the clutch inspection cover screws and thread the Vacuum Gauge (part No. HD 96950-68) into the screw hole.

2. Start the engine and allow it to idle. The vacuum gauge should read 9 inches of water vacuum (minimum).

3. Locate the 3/8 in. (9.5 mm) vent hose connected between the chain case and the tee connector. Pinch this hose closed and bring the engine idle speed up to 2,000 rpm. The vacuum gauge should now read 25 inches of water vacuum.

4. If the vacuum gauge shows a lower reading, there is an air leak into the primary housing; proceed to Step 5.

5. Pinch all of the oil lines running to the primary housing. Pinch the hoses as close to the housing as possible.

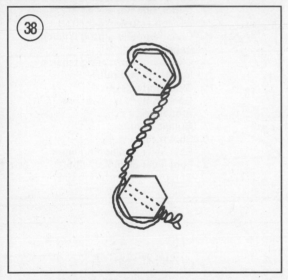

6. Pressurize the housing with 10 psi (0.7 kg/cm²) of compressed air. Now listen for air leaks at the following locations:

a. All O-ring and gasket surfaces.
b. All hose and oil seal fittings.
c. Starter drive and solenoid mounting areas.
d. Timing inspection hole.
e. Transmission filler hole.
d. Along the primary chain case housing and cover (possible cracks or casting defects).

7. Leaking areas must be repaired before putting the bike back into service.

## TRANSMISSION

### Removal/Installation

The transmission is installed in a separate housing behind the engine (**Figure 39**).

1. Drain the transmission oil as described in Chapter Three.

2. Remove the battery and battery carrier as described in Chapter Three.

3. Remove the passenger grab strap and seat.

4. Remove the master cylinder reservoir mount bracket at the transmission end cover.

5. Remove the starter as described in Chapter Seven.

6. Remove the primary housing as described in this chapter.

7. Disconnect the transmission shifter rod from the shifter linkage (A, **Figure 40**).

8A. *1966-1969:* Disconnect the clutch control rod at the clutch release lever by loosening the locknut at the pedal (foot control models) or at the booster connection (hand control models). See **Figure 41** or **Figure 42**. Then turn the rod out until the length increases enough to slide the flat portion of rod out of the clutch release lever slot.

8B. *1970-on:* Disconnect the clutch cable from the release lever (B, **Figure 40**) at the transmission.

9. Disconnect the wiring at the solenoid.

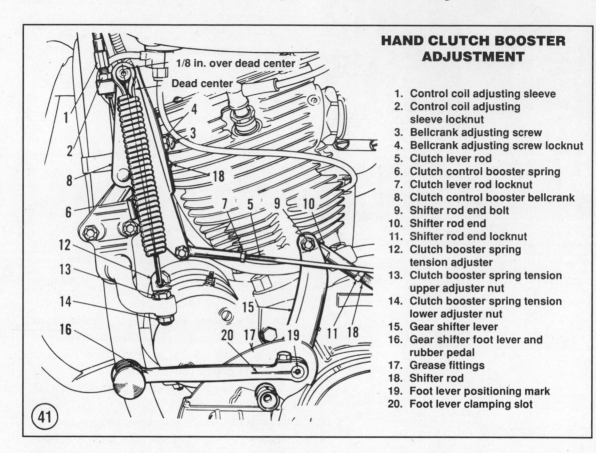

## HAND CLUTCH BOOSTER ADJUSTMENT

1. Control coil adjusting sleeve
2. Control coil adjusting sleeve locknut
3. Bellcrank adjusting screw
4. Bellcrank adjusting screw locknut
5. Clutch lever rod
6. Clutch control booster spring
7. Clutch lever rod locknut
8. Clutch control booster bellcrank
9. Shifter rod end bolt
10. Shifter rod end
11. Shifter rod end locknut
12. Clutch booster spring tension adjuster
13. Clutch booster spring tension upper adjuster nut
14. Clutch booster spring tension lower adjuster nut
15. Gear shifter lever
16. Gear shifter foot lever and rubber pedal
17. Grease fittings
18. Shifter rod
19. Foot lever positioning mark
20. Foot lever clamping slot

1/8 in. over dead center
Dead center

(41)

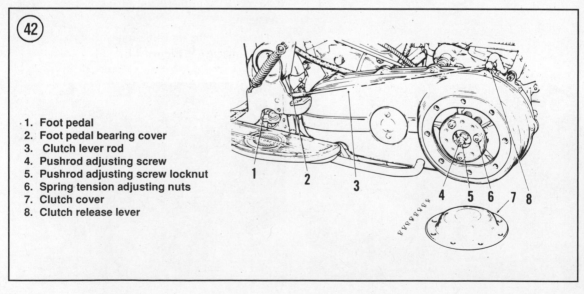

(42)

1. Foot pedal
2. Foot pedal bearing cover
3. Clutch lever rod
4. Pushrod adjusting screw
5. Pushrod adjusting screw locknut
6. Spring tension adjusting nuts
7. Clutch cover
8. Clutch release lever

10. Disconnect the speedometer drive cable and housing at the transmission (if so equipped).

11. Disconnect the neutral indicator switch wire at the transmission.

12. Remove the drive chain or drive belt as described in Chapter Nine.

> *NOTE*
> *The transmission can be removed with its mounting plate attached.*

13. Remove the transmission mounting plate-to-frame bolts.

14. Remove the starter relay and wiring.

15. Remove the rear brake line bracket.

16. Remove the transmission-to-frame mounting bolt from underneath the right side.

17. Remove the master cylinder alignment plate, if necessary.

18. On FXWG models, remove the rear brake line clip from the transmission end cover.

19. Remove the transmission and mounting plate from the left-hand side.

20. Installation is the reverse of these steps. Note the following.

21. Install the transmission assembly in the frame and install the mounting hardware. Tighten the mounting plate bolt to 30-33 ft.-lb. (41.4-45.5 N•m).

22. When installing the footrest and brake pedal assembly, make sure the brake pushrod seats in the rear brake master cylinder. See Chapter Ten.

23. Refill the transmission with the correct type and quantity oil as described in Chapter Three.

24. Check primary chain or primary belt as described in this chapter.

25. Adjust the primary chain or primary belt as described in Chapter Three.

26. Adjust the rear drive chain or drive belt as described in Chapter Three.

## Countershaft Disassembly

Refer to **Figure 43** (1966-early 1976) or **Figure 44** (1977-on).

1. Remove the following assemblies as described in this chapter:
   a. Transmission.
   b. Shifter cover.
   c. Side cover.
   d. Shift forks.

2A. *1966-1979:* Bend the lockwasher tabs back. Then remove the nut (**Figure 45**), lockwasher (**Figure 46**) and lockplate (**Figure 47**).

2B. *1980-on:* Referring to **Figure 48**, remove the retaining plate screws and remove the retaining plate.

3. While holding the gear cluster (**Figure 49**) with one hand, withdraw the countershaft through the side cover end (**Figure 50**).

4. Lift the gear cluster out of the housing (**Figure 51**).

5. Remove the thrust washer from the transmission case (B, **Figure 52**).

6. If necessary, disassemble countershaft as follows.

7. Remove the following parts in order:
   a. First gear (**Figure 53**).
   b. Bushing (if necessary).
   c. Washer (**Figure 54**).
   d. Shifter clutch (**Figure 55**).

8. Using a pointed tool, carefully pry the retaining ring from the gear cluster (**Figure 56**). Then remove the following parts in order:
   a. Washer (**Figure 57**).
   b. Second gear (**Figure 58**).
   c. Bushing (**Figure 59**).

9. Remove the thrust washer (12, **Figure 44**).

> *NOTE*
> *When removing the bearing rollers on 1966-1976 models, keep each roller set separate. If the roller sets are intermixed, or any roller lost, both sets must be replaced.*

10A. *1966-1976*: Referring to **Figure 43**, remove bearing rollers as follows:
   a. Remove bearing rollers (15) and roller retainer (16) from recess in shaft end of countershaft gear (22). Pry out the lock ring (17).
   b. Remove roller thrust washer (18), roller bearings (19), washers (20), and lock ring (21) from end of countershaft gear (22).

> *NOTE*
> *On 1966-1976 models, 22 bearing rollers are used in each bearing.*

10B. *1977-on*: The countershaft bearings on these models are single unit needle bearings. Replace the bearing as described under *Inspection*.

## Inspection

1. Examine gears for worn or chipped teeth, pitting, scoring or other damage. See **Figure 60** and **Figure 61**.

2. Examine shifter clutch (**Figure 62**) for chips and rounded edges or wear.

3. Check gear dogs for wear or rounded edges (**Figure 63**).

4. Examine gear and shaft splines (**Figure 64**) for wear or rounded edges.

5. Slip gears on shafts and check for free movement without appreciable play.

6. Replace worn or damaged thrust washers.

7A. *1966-1976:* Check the countershaft bearing rollers for wear or damage. If necessary, replace the bearings as a set. Bearing rollers (15 and 19) are available in standard size and oversizes of 0.0004 to 0.0008 in. (0.010-0.020 mm).

7B. *1977-on:* Check the countershaft needle bearings (**Figure 65**) for wear or roughness. If worn or damaged, they must be replaced by a Harley-Davidson dealer as special tools and procedures are required.

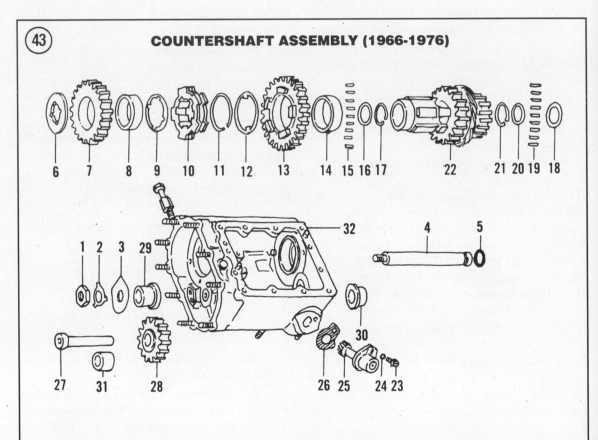

**COUNTERSHAFT ASSEMBLY (1966-1976)**

1. Countershaft nut
2. Lockwasher
3. Lock plate
4. Countershaft (1966-early 1976)
5. O-ring (1966-early 1976)
6. Countershaft gear end washer
7. Low gear
8. Low gear bushing
9. Washer
10. Shifter clutch
11. Spring lock ring
12. Gear retaining ring
13. Countershaft second gear
14. Second gear bushing
15. Bearing rollers
16. Roller retainer washer
17. Lock ring
18. Roller thrust washer
19. Roller bearing
20. Retaining washer
21. Lock ring
22. Countershaft gear
23. Screw
24. Washer
25. Speedometer drive unit
26. Gasket
27. Idler gear shaft
28. Idler gear
29. Collar
30. Collar
31. Idler gear bushing
32. Transmission housing

8. Inspect the transmission housing bushings and oil seals as described in this chapter.

### Countershaft Assembly/Installation

1. Coat all parts with engine oil prior to assembly.

2. *1966-1976:* Referring to **Figure 43**, install the bearing rollers as follows:

   a. Install new countershaft lock rings (17 and 21). Make sure the rings seat in their grooves.

   b. Install new roller retainer washers (16 and 20).

   c. Coat the bearing rollers (15 and 19) with grease and install them. If the bearings are being reused, make sure to install them in their original positions.

   d. Install the thrust washer (18) into its recess. See **Figure 66**.

3. Slide on the bushing and install second gear (**Figure 58**).

4. Install the washer (**Figure 57**).

5. Install a new circlip (**Figure 56**). Make sure the circlip seats completely in the gear cluster groove.

6. Install the shifter clutch (**Figure 55**), washer (**Figure 54**), bushing and first gear (**Figure 53**).

7. *1977-on:* Install the thrust washer (A, **Figure 52**) into its recess.

8. Coat the countershaft end thrust washer (B, **Figure 52**) with grease and install it in the transmission case.

9. Install a new O-ring on the countershaft.

10. Install the gear cluster in the transmission case (**Figure 51**). Hold it in position with one hand.

11. Insert the countershaft through the gear cluster from the sprocket side of the transmission case. The O-ring should be on the sprocket side. See **Figure 50**.

12. Measure the gear end play between the washer and the countershaft gear with a feeler gauge. Correct end play is listed in **Table 5** or **Table 6**. If the end play is incorrect, replace the washer (B, **Figure 52**) with a suitable size washer. Washers are available in the following sizes: 0.074, 0.078, 0.082, 0.085, 0.090, 0.095 and 0.100 in. (1.88, 1.98, 2.08, 2.16, 2.29, 2.41 and 2.54 mm).

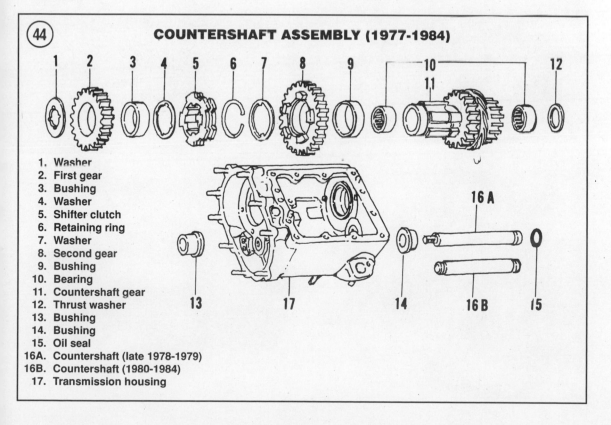

**(44) COUNTERSHAFT ASSEMBLY (1977-1984)**

1. Washer
2. First gear
3. Bushing
4. Washer
5. Shifter clutch
6. Retaining ring
7. Washer
8. Second gear
9. Bushing
10. Bearing
11. Countershaft gear
12. Thrust washer
13. Bushing
14. Bushing
15. Oil seal
16A. Countershaft (late 1978-1979)
16B. Countershaft (1980-1984)
17. Transmission housing

*NOTE*
*If the main shaft does not require re-*
*moval, perform Step 13. If main shaft*
*removal is required, remove it now as*
*described in this chapter.*

13A. *1966-1979:*

a. Install the retaining plate (**Figure 67**) so that
   the flat sides align with the main shaft retaining plate.

b. Install the lockwasher and countershaft nut
   (**Figure 68**).

c. Tighten the nut to 55-60 ft.-lb. (75.9-82.8
   N•m).

d. Bend the lockwasher arms over the nut to lock
   it.

13B. *1980-on:* Install the retaining plate (**Figure
48**). Tighten the screws to 7-9 ft.-lb. (9.6-12.4 N•m).

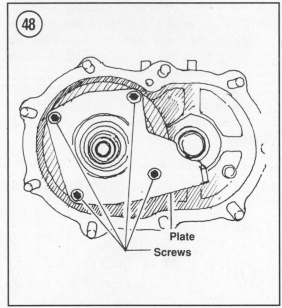

Plate
Screws

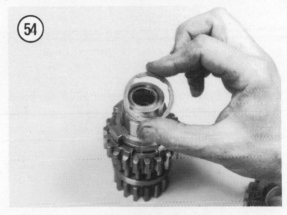

A          B

5

## Main Shaft Removal/Disassembly

Refer to **Figure 69** (1966-early 1978) or **Figure 70** (late 1978-on).

1. Remove the transmission case as described in this chapter.

2. Remove the countershaft as described in this chapter.

3. *1966-early 1978:* Referring to **Figure 69**, perform the following:

   a. Remove the 4 retaining plate screws (1).

   b. Remove the oil deflector (2) and retaining plate (**Figure 71**).

*NOTE*
*On 1980 and later models, the main-shaft retaining plate (**Figure 48**) was removed during countershaft removal.*

4. Using a brass or rawhide mallet, drive the main shaft (**Figure 72**) out of the transmission case through the side cover end until second gear almost contacts the case.

5. Pry the circlip (**Figure 73**) between the washer and shifter clutch out of its groove and slide it on the main shaft splines. See **Figure 74**.

6. Slide the main shaft out of the case while at the same time sliding third gear (**Figure 75**), washer (**Figure 76**), circlip (**Figure 77**) and the shifter clutch (**Figure 78**) off the shaft. Then remove the parts through the case opening.

7. If necessary, remove the main drive gear from the transmission case as described in this chapter.

8. If removal of the bearing (A, **Figure 79**) and the first/second gear combination (B, **Figure 79**) is required, refer all service to a Harley-Davidson dealer or machine shop, as a hydraulic press is required.

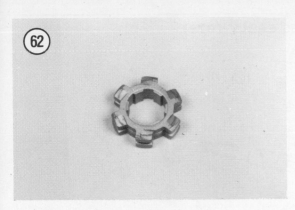

### Main Shaft Inspection

1. Examine gears (**Figure 60**) for worn or chipped teeth, pitting, scoring or other damage.

2. Examine dog clutches (**Figure 62**) for chips and rounded edges or wear.

3. Slip gears on shafts and check for free movement without appreciable play.

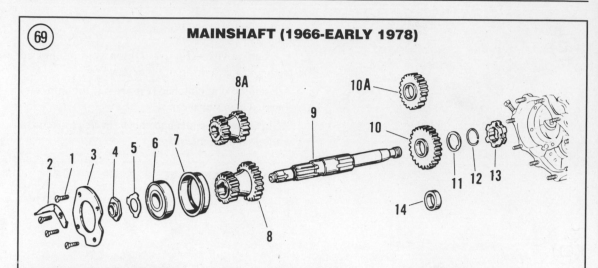

**MAINSHAFT (1966-EARLY 1978)**

1. Bearing housing retaining plate screw
2. Oil deflector
3. Retaining plate
4. Ball bearing nut
5. Ball bearing washer
6. Main shaft bearing
7. Main shaft bearing housing
8. Low and second gear
8A. Low and reverse gear (hand shift)
9. Main shaft
10. Third gear
10A. Main shaft second gear
11. Retaining washer
12. Lock ring
13. Shifter clutch
14. Third gear bushing (hand shift)

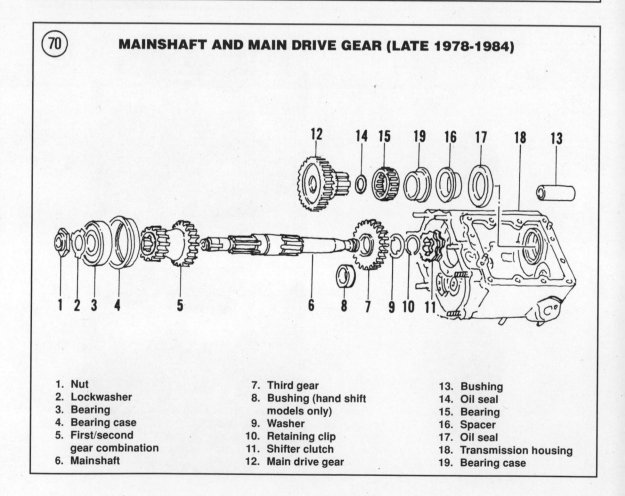

**MAINSHAFT AND MAIN DRIVE GEAR (LATE 1978-1984)**

1. Nut
2. Lockwasher
3. Bearing
4. Bearing case
5. First/second gear combination
6. Mainshaft
7. Third gear
8. Bushing (hand shift models only)
9. Washer
10. Retaining clip
11. Shifter clutch
12. Main drive gear
13. Bushing
14. Oil seal
15. Bearing
16. Spacer
17. Oil seal
18. Transmission housing
19. Bearing case

4. Replace worn or damaged thrust washers.

5. Inspect all main shaft bushings. Worn bushings should be replaced by a Harley-Davidson dealer as special tools and procedures are required.

### Main Shaft Assembly/Installation

1. Coat all parts with engine oil prior to assembly.

2. Install the main drive gear in the transmission case as described in this chapter.

3. If the main shaft nut was loosened, place the main shaft assembly in a vise with soft jaws. Tighten the nut to 50-60 ft.-lb. (69-83 N•m). Bend the lock-washer tab over the nut to lock it.

4. Install the main shaft into the transmission and slide it so that second gear barely contacts the case.

5. Slide the following parts on the main shaft:

    a. Bushing.

    b. Third gear (**Figure 75**).

c. Washer (**Figure 76**).

d. New circlip (**Figure 77**).

> *NOTE*
> *Make sure the circlip seats in the main shaft groove completely.*

e. Shifter clutch (**Figure 78**).

> *NOTE*
> *Install the shifter clutch so that the word HIGH on one side faces toward the main drive gear.*

6. Lightly tap the main shaft into the transmission case until the bearing housing flange seats against the case.

7. Install the countershaft as described in this chapter.

8A. *1966-1979:*

a. Install the main shaft retaining plate (**Figure 71**) and oil deflector. Tighten the screws to 7-9 ft.-lb. (9.6-12.4 N•m).

b. Install the countershaft retaining plate (**Figure 67**) so that the flat side aligns with the main shaft retaining plate.

c. Install the lockwasher and countershaft nut (**Figure 68**).

d. Tighten the nut to 55-60 ft.-lb. (75.9-82.8 N•m).

e. Bend the lockwasher arms over the nut to lock it.

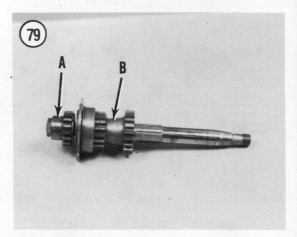

8B. *1980-on:* Install the retaining plate (**Figure 80**). Tighten the screws to 7-9 ft.-lb. (9.6-12.4 N•m).

9. Install the transmission case as described in this chapter.

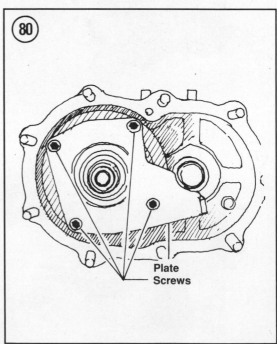

Plate Screws

## Main Drive Gear Removal/Installation

Refer to **Figure 70** (late 1978-on) or **Figure 81** (1966-early 1978).

1. Remove the countershaft and main shaft as described in this chapter.

2. If the countershaft sprocket was not removed during transmission case removal, remove it as follows:

   a. Hold the sprocket with a chain wrench or a universal holding tool.

   b. On early models, pry the lockwasher tabs away from the sprocket nut. On later models, remove the set screw that locks against the sprocket nut.

> *NOTE*
> *The sprocket nut uses left-hand threads. Turn the nut clockwise to remove it.*

   c. Loosen the sprocket nut.

   d. Remove the lockwasher, if used.

   e. Remove the oil deflector (3, **Figure 81**) if used.

   f. Remove the sprocket.

3. Measure the main drive gear end play with a dial indicator. Correct end play is listed in **Table 5** or **Table 6**. If the end play is incorrect, replace the main drive gear.

4. Push the main drive gear (**Figure 82**) into the case, then remove it through the shifter cover opening. See **Figure 83**. On 1966-1977 models, remove the roller bearing thrust washer (6, **Figure 81**).

5. Remove main drive gear oil seal (**Figure 84**) by carefully prying it out of the transmission housing. On 1966-1977 models, remove the oil seal cork washer (9, **Figure 81**).

6. Remove the main drive gear spacer and bearing case (1978-on).

> *NOTE*
> *When removing the main drive gear bearing rollers for 1966-1976, keep track of the 44 roller bearings (7, **Figure 81**). If any roller is lost, the set must be replaced.*

7. *1966-1977:* Remove all 44 roller bearings (**Figure 85**) from bearing race in housing. Store bearings in a separate container.

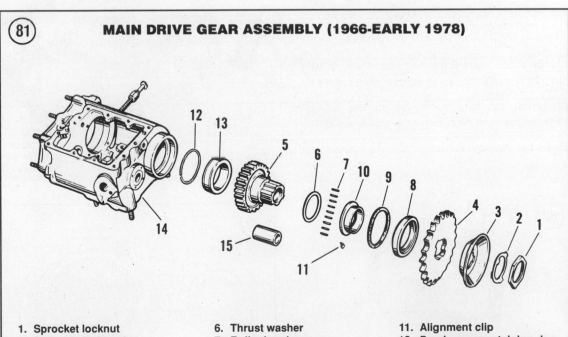

**(81)  MAIN DRIVE GEAR ASSEMBLY (1966-EARLY 1978)**

1. Sprocket locknut
2. Sprocket lockwasher
3. Oil deflector
4. Chain sprocket
5. Main drive gear
6. Thrust washer
7. Roller bearings
8. Main drive gear oil seal
9. Oil seal cork washer
10. Main drive gear spacer
11. Alignment clip
12. Bearing race retaining ring
13. Bearing race
14. Gearbox
15. Main drive gear bushing

8. Examine gears for worn or chipped teeth, pitting, scoring or other damage.

9. Examine gear and shaft splines for wear or rounded edges.

10. Replace worn or damaged thrust washer on 1966-1976 models.

11. Remove the drive gear seal by prying it out with a sharp tool or a small screwdriver. Lightly oil a new drive gear seal. Then install the seal into the end of the drive gear using a piece of pipe with a 1 in. inside diameter and 1 3/16 in. outside diameter.

12A. *1966-1976:* Check the countershaft bearing rollers for wear or damage. If necessary, replace the bearings and race as a set. A press is required to remove and install the bearing race. Refer service to a Harley-Davidson dealer or machine shop.

12B. *1977-on:* Check the countershaft needle bearing (15, **Figure 70**) for wear or roughness. If worn or damaged, it must be replaced by a Harley-Davidson dealer as a press is required.

> *CAUTION*
> *Attempting to install new bearings without the use of a press can damage the bearing race or bearing.*

13. Install the main drive gear spacer and bearing case (1977-on) as shown in **Figure 70** or **Figure 81**.

14. Install the main drive gear oil seal into the transmission case bore. Drive the seal in squarely.

15. Install the main drive gear into the transmission housing.

## Main Drive Gear Oil Seal Replacement (With Transmission Installed)

The main drive gear oil seal (**Figure 86**) can be replaced with the transmission installed in the frame.

1. Remove the countershaft sprocket as described in Chapter Nine.

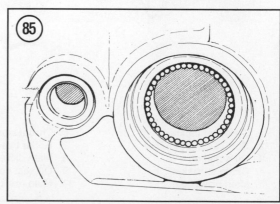

2. Pry the oil seal from the transmission as shown in **Figure 87**.

3. Carefully install the new oil seal over the shaft (**Figure 88**) and align it with the seal bore in the transmission case.

4. Install the new seal by driving it squarely into the transmission case.

5. Install all parts previously removed.

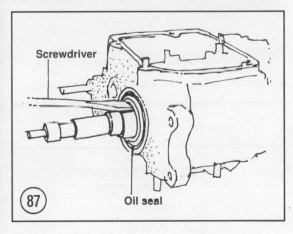

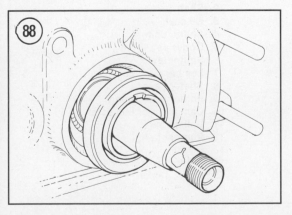

**Transmission Housing**

Remove the transmission mounting plate bolts (**Figure 89**) and remove the plate. Reverse to install. Tighten the nuts securely.

## HAND SHIFTER COVER
### (1966-EARLY 1979)

### Removal/Installation

Refer to **Figure 90** (1966-1969) or **Figure 91** (1970-early 1979).

1. Remove the transmission housing from the frame as described in this chapter.

*NOTE*
*When removing the shift cover screws, note that one of the screws is a transmission vent. This screw is installed in the hole nearest the dowel pin on the right hand side of the transmission. If the wrong type screw is installed in its position, transmission oil may be forced into the clutch.*

2. Remove the shift cover screws. Identify the transmission vent hole with a marker to insure the following.

3. Installation is the reverse of these steps. Note the following.

4. Install the shifter cover with a new gasket. Coat the gasket with Perfect Seal No. 4 or equivalent.

5. Install the shift cover screws. The 2 long screws are installed in the holes next to the bulge in the cover over the shift gear. Install the transmission vent screw in its original position.

6. Coat all shift cover screws, except the transmission vent screw, with Loctite thread sealant.

7. Adjust the shift linkage as described in Chapter Three.

### Disassembly/Assembly

Refer to **Figure 90** (1966-1969) or **Figure 91** (1970-early 1979).

1. *1970-early 1979:* Remove the neutral switch and cam follower mechanism.

2. Before disassembling the shifter cover, check the shifter cam end play as follows:

    a. Measure the shifter cam end play with a feeler gauge or dial indicator.

    b. Correct end play is 0.0005-0.0065 in. (0.013-0.165 mm).

    c. If end play is excessive, install a shim washer on the shaft.

    d. If end play is too tight, file down the boss on the inside of the shifter cover.

3. Remove the shaft lock screw.

4. Insert a drift into the shaft hole. Then tap the shaft out of the cover.

5. Remove the shifter cam from inside cover.

6. Remove the cotter pin from the shift lever shaft. Discard the cotter pin.

7. Wedge a screwdriver or chisel between the shifter gear and the shifter case. Lightly tap the tool and force the gear from the shaft.

8. Withdraw the shift lever shaft from the shifter case. Discard the leather washer.

9. Remove the cam plunger cap screw. Then remove the ball spring and ball.

10. Assembly is the reverse of these steps. Note the following.

11. Install required shim washers or file down boss inside shifter cover to obtain correct shifter cam end play as described in Step 1.

12. Time the shifter gear to the shifter cam gear as follows:

    a. Install shifter gear spring and shifter gear with spring over gear hub and timing mark between gear teeth facing bushing in cover.

    b. Install shifter cam so that notch in tooth on its gear aligns with timing mark on shifter gear.

    c. Install shifter lever so that lever points toward the left screw hole in cover.

    d. Install shifter cam so that its timing mark aligns with the shifter gear timing mark.

13. Install a new leather washer and shifter lever cotter pin.

### Inspection

1. Clean all parts in solvent, except the oil seal.

2. Inspect all parts for wear or damage. Replace parts as required.

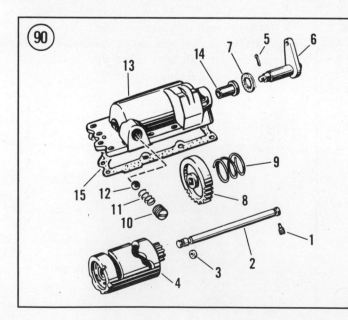

**HAND SHIFTER COVER
(1966-1969)**

1. Shaft lock screw
2. Shaft
3. Oil seal
4. Shifter cam
5. Cotter pin
6. Shifter lever
7. Leather washer
8. Shifter gear
9. Shifter gear spring
10. Cam plunger cap screw
11. Ball screw
12. Plunger ball
13. Cover
14. Shifter lever bushing
15. Cover gasket

3. Inspect the shifter cam slots and plunger ball seats in cam. They must be unworn and have sharp edges. Replace the shifter cam if worn.

4. Replace the shaft oil seal if worn or damaged. When installing the new seal, place seal in the wider shaft groove.

*WARNING*
*Parts will be hot enough to cause burns during the next steps. Protect yourself accordingly.*

5. If the shifter lever bushing is worn or damaged, replace it as follows:

   a. Thread a 5/8-18 in. tap into the bushing approximately 1/2 in. (12.7 mm) deep.
   b. Remove the tap.
   c. Heat the shifter cover bushing area to approximately 300° F.
   d. Quickly install a 5/8-18 bolt approximately 2 in. (50.8 mm) long in the bushing and clamp the bolt in a vise so that the shifter cover faces down.
   e. Tap the shifter cover with a plastic mallet until the cover is driven off of the bushing.
   f. To install a new bushing, place it in a freezer for approximately 1/2 hour.
   g. Reheat the shifter cover bushing area and carefully tap the new bushing into the cover with a suitable size drift.

## FOOT SHIFTER
### (1966-EARLY 1979)

**Removal/Installation**

The shifter cover can be removed with the transmission installed on the motorcycle. If transmission repairs are also required, remove the transmission/shifter cover assemblies as one unit. Refer to **Figure 92**.

1. Remove the battery and battery carrier.

2. Drain the oil tank as described in Chapter Three. Then remove the oil tank.

3. Disconnect the transmission shifter rod from the shifter linkage (A, **Figure 93**).

4A. *1966-1969:* Disconnect the lever control rod at the clutch release lever by loosening the locknut at the pedal. Then turn the rod out until the length increases enough to slide the flat portion of rod out of the clutch release lever slot. See **Figure 94**.

4B. *1970-early 1979:* Disconnect the clutch cable from the release lever (B, **Figure 93**) at the transmission.

*CAUTION*
*When removing the shift cover screws, note that one of the screws is a transmission vent (A, **Figure 95**). If the wrong type screw is installed in its po-*

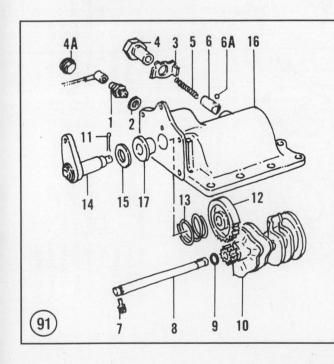

**(91)**

## HAND SHIFT COVER
### (1970-EARLY 1979)

1. Neutral indicator switch
2. Washer
3. Lockwasher (late 1976-early 1979)
4. Cam follower retainer (late 1976-early 1979)
4A. Cam follower cap screw (1970-early 1976)
5. Spring
6. Cam follower (late 1976-early 1979)
6A. Plunger ball (1970-early 1976)
7. Camshaft lock screw
8. Camshaft
9. Oil seal
10. Shifter cam
11. Cotter pin
12. Shifter gear
13. Shifter gear spring
14. Shifter lever
15. Leather washer
16. Shifter cover
17. Bushing

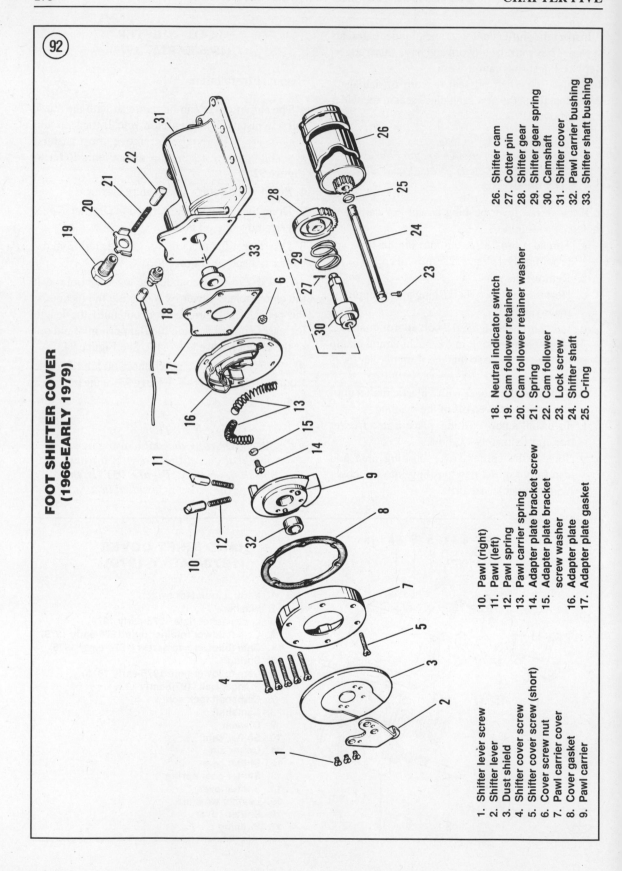

**FOOT SHIFTER COVER (1966-EARLY 1979)**

1. Shifter lever screw
2. Shifter lever
3. Dust shield
4. Shifter cover screw
5. Shifter cover screw (short)
6. Cover screw nut
7. Pawl carrier cover
8. Cover gasket
9. Pawl carrier
10. Pawl (right)
11. Pawl (left)
12. Pawl spring
13. Pawl carrier spring
14. Adapter plate bracket screw
15. Adapter plate bracket screw washer
16. Adapter plate
17. Adapter plate gasket
18. Neutral indicator switch
19. Cam follower retainer
20. Cam follower retainer washer
21. Spring
22. Cam follower
23. Lock screw
24. Shifter shaft
25. O-ring
26. Shifter cam
27. Cotter pin
28. Shifter gear
29. Shifter gear spring
30. Camshaft
31. Shifter cover
32. Pawl carrier bushing
33. Shifter shaft bushing

*sition, transmission oil may be forced into the clutch.*

5. Remove the shift cover screws. Identify the transmission vent hole with a marker to insure correct installation.

6. Remove the shift cover (B, **Figure 95**).

7. Installation is the reverse of these steps. Note the following.

8. Install the shifter cover with a new gasket. Coat the gasket with Perfect Seal No. 4 or equivalent.

9. Coat all shift cover screws, except the transmission vent screw, with Loctite thread sealant.

10. Adjust the shift linkage as described in Chapter Three.

### Disassembly

Refer to **Figure 92** for this procedure.

1. Remove the shift lever screws and shift lever (**Figure 96**).

2. Remove the dust cover (**Figure 97**).

3. Remove the neutral switch (18, **Figure 92**).

4. Pry the lockwasher tab (20) away from the cam follower retainer (19). Then remove the retainer, lockwasher, spring (21) and cam follower (22).

> *WARNING*
> *The pawl carrier springs (12, **Figure** 92) are under pressure, and may fly out when the pawl carrier cover (7, **Figure** 92) is removed.*

5. Remove the pawl carrier cover screws. The bottom cover screw is secured by a nut at the back of the adapter plate.

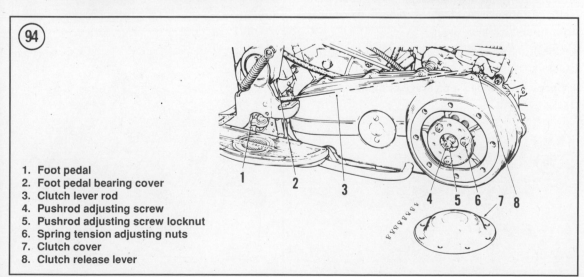

1. Foot pedal
2. Foot pedal bearing cover
3. Clutch lever rod
4. Pushrod adjusting screw
5. Pushrod adjusting screw locknut
6. Spring tension adjusting nuts
7. Clutch cover
8. Clutch release lever

6. Remove the pawl carrier cover (**Figure 98**).

7. Remove the pawls (10 and 11, **Figure 92**) and springs (12, **Figure 92**).

8. Remove the adapter plate (16) and pawl carrier springs (13).

> *NOTE*
> *When removing the shifter cam shaft in Step 9, remove the shaft from the opposite end of the shaft oil seal. See* **Figure 99**.

9. Remove the shifter cam lock screw (23). Then tap the shifter cam shaft out of the cam and remove the cam (**Figure 99**).

10. Remove the shifter shaft cotter pin (27). Then tap the shifter shaft (30) out of the cover.

11. Remove the shifter gear (28) and spring (29).

## Inspection

1. Thoroughly clean all parts (except neutral switch) in solvent, then blow dry.

> *NOTE*
> *When parts have been disassembled and cleaned, visually inspect them for any signs of wear, cracks, breakage or other damage. If there is any doubt as to the condition of any part, replace it with a new one.*

2. Replace any circlips and retaining rings that were removed during disassembly, as removal sometimes deforms and weakens them.

3. Check the shift cam slots for wear or grooves. Excessive wear will result in difficult shifting.

4. Check pawl stops for breakage or minute surface cracks.

## Assembly

> *NOTE*
> *The shifter shaft and shifter gear must be timed correctly as described in Steps*

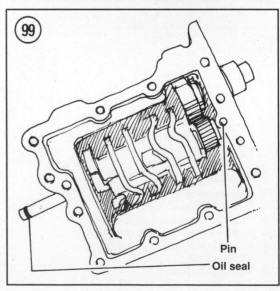

Pin
Oil seal

*1-4 or the transmission will not shift into all gears.*

1. There is a timing cut (2 cuts on 1970 and later models) between the 2 center teeth on one side of the shifter gear. This mark must be aligned with the corner of the squared end of the shifter shaft which is just to the left of its last ratchet tooth. See **Figure 100**.

2. Position the shifter gear (28) and spring (29) into the shifter cover. Install the shifter gear so that the timing marks face as shown in **Figure 101**.

3. Install the shifter shaft into the case. Align the timing mark on the shifter gear with the first slot on the shifter shaft. See **Figure 101**.

4. Support the shifter gear with a 5/8 in. wrench as shown in **Figure 102**. Then tap the shifter shaft into the shifter gear.

5. Install a new cotter pin (27) in the end of the shifter shaft.

6. Install the shifter cam as follows:

   a. Examine the shifter cam gear teeth. Note that one tooth is shorter than the rest or that a timing mark is stamped on the top of one gear tooth. This tooth is the timing mark.

   b. Place the shifter cam into the shifter cover.

   c. Engage the shifter cam gear with the shifter gear by aligning the shifter cam timing mark with the shifter gear timing mark. See **Figure 103**.

   d. Install a new camshaft O-ring (25, **Figure 92**).

   e. Install the cam shaft through the shifter cover and shifter cam until the shaft's slotted groove is aligned with the set screw hole. Install the set screw until it bottoms on the cam shaft.

7. Install the cam follower (22), spring (21), retainer washer (20) and cam follower retainer (19). Tighten the retainer and lock it with one lockwasher tab.

8. Install a new adapter plate gasket (17).

9. Install the adapter plate (16) and screws. Coat the screws with Loctite Lock N' Seal. Install the screws finger tight at this time.

10. Rotate the shifter cam into first gear. When in first gear, the cam follower will rest on the first detent.

11. Adjust the adapter plate (16) as follows:

   a. *1966-early 1978:* Align the adapter plate timing notch with the notch between the 2 shifter gear teeth. See **Figure 104**.

   b. *Late 1978-early 1979:* Align the adapter plate timing notch with the first shifter gear tooth notch. See **Figure 105**.

   c. Tighten the adapter plate screws (14) to 6-9 ft.-lb. (8.3-12.4 N•m).

   d. Rotate the shifter cam into all 4 gears and check the adapter plate timing notch at each gear position. The notch should align between 2 gear teeth on each gear shaft (except in neutral). When the cam is in neutral, the adapter plate timing notch will line up with the first

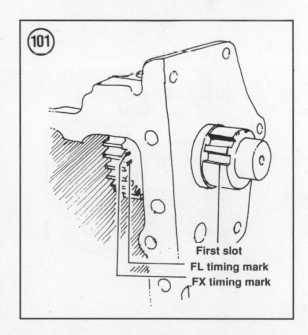

**100 TIMING SHIFTER GEAR**

FL timing mark
FX timing mark

Shifter gear
Shifter shaft

**101**

First slot
FL timing mark
FX timing mark

gear tooth. If the shifter gear and shifter cam do not align correctly, remove the cam and reassemble.

12. Grease the pawl carrier springs (13) and install them on the adapter plate.

13. Grease the shifter shaft end.

14. Lubricate the right and left pawls with light machine oil.

15. Install the pawl springs (12) and pawls (10 and 11) into the pawl carrier (9). Make sure to align the pawl grooves with pawl carrier pins.

16. Align the pawl carrier tab with the 2 pawl carrier springs and install the pawl carrier.

17. Lubricate the back of the pawl carrier with grease.

18. Install a new pawl carrier cover gasket. Then install the pawl carrier cover. Align the notch in the cover with the notch in the adapter plate.

19. Apply Loctite Lock N' Seal to the cover mounting screws and install the screws. Install the short screw in the bottom hole and secure it with the nut.

20. Install the dust shield. Then align the shift lever with the dust shield and install it. Coat the mounting screws with Loctite Lock N' Seal. Tighten the screws securely.

21. Apply Loctite Pipe Sealant With Teflon onto the neutral switch threads. Install the switch and tighten securely.

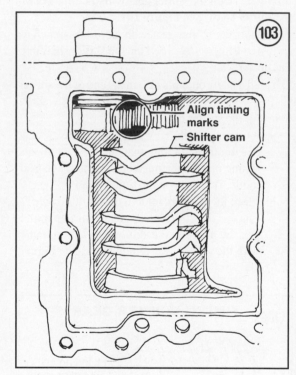

Align timing marks
Shifter cam

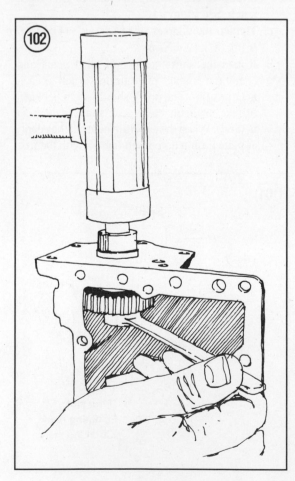

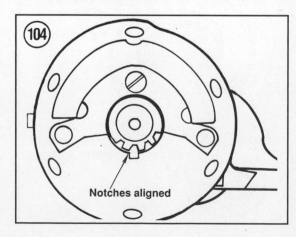

Notches aligned

## SHIFTER COVER
## (LATE 1979-1984)

### Removal

The shifter cover can be removed with the transmission installed on the motorcycle. If transmission repairs are also required, remove the transmission and shifter cover as one unit. This procedure describes removal of the shifter cover only.

Refer to **Figure 106** for this procedure.

1. Remove the battery and battery cover.

2. Drain the oil tank as described in Chapter Three. Then remove the oil tank.

3. Remove the bolts and washers attaching the shift cover to the transmission; the bolt from the hole indicated by an arrow in **Figure 106** can be loosened, but not removed. To remove the bolt, the shifter shaft cover (11) must be removed first.

4. Lift the shifter cover off of the transmission housing. Discard the shifter cover gasket.

5. Installation is the reverse of these steps. Note the following.

6. Install a new shifter cover gasket during installation.

7. Lightly coat the shift cover mounting bolts with Loctite Lock N' Seal. Tighten the mounting bolt to 13-16 ft.-lb. (17.9-22.1 N•m).

8. Adjust the shift linkage as described in Chapter Three.

### Disassembly

Refer to **Figure 106** for this procedure.

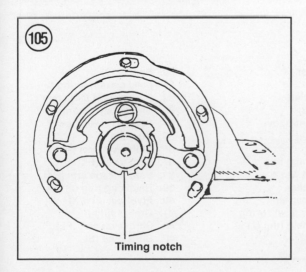

(105)

**Timing notch**

1. Remove the neutral indicator switch (5) and washer.

2. Remove the shifter shaft cover bolts (7) and washers.

3. Remove the shift lever bolt (9) and washer.

4. Remove the shifter linkage assembly from the transmission cover.

5. Remove remaining shifter cover bolt.

6. If necessary, remove the top plug (17) as follows:

    a. Drill a 1/4 in. (6.35 mm) hole through the top plug. Drill only far enough to penetrate completely through the top plug.

    b. Pry the top plug off of the shift cover with a punch inserted through the drilled hole.

    c. Discard the top plug; a new plug must be installed during reassembly.

7. Remove the shift cam retaining ring (23) and washer through the top plug hole opening and remove the shift cam (25) and pawl assembly.

8. Disassemble the shifter cover as follows:

    a. Remove the cam follower (1, **Figure 107**) and spring (2, **Figure 107**) from the cam follower body.

    b. Pry back the lockwasher tabs and remove the cam follower body bolts (4, **Figure 107**).

    c. Lift the cam follower body (5, **Figure 107**) out of the shift cover.

    d. Remove the screws, pawl stops and remove the long springs (8, **Figure 107**).

### Inspection

1. Thoroughly clean all parts (except neutral switch) in solvent, then blow dry.

> *NOTE*
> *When parts have been disassembled and cleaned, visually inspect them for any signs of wear, cracks, breakage or other damage. If there is any doubt as to the condition of any part, replace it with a new one.*

2. Replace any circlips and retaining rings that were removed during disassembly, as removal sometimes deforms and weakens them.

3. Check the shift cam slots for wear or grooves. Excessive wear will result in difficult shifting.

4. Check pawl stops for breakage or surface cracks.

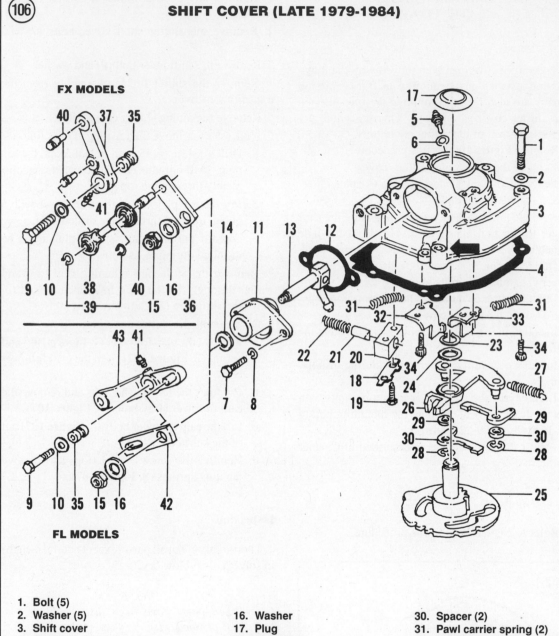

**106** **SHIFT COVER (LATE 1979-1984)**

FX MODELS

FL MODELS

1. Bolt (5)
2. Washer (5)
3. Shift cover
4. Gasket
5. Neutral indicator switch
6. Washer
7. Bolt (2)
8. Lockwasher (2)
9. Bolt
10. Washer (2)
11. Shifter shaft cover
12. Gasket
13. Shifter shaft
14. Oil seal
15. Nut
16. Washer
17. Plug
18. Lockplate
19. Bolt (2)
20. Plunger body
21. Plunger
22. Spring
23. Retaining ring
24. Thrust washer
25. Shift cam
26. Pawl carrier
27. Shifter pawl spring
28. Retaining ring (2)
29. Pawl (2)
30. Spacer (2)
31. Pawl carrier spring (2)
32. Shifter pawl stop, rear
33. Shifter pawl, front
34. Socket head screw
35. Bushing
36. Shift lever arm (FX)
37. Shift lever (FX)
38. Shift linkage arm (FX)
39. Retaining ring (2) (FX)
40. Pivot pin (3) (FX)
41. Grease fitting
42. Shift lever arm (FL)
43. Shift lever (FL)

## Assembly

1. Lubricate the pawl stop springs (8, **Figure 107**) with multi-purpose grease.

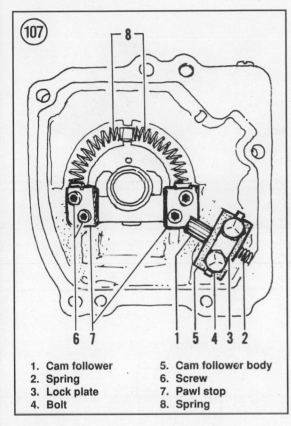

**107**

1. Cam follower
2. Spring
3. Lock plate
4. Bolt
5. Cam follower body
6. Screw
7. Pawl stop
8. Spring

**108**

Pawl carrier

Shift cam

2. Install the pawl stops (7, **Figure 107**) and secure with the Allen bolts.

3. Install the pawl stop springs (8, **Figure 107**).

4. Install the cam follower body (5, **Figure 107**), lockplate and bolts. Tighten the bolts and bend the lockplate tabs over the bolts to lock them.

5. Install the spring (2, **Figure 107**) and cam follower (1, **Figure 107**) into the cam follower body.

6. Coat the neutral switch threads with Loctite Pipe Sealant With Teflon. Install the neutral switch and washer and tighten to 5-10 ft.-lb. (6.9-13.8 N•m).

7. Refer to **Figure 108**. Slide the pawl assembly onto the shift cam. Engage the pawls with the shift cam gear teeth.

8. Install the shift cam and pawl assembly into the shift cover. Position the tab on the pawl assembly between the pawl stop springs.

9. Install the shift cam washer and a new circlip.

10. Coat a new top plug (17, **Figure 106**) with Seal-All. Then place the top plug in the cover and seat it with a ball peen hammer.

11. Assemble the shift linkage using a new circlip. Tighten all linkage bolts securely.

## SHIFT FORKS

### Removal/Disassembly

Refer to **Figure 109** for this procedure.

1. Remove the shift cover as described in this chapter.

2. Remove the shifter finger rollers (**Figure 110**) from each shifter finger.

3A. *1966-1976:* Remove the lock screw (**Figure 111**) securing the shifter fork shaft.

3B. *1977-on:* Remove the circlip (**Figure 112**) from the end of the shifter fork shaft.

4. Remove the shifter fork shaft (**Figure 113**) from the case by tapping it out with a drift.

5. Remove the shifter forks (**Figure 114**).

*NOTE*
*Do not disassemble the forks unless replacement of a part or parts is necessary.*

*NOTE*
*When disassembling the shifter fork assemblies, label each sub-assembly as they are not interchangeable (**Figure 115**).*

6. Loosen and remove the shift finger nuts (**Figure 116**) and remove them. Then disassemble the shifter forks in the order shown in **Figure 109**.

7. Repeat Step 4 for the opposite shift finger assembly.

**Inspection**

1. Clean all parts in solvent.

2. Inspect each shifter fork (**Figure 116**) for signs of wear or cracking. Make sure the forks slide smoothly on the shifter fork shaft. Replace any worn forks.

3. Check for any arc-shaped wear or burn marks on the shifter forks (**Figure 117**). If these are apparent, the shifter fork has come in contact with the gear, indicating that the fingers are worn beyond use and the fork must be replaced.

4. Roll the shifter fork shaft on a flat surface such as a piece of plate glass and check for bends. If the shaft is bent, it must be replaced.

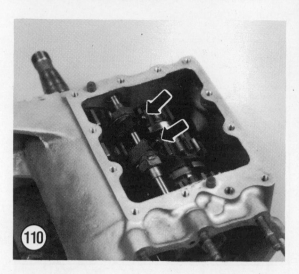

**SHIFTER FORKS**

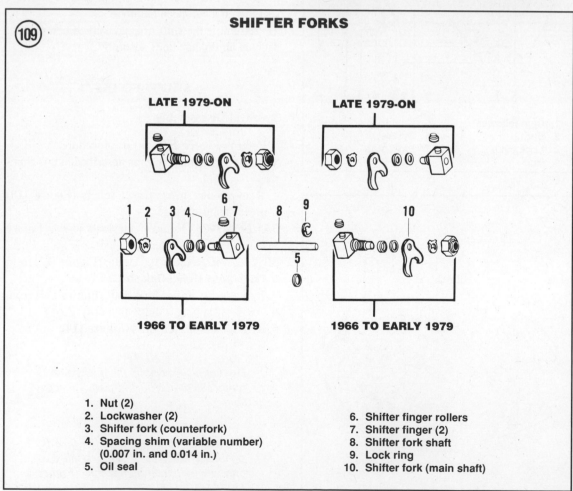

1. Nut (2)
2. Lockwasher (2)
3. Shifter fork (counterfork)
4. Spacing shim (variable number)
   (0.007 in. and 0.014 in.)
5. Oil seal
6. Shifter finger rollers
7. Shifter finger (2)
8. Shifter fork shaft
9. Lock ring
10. Shifter fork (main shaft)

5. Install each shift finger on the shift shaft (**Figure 118**). The shift fingers should slide smoothly without any signs of binding.

## Assembly

*NOTE*
*If any part was replaced, the shifter clutch clearance must be checked and adjusted. Because this procedure requires special tools, refer clearance check to a Harley-Davidson dealer.*

1. Coat all bearing and sliding surfaces with assembly oil.

2. Install the spacer(s), shifter fork, lockwasher and nut on the shift finger. Tighten the nut to 10-12 ft.-lb. (13.8-16.5 N•m). Bend the lockwasher tab to lock against the nut.

*CAUTION*
*Do not exceed the torque specifications in Step 2 as this could cause the shift finger to bind on the shift shaft.*

3. Position the shifter fork assemblies in the transmission case (**Figure 114**). Engage the shifter forks with their respective gears.

4. Slide the shift shaft through the transmission case and through the shifter fork assemblies. Secure the shaft with the lockscrew (**Figure 111**) or a new circlip (**Figure 119**).

5. Install the shift rollers (**Figure 110**) onto the shift fingers.

6. If any part was replaced, have the shift fork adjustment checked by a Harley-Davidson dealer.

## KICKSTARTER AND STARTER CLUTCH

### Removal/Installation

Refer to **Figure 120** (1966-1969) or **Figure 121** (1971-1984).

1. Disconnect the negative battery cable.

2. Drain the transmission oil as described in Chapter Three.

3. Remove the exhaust system as described in Chapter Seven.

4. Remove the rear master cylinder. See Chapter Ten.

*NOTE*
*Do not disconnect the brake hydraulic*
*line or hose or the brakes will have to*
*be bled.*

5. Remove the clutch cover (**Figure 122**) on the primary cover.

6. See **Figure 123**. Loosen the pushrod locknut and turn the adjusting screw counterclockwise to remove all tension on the pushrod.

7. Remove the kickstarter pedal.

8. Remove the starter bracket (A, **Figure 124**).

9. Remove the side cover mounting bolts and remove the side cover (B, **Figure 124**) and gasket.

10. If necessary, disassemble the side cover as described in this chapter.

11. Installation is the reverse of these steps. Note the following.

12. Install a new side cover gasket.

13. Pull the pushrod assembly partway out as shown in **Figure 125**.

14. Position the release lever to the left of the cover as shown in A, **Figure 126**.

15. Align the pushrod oil slinger (**Figure 127**) with the release lever mechanism (B, **Figure 126**) and install the side cover. The lever should face as shown in **Figure 128**.

16. Install the side cover mounting bolts and tighten them to 13-16 ft.-lb. (17.9-22.1 N•m).

17. Refill the transmission with the correct type and quantity of oil as described in Chapter Three.

18. Adjust the clutch as described in Chapter Three.

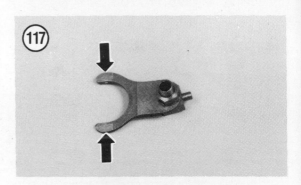

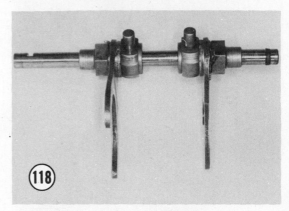

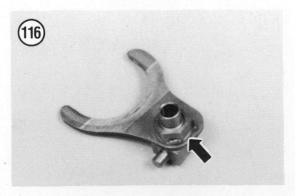

## Disassembly

Refer to **Figure 120** or **Figure 121** for this procedure.

1. Clamp the end of the kickstarter shaft in a vise with soft jaws. Then from inside the cover, pry back the kickstarter shaft lockwasher and remove the nut (A, **Figure 129**).

2. Using a universal type claw puller (**Figure 130**), remove the kickstarter gear. See B, **Figure 129**.

3. Support the side cover in a vise or on wood blocks. Do not block the kickstarter shaft or spring. Drive the kickstarter shaft out of the cover using a plastic hammer. Remove the thrust washer (11, **Figure 120** or 27, **Figure 121**).

4. Remove the release lever nut and washer.

5. Pull the release lever off of the shaft with a universal type claw puller.

6. Remove the clip. Then remove the shaft from the cover.

7. Remove the release finger and washer.

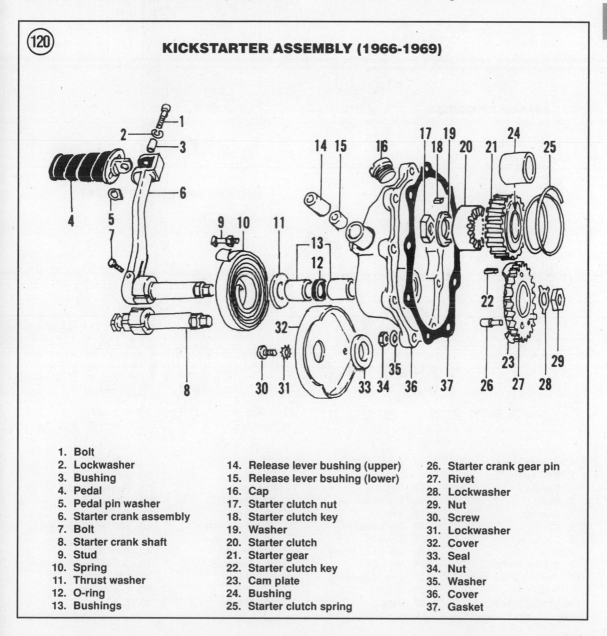

**KICKSTARTER ASSEMBLY (1966-1969)**

| | | |
|---|---|---|
| 1. Bolt | 14. Release lever bushing (upper) | 26. Starter crank gear pin |
| 2. Lockwasher | 15. Release lever bsuhing (lower) | 27. Rivet |
| 3. Bushing | 16. Cap | 28. Lockwasher |
| 4. Pedal | 17. Starter clutch nut | 29. Nut |
| 5. Pedal pin washer | 18. Starter clutch key | 30. Screw |
| 6. Starter crank assembly | 19. Washer | 31. Lockwasher |
| 7. Bolt | 20. Starter clutch | 32. Cover |
| 8. Starter crank shaft | 21. Starter gear | 33. Seal |
| 9. Stud | 22. Starter clutch key | 34. Nut |
| 10. Spring | 23. Cam plate | 35. Washer |
| 11. Thrust washer | 24. Bushing | 36. Cover |
| 12. O-ring | 25. Starter clutch spring | 37. Gasket |
| 13. Bushings | | |

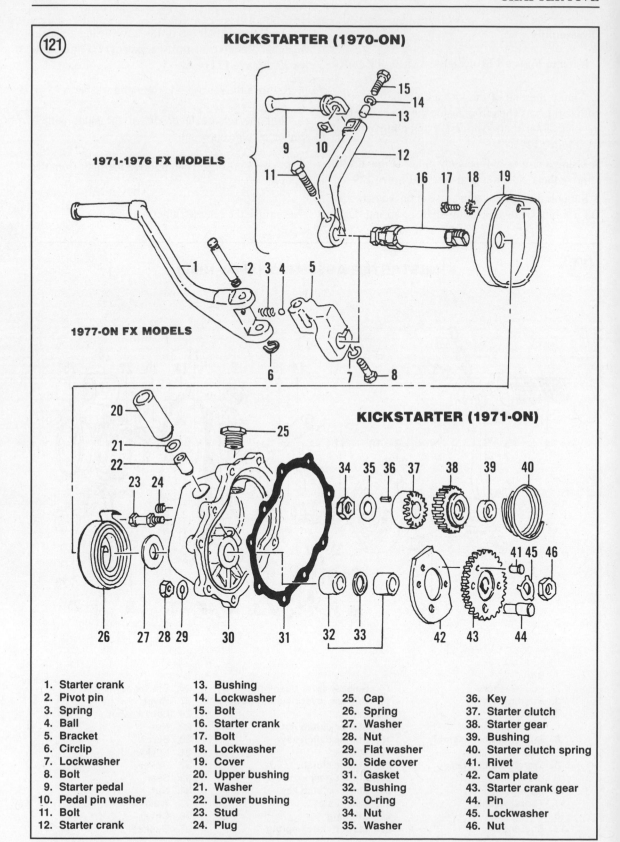

**KICKSTARTER (1970-ON)**

**1971-1976 FX MODELS**

**1977-ON FX MODELS**

**KICKSTARTER (1971-ON)**

| 1. Starter crank | 13. Bushing | 25. Cap | 36. Key |
| 2. Pivot pin | 14. Lockwasher | 26. Spring | 37. Starter clutch |
| 3. Spring | 15. Bolt | 27. Washer | 38. Starter gear |
| 4. Ball | 16. Starter crank | 28. Nut | 39. Bushing |
| 5. Bracket | 17. Bolt | 29. Flat washer | 40. Starter clutch spring |
| 6. Circlip | 18. Lockwasher | 30. Side cover | 41. Rivet |
| 7. Lockwasher | 19. Cover | 31. Gasket | 42. Cam plate |
| 8. Bolt | 20. Upper bushing | 32. Bushing | 43. Starter crank gear |
| 9. Starter pedal | 21. Washer | 33. O-ring | 44. Pin |
| 10. Pedal pin washer | 22. Lower bushing | 34. Nut | 45. Lockwasher |
| 11. Bolt | 23. Stud | 35. Washer | 46. Nut |
| 12. Starter crank | 24. Plug | | |

(122)

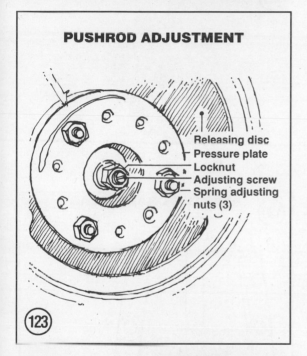

**PUSHROD ADJUSTMENT**

Releasing disc
Pressure plate
Locknut
Adjusting screw
Spring adjusting nuts (3)

(123)

(124)

A

B

## Inspection

1. Clean all components thoroughly with solvent. Remove any gasket residue from the cover-to-transmission machined surfaces. Check the threads in the transmission case to be·sure they are clean. If dirty or damaged, use a tap to true up the threads and remove any deposits.

(125)

(126)

A

B

(127)

5

2. Visually check the shaft surfaces for cracks, deep scoring, excessive wear or heat discoloration.

3. Check the kickstarter gear for worn or damaged gear teeth.

4. If oil leaks out of the side cover along the kickstarter shaft, the oil seal must be replaced by removing the front bushing with a blind hole bearing remover and slide hammer. Have the new bushings installed by a Harley-Davidson dealer or machine shop, as a press is required.

5. Check the kickstarter spring. If it is damaged or broken, replace it by performing the following:

   a. Remove the kickstarter shaft as described in *Disassembly.*

   b. Lift the outer spring hook off of the shaft spring stop.

   c. Tap the spring off the kickstarter shaft with a punch and hammer.

   d. Install the new springs so that the outer spring hook faces to the left-hand side when looking at the kickstarter crank end. See **Figure 131**.

### Assembly

Refer to **Figure 120** or **Figure 121** for this procedure.

1. Assemble the release finger as follows:

   a. Install the washer and release finger into the side cover.

   b. Insert the release lever shaft into the side cover and through the release finger and washer. Secure the shaft with a new circlip.

   c. Install the release lever. Then install the lockwasher and the nut. Tighten the nut until the release lever bottoms on the shaft.

2. Install the washer on the kickstarter shaft so that the chamfered side of the washer faces the spring.

3. See **Figure 132**. Install the kickstarter gear as follows:

   a. Turn the kickstarter shaft so that the flat side is in the 12 o'clock position.

   b. Slide the kickstarter gear onto the shaft. When installed correctly, the kickstarter gear dowel pin is in the 7 o'clock position.

4. Hold the kickstarter shaft in a vise with soft jaws as during disassembly.

5. Engage the end of the kickstarter spring with the stud on the side cover.

6. Press the kickstarter gear onto the kickstarter shaft.

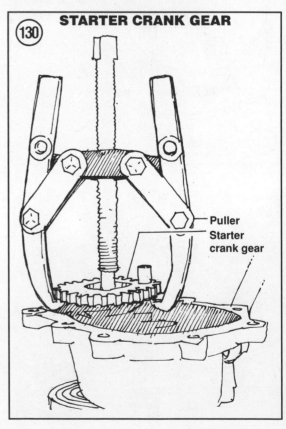

**STARTER CRANK GEAR**

Puller
Starter crank gear

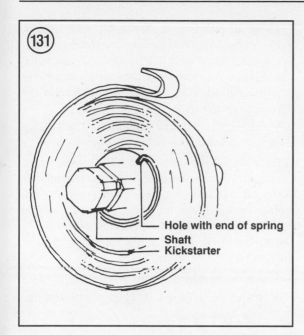

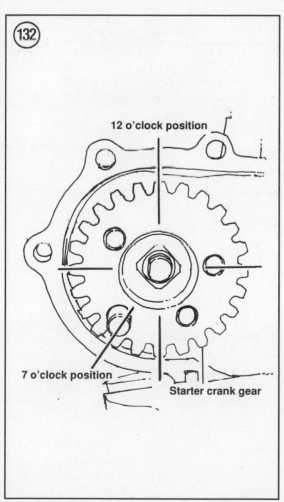

7. Install the kickstarter gear lockwasher and nut. Tighten the nut to 50-60 ft.-lb. (69-82.8 N•m) (1966-early 1978) or 30-40 ft.-lb. (41.4-55.2 N•m) (late 1978-on). Bend the lockwasher tab over the nut to lock it.

### Starter Clutch Removal/Inspection/Installation

Refer to **Figure 120** or **Figure 121** for this procedure.

1. Remove the kickstarter side cover as described under *Removal/Installation* in this chapter.

2. Remove the pushrod (**Figure 125**).

3. Lock the transmission into 2 gears at once.

4. Bend the lockwasher tab away from the starter clutch nut (**Figure 133**).

5. Remove the starter clutch with the Harley-Davidson starter clutch puller (HD-95650-42). See **Figure 134**.

6. Remove the Woodruff key, starter clutch gear and spring.

7. Check the starter gear and starter clutch teeth. Teeth should be sharp and show no signs of cracking, scoring or excessive wear. Replace starter clutch and starter gear if teeth are rounded or if the kickstarter has been slipping.

8. Visually check the starter clutch for cracks, deep scoring and excessive wear.

9. Slide the starter gear on the main shaft and check side play by hand. If gear is loose, have the bushing replaced by a Harley-Davidson dealer or machine shop. Do not attempt to drive a new bushing into the gear.

10. Installation is the reverse of these steps. Note the following:

   a. Lubricate the main shaft with engine oil.

   b. Install the spring and starter gear onto main shaft.

   c. Push the starter gear toward the transmission and release it. The spring should be able to push the gear forward. If the gear is tight, the gear bushing should be reamed. Refer this service to a Harley-Davidson dealer.

   d. Press the starter clutch onto the main shaft.

   e. Tighten the starter clutch nut to 34-42 ft.-lb. (46.9-57.9 N•m). Bend the lockwasher tab over to lock the nut.

## SIDE COVER
## (ELECTRIC START)

### Removal/Disassembly

1. Refer to **Figure 135** for this procedure. Disconnect the negative battery cable.

2. Drain the transmission oil as described in Chapter Three.

3. Remove the exhaust system as described in Chapter Seven.

4. Remove the rear master cylinder reservoir. See Chapter Twelve.

5. Remove the clutch cover (**Figure 122**) on the primary cover.

6. See **Figure 123**. Loosen the pushrod locknut and turn the adjusting screw counterclockwise to remove all tension on the pushrod.

7. Remove the starter bracket.

8. Remove the side cover mounting bolts and remove the side cover and gasket.

9. If necessary, disassemble the side cover as described in this chapter.

10. Installation is the reverse of these steps. Note the following.

11. Install a new side cover gasket.

12. Pull the pushrod assembly partway out as shown in **Figure 125**.

13. Position the release lever to the left of the cover as shown in A, **Figure 126**.

14. Align the pushrod oil slinger (**Figure 127**) with the release lever mechanism (B, **Figure 126**) and install the side cover. The lever should face as shown in **Figure 128**.

15. Install the side cover mounting bolts and tighten them to 13-16 ft.-lb. (17.9-22.1 N•m).

16. Refill the transmission with the correct type and quantity of oil as described in Chapter Three.

17. Adjust the clutch as described in Chapter Three.

### Disassembly/Inspection/Assembly

Refer to **Figure 135** for this procedure.

1. Remove the release lever nut and washer.

2. Pull the release lever off of the shaft with a universal type claw puller.

3. Remove the clip. Then remove the shaft from the cover.

4. Remove the release finger and washer.

5. Clean all components thoroughly with solvent. Remove any gasket residue from the cover-to-transmission machined surfaces. Check the threads in the transmission case to be sure they are clean. If dirty or damaged, use a tap to true up the threads and remove any deposits.

6. Visually check the shaft surfaces for cracks, deep scoring, excessive wear or heat discoloration.

7. Slide the release lever shaft into the side cover. Check the shaft-to-bushing wear by moving the shaft back and forth. If there is excessive shaft wear, replace the upper and lower bushings with a blind hole bearing remover and slide hammer. Have a Harley-Davidson dealer or machine shop press in the new bushing.

8. Assemble the release finger as follows:

   a. Install the washer and release finger into the side cover.

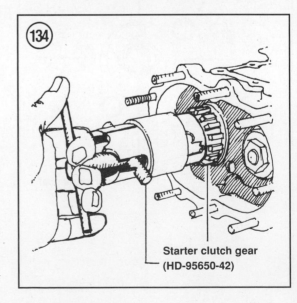

**Starter clutch gear**
**(HD-95650-42)**

b. Insert the release lever into the side cover and through the release finger and washer. Secure the shaft with a new circlip.

c. Install the release lever. Then install the lockwasher and the nut. Tighten the nut until the release lever bottoms on the shaft.

## ELECTRIC STARTER DRIVE

### Removal (Type I)

The Type I electric starter drive (**Figure 136**) is used on the following models:

    a. 1966-1982 FL.

    b. All FX chain driven models.

1. Disconnect the negative battery cable.

2. Remove the starter as described in Chapter Seven.

3. Remove the primary cover as described in this chapter.

4. Remove the drive gear housing bolts (1, **Figure 136**) and remove the drive gear housing (12).

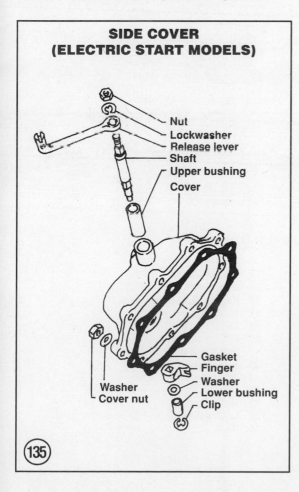

**SIDE COVER (ELECTRIC START MODELS)**

Nut
Lockwasher
Release lever
Shaft
Upper bushing
Cover
Gasket
Finger
Washer
Washer
Lower bushing
Cover nut
Clip

(135)

5. Remove the oil deflector (3) and gasket (4).

*NOTE*
*The gasket (4) was not used on 1980 and earlier models. However, Harley-Davidson recommends installing gasket on reassembly to prevent primary case seal leaks.*

6. Working from the left-hand side, disengage the shifter lever finger (7) from the shifter collar (18). Then remove the pinion gear and shaft assembly (5).

7. To remove the shifter lever, perform the following:

    a. Remove the battery and battery carrier as described in Chapter Three.

    b. Remove the oil tank mounting brackets.

    c. Remove the solenoid as described in Chapter Seven.

    d. Remove the shifter lever screw (6) and remove the shifter lever.

### Inspection

1. Check the drive gear for worn, chipped or broken teeth. Replace the gear if necessary.

*NOTE*
*If the drive gear is worn, check the starter gear as described in Chapter Seven.*

2. Check the drive gear thrust washer (9), if used, for damage or cupping. Replace the washer if necessary.

3. Check the drive gear housing needle bearing for wear or damage. Rotate the bearing with your fingers and check for noise, roughness or looseness. If the bearing's condition is doubtful, replace it. Have the bearing pressed out and a new one pressed in by a Harley-Davidson dealer or machine shop. Do not attempt to drive in the bearing.

4. Lubricate the needle bearing with a high temperature grease such as Lubriplate 110.

5. Install the drive gear and thrust washer (if used) in the drive housing.

6. Inspect the pinion gear needle bearing (15) installed in the primary cover as described in Step 3. See **Figure 137**. Also check the pinion shaft collar (16) bearing surface. If bearing or collar is worn, replace them both. Replace the bearing as described in Step 3.

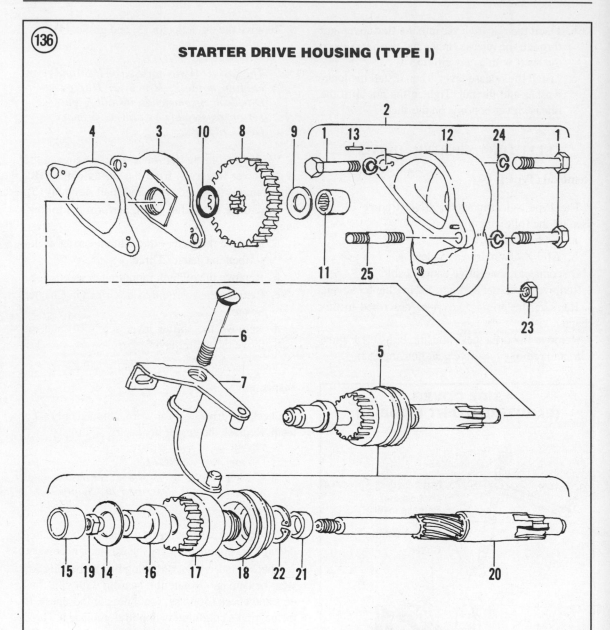

**STARTER DRIVE HOUSING (TYPE I)**

1. Bolt
2. Drive gear housing assembly
3. Oil defector
4. Gasket
5. Pinion gear and shaft assembly
6. Shifter lever screw
7. Shifter lever
8. Drive gear
9. Thrust washer
10. O-ring
11. Bearing (in drive gear housing)
12. Drive gear housing
13. Locating pin
14. Thrust washer
15. Bearing (in primary cover)
16. Pinion shaft collar
17. Pinion gear
18. Shifter collar
19. Pinion shaft nut
    (left-hand thread)
20. Pinion shaft
21. Spacer
22. Circlip
23. Nut
24. Lockwasher
25. Stud

7. Check the pinion gear for worn, chipped or broken teeth. Replace the gear if necessary.

*NOTE*
*If the pinion gear is worn, check the clutch ring gear as described in this chapter under **Clutch Inspection**.*

8. Check the shifter collar groove and the shifter lever fingers for wear. Replace both parts if either is worn.

9. To disassemble the pinion gear shaft assembly (5), perform the following:

*NOTE*
*The pinion shaft nut uses left-hand threads.*

a. Secure the pinion shaft in a vise with soft jaws.
b. Remove the pinion shaft nut (19).
c. Remove the washer (14) and pinion shaft collar (16).
d. Remove the slide pinion gear (17) and shifter collar (18) as one unit.
e. Remove the spacer (21).
f. If necessary, remove the circlip (22) and separate the pinion gear and shifter collar.
g. Inspect components as described in this section.

10. Assemble the pinion gear and shaft assembly by reversing Step 9. Note the following:
a. Inspect all parts for wear and damage as described in this procedure. Replace parts as necessary.
b. Install the circlip (22).
c. Lubricate all parts with a high temperature grease.

## Installation

1. Assemble the pinion gear and shaft assembly as described under *Inspection*.
2. Install the pinion shaft assembly. Engage the shifter lever fingers (7, **Figure 136**) with the shifter collar drum (18).
3. Install the shifter lever assembly into the inner primary case. Lubricate the shifter lever screw with a high temperature grease and install it through the shifter lever. Tighten the screw securely.
4. Install the drive gear housing (with gear). See **Figure 138**.
5. Install a new O-ring (10, **Figure 136**) in the oil deflector.
6. Install the oil deflector onto the drive gear housing.
7. Install a new gasket (4) on the oil deflector.
8. Install the primary cover as described in this chapter.
9. Install the solenoid and starter as described in Chapter Seven.

## Removal (Type II)

The Type II electric starter drive (**Figure 139**) is used on all belt drive models.
1. Disconnect the negative battery cable.
2. Remove the starter as described in Chapter Seven.

3. Remove the primary cover as described in this chapter.

4. Remove the outer drive gear housing bolts (1 and 2, **Figure 139**).

5. Remove the outer drive housing (3), gasket (4) and drive gear (5).

6. Remove the inner drive housing bolts (6) and remove the housing (7) and gasket (8).

7. Working from the left-hand side, disengage the shifter lever fingers (10) from the shifter collar (17). Then remove the pinion gear and shaft assembly (16).

8. To remove the shifter lever, perform the following:

a. Remove the battery and battery carrier as described in Chapter Three.

b. Remove the oil tank mounting brackets.

c. Remove the solenoid as described in Chapter Seven.

d. Remove the shifter lever screw (9) and remove the shifter lever.

**Inspection**

1. Check the drive gear (5) for worn, chipped or broken teeth. Replace the gear if necessary.

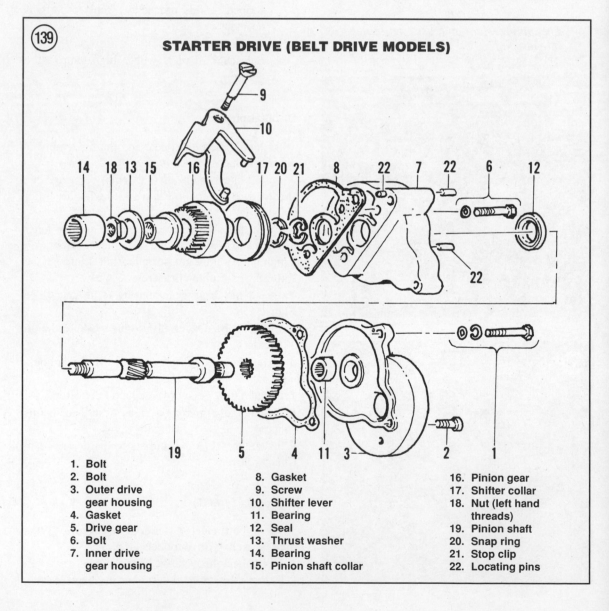

**(139)  STARTER DRIVE (BELT DRIVE MODELS)**

1. Bolt
2. Bolt
3. Outer drive gear housing
4. Gasket
5. Drive gear
6. Bolt
7. Inner drive gear housing
8. Gasket
9. Screw
10. Shifter lever
11. Bearing
12. Seal
13. Thrust washer
14. Bearing
15. Pinion shaft collar
16. Pinion gear
17. Shifter collar
18. Nut (left hand threads)
19. Pinion shaft
20. Snap ring
21. Stop clip
22. Locating pins

*NOTE*
*If the drive gear is worn, check the starter gear as described in Chapter Seven.*

2. Inspect the inner drive gear housing seal (12) for wear or damage. If necessary, remove the seal by prying it out of the housing with a screwdriver. Install the new seal with a large socket placed on the outside seal surface. Install the seal with the lip side facing toward the drive gear.

3. Grasp the locating pins (22) on both sides of the inner drive gear housing (7). The pins should be tight. If not, check the pin locations to make sure the housing is not cracked.

4. Check the inner drive gear housing needle bearing (14) for wear or damage. Rotate the bearing with your fingers and check for noise, roughness or looseness. If the bearing's condition is doubtful, replace it. Have the bearing pressed out and a new one pressed in by a Harley-Davidson dealer or machine shop. Do not attempt to drive in the bearing.

5. Lubricate the needle bearings with a high temperature grease.

6. Inspect the pinion gear needle bearing (11) installed in the outer drive gear housing as described in Step 4. Also check the pinion shaft collar (15) bearing surface. If bearing or collar is worn, replace them both. Replace the bearing as described in Step 4.

7. Check the pinion gear for worn, chipped or broken teeth. Replace the gear if necessary.

*NOTE*
*If the pinion gear is worn, check the clutch ring gear as described in this chapter under* **Clutch Inspection**.

8. Check the shifter collar groove (17) and the shifter lever fingers (10) for wear. Replace both parts if any one part is worn.

9. To disassemble the pinion gear shaft assembly (16), perform the following:

*NOTE*
*The pinion shaft nut uses left-hand threads.*

a. Secure the pinion shaft in a vise with soft jaws.
b. Remove the pinion shaft nut (18).
c. Remove the washer (13) and pinion shaft collar (15).

d. Remove the slide pinion gear (16) and shifter collar (17) as one unit.
e. If necessary, remove the snap ring (20) and separate the pinion gear and shifter collar.
f. On 1983 and earlier FXSB models, remove the stop clip (21).
g. Inspect components as described in this section. On 1983 and earlier FXSB models, replace the pinion shaft if the stop clip groove is worn.

10. Assemble the pinion gear and shaft assembly by reversing Step 9. Note the following:

a. Inspect all parts for wear and damage as described in this section. Replace parts as necessary.
b. Install new snap ring (20).
c. On 1983 and earlier FXSB models, install a new stop clip (21).
d. Lubricate all parts with a high temperature grease.

**Installation**

1. Assemble the pinion gear and shaft assembly as described under *Inspection*.

2. Install the shifter lever assembly into the inner primary case. Lubricate the shifter lever screw with a high temperature grease and install it through the shifter lever. Tighten the screw securely.

3. Install the pinion shaft assembly. Engage the shifter lever fingers (10) with the shifter collar (18).

4. Glue a new gasket (8) onto the inner drive gear housing (7).

5. Install the inner drive gear housing over the pinion shaft assembly and install onto the primary case. Install the mounting bolts and tighten securely.

6. Lubricate the drive gear with high temperature grease and slide it onto the pinion shaft.

7. Cement a new gasket (4) onto the outer drive gear housing. Then install the housing and tighten the mounting bolts securely.

8. Install the starter as described in Chapter Seven.

9. Install the primary cover as described in this chapter.

10. Install the solenoid as described in Chapter Seven.

## SHIFTER ADJUSTMENT

Shifter adjustment is required to compensate for component wear or whenever the transmission has been removed.

### Hand Shift Linkage Adjustment (1966-early 1978)

1. Move the shift lever into 3rd gear.
2. Disconnect the shifter rod from the shifter lever.
3. Carefully move lever on transmission until shifter spring plunger inside transmission drops into its detent. This position can be determined by feel.
4. Adjust clevis on shift rod until it aligns exactly with hole in shifter cover.
5. Reconnect the shifter rod to the shifter lever.

### Foot Shift Linkage Adjustment (1966-early 1978 FL and 1966-1984 FX Except FXWG)

1A. *FL models:* Determine that notch or mark in end of foot shifter shaft aligns with clamp slot in shifter lever. If necessary, adjust length of rod by removing shifter rod end bolt, then turning rod end as required.
1B. *FX models:* Shift linkage should operate so that the shift pedal can travel to its full limit without interference when making gear changes. If necessary, adjust threaded ends on rod after removing retainer clip (**Figure 140**).
2. Recheck the adjustment.

### Shift Linkage Adjustment (Late 1978-1984 FL and FXWG)

Refer to **Figure 141** for this procedure.
1. The shift linkage should operate so that the shift pedal can travel to its full limit without interference when making gear changes. If necessary, adjust as follows.
2. On FL models, remove the left-hand side exhaust pipe. See Chapter Six.
3. Remove the cotter pin and clevis pin securing the shift rod to the shift lever.

4. Loosen the shifter rod locknut. Then adjust the shifter rod so that the shift pedal can travel correctly as described in Step 1. Reinstall the clevis pin and a new cotter pin.

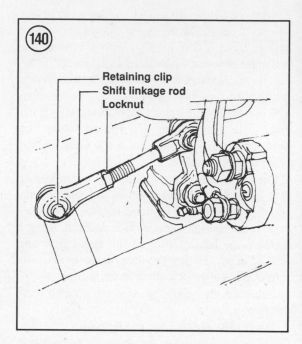

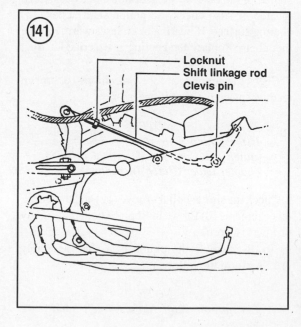

### Table 1 CLUTCH SPECIFICATIONS

| Item | in. | mm |
|---|---|---|
| Roller bearing fit | | |
|   1966-early 1978 | 0.002-0.003 | 0.051-0.076 |
| Spring free length | | |
|   1966-1967 | 1 31/64 | 37.70 |
|   1968-on | 1 45/64 | 43.26 |
| Spring adjust ment | | |
|   1966-1967 | 31/32 | 24.61 |
|   1968-on | 1 1/32 | 26.19 |
| Lining thickness | 1/32 | 0.80 |

### Table 2 CLUTCH TIGHTENING TORQUE

| Item | ft.-lb. | N•m |
|---|---|---|
| Compensating sprocket nut | | |
|   Late 1978-on | 80-100 | 110.4 |
| Transmission sprocket locknut | | |
|   1966-1969 | See text | |
|   1970-early 1978 | 50-60 | 69-82.8 |
|   Late 1978-early 1983 | 105-120 | 144.9-165.6 |
|   Late 1986-on | 80-90 | 110.4-124.2 |
| Primary cover screws | | |
|   Late 1978-on | 18-22 | 24.8-30.4 |

### Table 3 PRIMARY CHAIN ALIGNMENT

| Dimension C | Spacer thickness |
|---|---|
| 0.2500-0.2812 in. (6.35-7.14 mm) | 0.060 in. (1.52 mm) |
| 0.2812-0.3125 In. (7.14-7.94 mm) | 0.090 in. (2.28 mm) |
| 0.3125-0.3427 in. (7.94-8.70 mm) | 0.120 in. (3.05 mm) |
| 0.3427-0.3750 in. (8.70-9.52 mm) | 0.150 in. (3.81 mm) |
| 0.3750-0.4063 In. (9.52-10.32 mm) | 0.180 in. (4.57 mm) |
| 0.4063-0.4375 in. (10.32-11.11 mm) | 0.210 in. (5.33 mm) |

### Table 4 PRIMARY BELT ALIGNMENT

| Dimension C | Spacer thickness |
|---|---|
| 0.324-0.304 in (8.23-7.72 mm) | 0.060 in. (1.52 mm) |
| 0.304-0.284 in. (7.72-7.21 mm) | 0.040 in. (1.02 mm) |
| 0.284-0.264 in. (7.21-6.71 mm) | 0.020 in. (0.51 mm) |
| 0.264-0.254 in. (6.71-6.45 mm) | 0.010 in. (0.25 mm) |
| 0.254-0.244 in. (6.45-6.20 mm) | None |

### Table 5 TRANSMISSION SPECIFICATIONS (1966-EARLY 1978)

| Item | Specification | |
| | in. | mm |
|---|---|---|
| Main shaft drive gear | | |
|   Roller bearing | 0.005-0.002 | 0.013-0.051 |
|   Inner bearing | 0.002-0.003 | 0.051-0.076 |
|   Drive gear end play | | |
|     1966-1969 | 0.003-0.013 | 0.076-0.330 |
|     1970-early 1978 | 0.0025-0.0135 | 0.063-0.343 |
| Main shaft | | |
|   Third gear end play | 0.000-0.017 | 0.00-0.43 |

(continued)

## Table 5 TRANSMISSION SPECIFICATIONS (1966-EARLY 1978) (continued)

| Item | Specification | |
|---|---|---|
| | in. | mm |
| Countershaft | | |
| Gear end play | | |
| 1966-1969 | 0.008-0.012 | 0.20-0.30 |
| 1970-early 1978 | 0.007-0.012 | 0.18-0.30 |
| Shifter clutch clearance | | |
| Low and second gear | | |
| 1966-1969 | 0.075 | 1.90 |
| 1970-early 1978 | 0.080-0.090 | 2.03-2.29 |
| Third and fourth gear | | |
| 1966-1969 | 0.100 | 2.54 |
| 1970-early 1978 | 0.100-0.110 | 2.54-2.79 |
| Sliding reverse gear | | |
| 1966-1969 | 0.055 | 1.40 |
| 1970-early 1978 | 0.060-0.070 | 1.52-1.78 |
| Gear backlash | 0.003-0.006 | 0.08-0.15 |
| Shifter cam end play | 0.0005-0.0065 | 0.013-0.165 |

## Table 6 TRANSMISSION SPECIFICATIONS (LATE 1978-1984)

| Item | Specification | | Wear limit | |
|---|---|---|---|---|
| | in. | mm | in. | mm |
| Mainshaft main drive gear | | | | |
| End play | | | | |
| Late 1978-1981 | 0.0025-0.0135 | 0.063-0.343 | 0.025 | 0.63 |
| 1982-on | 0.010-0.025 | 0.254-0.635 | 0.025 | 0.63 |
| Mainshaft bushing | 0.0018-0.0032 | 0.046-0.081 | 0.004 | 0.10 |
| Mainshaft | | | | |
| Third gear end play | 0.000-0.017 | 0.00-0.43 | 0.020 | 0.51 |
| Fit on shaft | 0.0012-0.0023 | 0.030-0.058 | 0.003 | 0.08 |
| Shift fork clutch gear spacing | 0.100-0.110 | 2.54-2.79 | 0.110 | 2.79 |
| Mainshaft low gear end bearing | | | | |
| Fit in housing | | | | |
| Loose | 0.0013 | 0.033 | 0.0020 | 0.051 |
| Press | 0.0001 | 0.003 | 0.0005 | 0.013 |
| Fit on shaft | | | | |
| Loose | 0.001 | 0.025 | 0.0015 | 0.038 |
| Press | 0.0007 | 0.018 | 0.001 | 0.025 |
| Housing in case | | | | |
| Loose | 0.0005 | 0.013 | 0.0009 | 0.023 |
| Press | 0.0010 | 0.025 | 0.0015 | 0.038 |
| Countershaft | | | | |
| Runout | – | – | 0.003 | 0.076 |
| Drive gear end bearing | 0.0025-0.020 | 0.063-0.51 | 0.002 | 0.051 |
| Low gear end bearing | 0.0005-0.0019 | 0.013-0.048 | 0.002 | 0.051 |
| Gear end play | 0.004-0.012 | 0.102-0.305 | 0.015 | 0.381 |
| Second gear | | | | |
| End play | 0.003-0.017 | 0.076-0.432 | 0.020 | 0.51 |
| Bushing on shaft | 0.000-0.0015 | 0.000-0.038 | 0.002 | 0.051 |
| Bushing in gear | 0.0005-0.0025 | 0.013-0.063 | 0.002 | 0.051 |
| Low gear | | | | |
| Bushing on shaft | 0.000-0.0015 | 0.000-0.038 | 0.002 | 0.051 |
| Bushing in gear | 0.0005-0.0025 | 0.013-0.063 | 0.002 | 0.051 |
| Shifter clutch | 0.080-0.090 | 2.032-2.286 | 0.090 | 2.29 |
| Gear backlash | 0.003-0.006 | 0.076-0.152 | 0.010 | 0.25 |
| Shifter cam (Late 1978-early 1979) | | | | |
| End play | 0.005-0.0065 | 0.127-0.165 | 0.007 | 0.178 |

# FUEL AND EXHAUST SYSTEMS

The fuel system consists of the fuel tank, shutoff valve and filter, carburetor and air cleaner.

This chapter includes service procedures for all parts of the fuel and exhaust system.

## AIR CLEANER

The air cleaner must be cleaned or replaced at the intervals specified in Chapter Three (or more frequently in dusty areas).

Service to the air cleaner element is described in Chapter Three.

## CARBURETORS

### Service

Major carburetor service (removal and cleaning) should be performed when poor engine performance and/or hesitation is observed. Carburetor jetting changes should be attempted only if you're experienced in this type of "tuning" work; a bad guess could result in costly engine damage or, at best, poor performance.

> *NOTE*
> *When tuning late model engines, check with local authorities for regulations concerning emissions.*

If after servicing the carburetors and making adjustments as described in this chapter, the motorcycle does not perform correctly (and assuming that other factors affecting performance are correct, such as ignition timing and condition, engine tuning, etc.), the motorcycle should be checked by a dealer or a qualified performance tuning specialist.

### Removal/Installation

1. Remove the air cleaner as described in Chapter Three.

2. Disconnect the fuel line at the carburetor.

> *NOTE*
> *Some factory installed hose clamps must be cut off with pliers and cannot be reused. Use a removable hose clamp during installation so that the fuel line can be removed more easily when performing future service or roadside troubleshooting.*

3. Disconnect the throttle cable(s) at the carburetor.

4. Loosen the choke cable set screw at the carburetor and slide it out of its holder.

5. Disconnect the vacuum line at the carburetor (if so equipped).

6. Remove the carburetor mounting nuts and washers and remove the carburetor.

*NOTE*
*Drain most of the gasoline from the carburetor assembly and place it in a clean heavy-duty plastic bag to keep it clean until it is worked on or reinstalled.*

*CAUTION*
*Cover the exposed manifold with shop rags or tape. This will prevent engine damage from small objects dropped undetected into the manifold.*

7. While the carburetor is removed, examine the intake manifold on the cylinder head for any cracks or damage that would allow unfiltered air to enter the engine. Any damaged parts should be replaced.

8. Install by reversing these removal steps. Install a new manifold gasket. Adjust the throttle grip as described in Chapter Three.

### Disassembly/Assembly (Model DC)

Model DC carburetors are used on 1966 models. Refer to **Figure 1** for this procedure.

1. Remove the support bracket (25), if so equipped.
2. Remove the throttle body screws and lockwashers (1) and remove the throttle body (42) assembly.
3. Remove the throttle body gasket (2).
4. Remove the idle hole plug (3) from the throttle body.
5. Remove the low speed needle valve (4), washer (5) and spring (6).
6. Remove the throttle shaft screws (7) and lockwashers. Then slide the throttle disc (8) out of the shaft.
7. Loosen the throttle lever clamping screw (9). Then remove the throttle lever stop screw (14) and spring (15).
8. Slide the throttle shaft (13) out of the main body.
9. Remove the float bowl screws and tap the float bowl (17) to remove it.
10. Remove the float bowl gasket (18).
11. Unscrew the float nut (19) at the top of the float (20). Then slide the float off of the float lever (23).
12. Remove the float valve and seat assembly (21).
13. Remove the float lever screw (22), lockwasher and float washer. Then remove the float lever and bracket assembly (23).

## DC CARBURETOR–1966

1. Throttle body screw and washer
2. Body gasket
3. Idle hole plug
4. Low speed needle valve
5. Low speed needle valve washer
6. Low speed needle valve spring
7. Throttle shaft screw
8. Throttle disc
9. Throttle lever clamping screw
10. Throttle lever
11. Throttle shaft spring
12. Throttle shaft washer
13. Throttle shaft
14. Throttle lever stop screw
15. Throttle lever stop screw spring
16. Bowl mounting screw
17. Bowl
18. Bowl gasket
19. Float nut
20. Float
21. Float valve and seat
22. Float lever screw and washers
23. Float lever and bracket assembly
24. Support bracket nut and lockwasher
25. Support bracket
26. Bowl nut
27. Bowl nut gasket
28. Idle tube assembly
29. Main nozzle
30. High-speed needle valve extension housing
31. High-speed needle valve
32. High-speed needle valve packing nut
33. High-speed needle valve packing
34. Carburetor jet
35. Drain plug and gasket
36. Idle passage tube
37. Throttle shaft screw
38. Vent clamp
39. Vent housing
40. Vent gasket
41. Idle bleed tube
42. Throttle body
43. Main housing

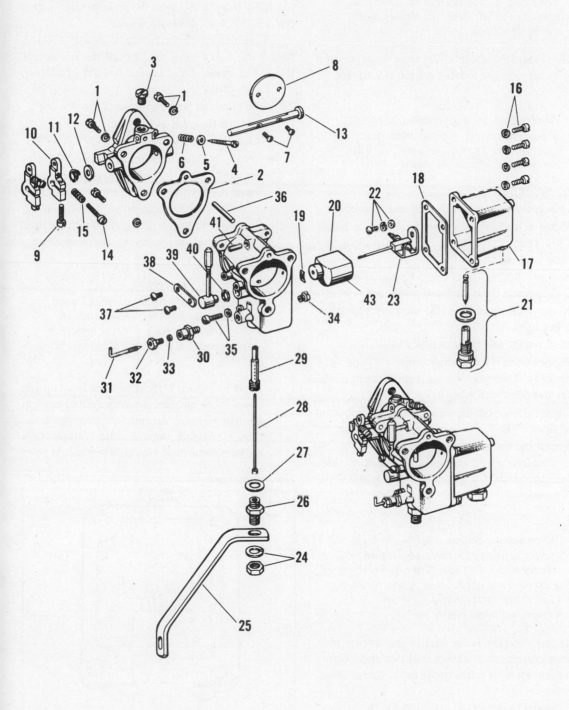

14. Remove the bowl nut (26) and gasket (27).

*NOTE*
*If the idle tube (28) comes out with the bowl nut, do not attempt to remove it. Usually the idle tube will remain in the carburetor body.*

15. Gently pull the idle tube (28) from the carburetor body by tugging and moving it from side to side.

*NOTE*
*Harley-Davidson recommends grinding a screwdriver to fit the main nozzle slot before removal in Step 16. The screwdriver slot in the main nozzle is 0.051 in. wide.*

16. Unscrew the main nozzle (29) and remove it.

17. Remove the high-speed needle extension housing (30). Then remove the needle (31), packing nut (32) and packing (33).

18. Remove the high-speed metering plug or fixed jet (34) opposite the high-speed needle valve hole.

19. Remove the drain plug (35) and gasket from the main housing.

20. Remove the idle passage tube (36).

21. Remove the throttle shaft screw (37) and vent clamp (38). Then remove the vent housing (39), gasket and idle bleed tube (41).

22. Clean and inspect the carburetor as described in this chapter.

23. Install the throttle shaft (13) and check for excessive clearance by moving the shaft side-to-side. Replace the shaft if worn approximately 0.002 in. (0.05 mm).

*NOTE*
*When using a drill to clean fuel holes in the following procedures, make sure to use the exact drill specified and to take extreme care not to enlarge the holes. Check the drill carefully for size and remove all burrs before use.*

24. If the idle port holes next to the throttle disc become crusted, and solvent will not clean them, open them up with a drill of the exact size as listed below.
    a. Model DC-1, DC-1L, DC-1M and DC-10; No. 70 drill.
    b. Model DC-2; No. 56 drill.

25. Clean the idle jet hole with a No. 57 (0.043 in., 1.092 mm) drill, and the angular hole it meets with a No. 52 (0.0635 in., 1.613 mm) drill.

26. Clean nozzle bleed holes with a No. 54 (0.055 in., 1.397 mm) drill. Clean the main passage with a No. 17 (0.173 in., 4.394 mm) drill.

27. Clean high speed needle seat holes with a No. 55 (0.052 in., 1.321 mm) drill for all models except DC-2. On model DC-2, use a No. 70 (0.028 in., 0.711 mm) drill.

28. Make sure that both carburetor vents are open.

29. Adjust the float rod as follows:
    a. Assemble float valve and seat (21).
    b. Install the float lever bracket screws (22) finger tight so that the bracket may be adjusted.
    c. Insert the float valve and seat assembly (21) partially into the float bowl. Then align the float lever fingers with the groove on the float valve. Screw the float valve assembly into the bowl.

*CAUTION*
*Never screw the float valve and seat assembly into bowl with bowl attached to the carburetor body. Fingers on lever will be damaged if they are not engaged properly.*

    d. Refer to **Figure 2**. Hold the float bowl upside down. Then measure the distance between top of float rod and the outer edge of the bowl flange opposite the fuel inlet fitting. This measurement must be made just as the float

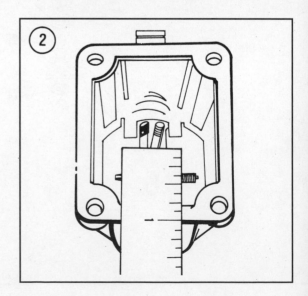

valve seats *lightly*. This point can be determined by moving the float lever up and down gently. The correct measurement is 1 in. ±1/64 in. (25.4 mm ± 0.397 mm).

e. If necessary, reposition the float lever bracket (23) to obtain the correct measurement.

f. When the adjustment is correct, tighten the bracket screw securely. Recheck adjustment after tightening screw.

g. Install the float (20) on the float lever (23) with its flat side facing up. Secure the float with the float nut (19).

30. Reassemble the carburetor in the reverse order of disassembly. Note the following.

### CAUTION
*When installing the high speed needle assembly into the main body, always back out the needle so that its point will not enter the valve hole in the main body when the body is tightened up.*

31. Assemble the high speed needle (31), packing (33) and packing nut (32) into the needle valve extension housing (30). Tighten the packing nut (32) just enough to keep the needle from turning freely.

32. With the carburetor main body inverted, drop in the idle passage tube (36), small end first. It may be necessary to jiggle the body until the idle tube locates itself in position. Press on the end of the tube until it seats and the bottom of the plug extends about 1/32 in. out of the main nozzle (29) passage.

33. Install the idle passage tube (36) in the main housing (43) with its chamfered end facing out.

## Disassembly/Assembly
## (HD Carburetor, 1967-1970)

Refer to **Figure 3** for this procedure.

1. Remove the idle (**Figure 4**) and intermediate (**Figure 5**) adjusting screws.

2. Remove the throttle valve screws and pull the throttle valve (51) out of its shaft.

3. Remove the accelerating pump lever screw (3). Then pull the throttle shaft out of the carburetor housing. Remove the compression spring (50), washers (49) and dust seals (48).

4. Remove the accelerating pump plunger assembly.

5. Remove the diaphragm cover (18).

6. Remove the diaphragm (17) and gasket (23). Separate the gasket from the diaphragm.

7. Remove the plug screw (21) from the diaphragm cover.

8. Remove the following parts in order:
   a. Inlet control lever screw (34).
   b. Inlet control lever pin (33).
   c. Inlet control lever (32).
   d. Inlet control lever tension spring (37).
   e. Inlet needle (35).

9. Using a 3/8 in. socket, remove the inlet needle and seat assembly (35).

10. Remove the inlet needle seat gasket (36) with a piece of wire.

11. Remove the main jet plug screw (44) from the side of the carburetor.

12. Remove the main jet (42) and its gasket.

13. Welch plugs (6, 7 and 8) are used to block the main nozzle, idle port and the economizer check ball. To remove any welch plug, perform the following:
   a. Drill a 1/8 in. (3.175 mm) diameter hole off center through the welch plug only. Do not drill deeper than the welch plug.
   b. Pry out the welch plug with a small punch. Make sure the casting surrounding the edges is not damaged.

### NOTE
*The sides of the choke valves are tapered 15° to conform to the throttle bore. Note the top and bottom sides of the choke valves before removal. They must be installed right side up.*

14. Remove the lower choke valve (15).

15. Slide the choke shaft (13) out of the carburetor. This in turn releases the upper choke valve (11) together with its spring (12), choke friction ball (9) and the friction ball spring (10).

16. Remove the choke shaft dust seal (14).

17. Clean and inspect the carburetor parts as described in this chapter.

18. Assembly is the reverse of these steps. Note the following.

19. Install new welch plugs by seating them with a flat end punch.

20. The inlet control lever tension spring (37) should seat into the body casting counterbore and locate on the inlet control lever protrusion. Adjust the inlet control lever (32) so that it is flush with the metering chamber floor by bending the diaphragm lever end as required. See **Figure 6**.

## HD CARBURETOR (1967-1970)

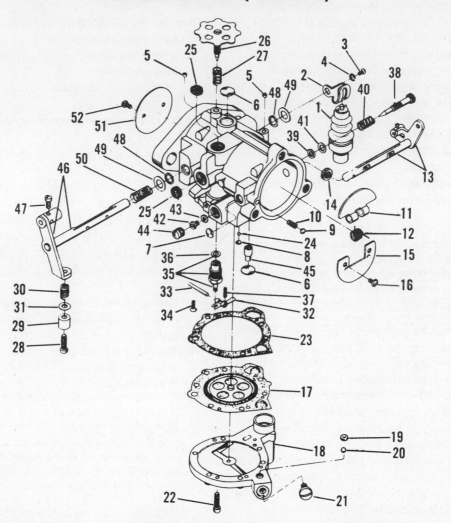

| | | |
|---|---|---|
| 1. Accelerating pump | 19. Accelerating pump check ball retainer | 37. Inlet control lever tension spring |
| 2. Accelerating pump lever | 20. Accelerating pump check ball | 38. Intermediate adjusting screw |
| 3. Accelerating pump lever screw | 21. Diaphragm cover plug screw | 39. Intermediate adjusting screw packing |
| 4. Accelerating pump lever screw lockwasher | 22. Diaphragm cover screws | 40. Intermediate adjusting screw spring |
| 5. Channel plug | 23. Diaphragm cover gasket | 41. Intermediate adjusting screw washer |
| 6. Welch plug | 24. Economizer check ball | 42. Main jet |
| 7. Welch plug | 25. Fuel filter screen | 43. Main jet gasket |
| 8. Welch plug | 26. Idle adjustment screw | 44. Main jet plug screw |
| 9. Choke shaft friction ball | 27. Idle adjustment screw spring | 45. Main nozzle check valve |
| 10. Choke shaft friction spring | 28. Throttle stop screw | 46. Throttle shaft assembly |
| 11. Choke valve (top) | 29. Throttle stop screw cup | 47. Throttle lever wire block screw |
| 12. Choke valve spring | 30. Throttle stop screw spring | 48. Dust seal |
| 13. Choke shaft assembly | 31. Throttle stop screw spring washer | 49. Washer |
| 14. Choke shaft dust seal | 32. Inlet control lever | 50. Throttle shaft spring |
| 15. Choke valve (bottom) | 33. Inlet control lever pin | 51. Throttle valve |
| 16. Choke valve screws | 34. Inlet control lever screw | 52. Throttle valve screws |
| 17. Diaphragm | 35. Inlet needle and seat | |
| 18. Cover | 36. Inlet needle seat gasket | |

21. Tighten the inlet seat assembly to 40-45 in.-lb. (4.6-5.2 N•m) and the accelerating pump channel plug to 23-28 in.-lb. (2.6-3.2 N•m).

22. Adjust the carburetor as described in Chapter Three.

### Disassembly/Assembly
### (Bendix Carburetor, 1971-1975)

The Bendix Model 16P12 carburetors are installed on 1971-1975 models. Refer to **Figure 7** for this procedure.

1. Remove the pump lever screw (1, **Figure 7**).

2. Remove the accelerator pump (3) as an assembly.

3. Remove idle tube (4) and its gasket (5).

4. Remove the jet and tube assembly (6) from the bottom of the float bowl. It may be necessary to give this assembly a firm tug to pull it from its bore. After pulling it out, remove its fiber washer (7) and O-ring (8).

5. Tap the float bowl (9) and remove it.

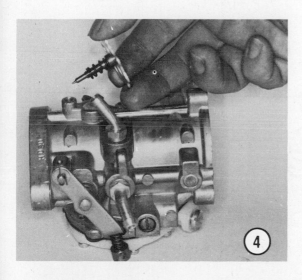

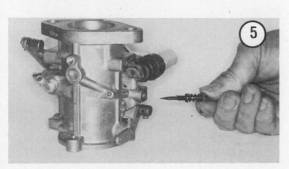

6. Carefully remove float pivot pin (11).

7. Remove float (12), float valve (14) and float spring (13) as an assembly.

8. Remove float bowl gasket (15).

9. Remove the idle mixture needle (16) and its spring (17).

10. Remove the throttle stop screw (18) and its spring (19).

*NOTE*
*The following steps are required only if removal of the choke and/or throttle valves is necessary.*

11. Remove the choke valve screws (21). Then slide the choke valve (20) from the choke shaft (22). Remove the choke shaft, together with the plunger (22A) and spring (22B).

12. Remove the seal retainer (23) and seal (24) from inside the choke shaft opening (only if replacement is required).

13. Remove the throttle valve screws (27). Then slide the valve (26) from the throttle shaft. Remove the throttle shaft, together with its lever (28).

14. Remove the throttle shaft seal retainers (30 and 31 only) and seals (only if replacement is required).

15. Clean and inspect the carburetor as described in this chapter.

16. Installation is the reverse of these steps. Note the following.

17. Make sure that the float needle clip is attached to float tang.

18. Make sure that the float spring is installed as shown in **Figure 8**. When installing the float bowl, bend end of spring up so that it rests against inside of float bowl.

19. Check and adjust float level, if necessary, before installing float bowl. See *Fuel Level Adjustment*.

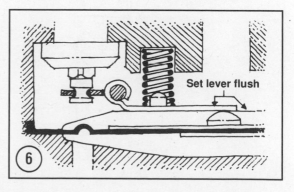

Set lever flush

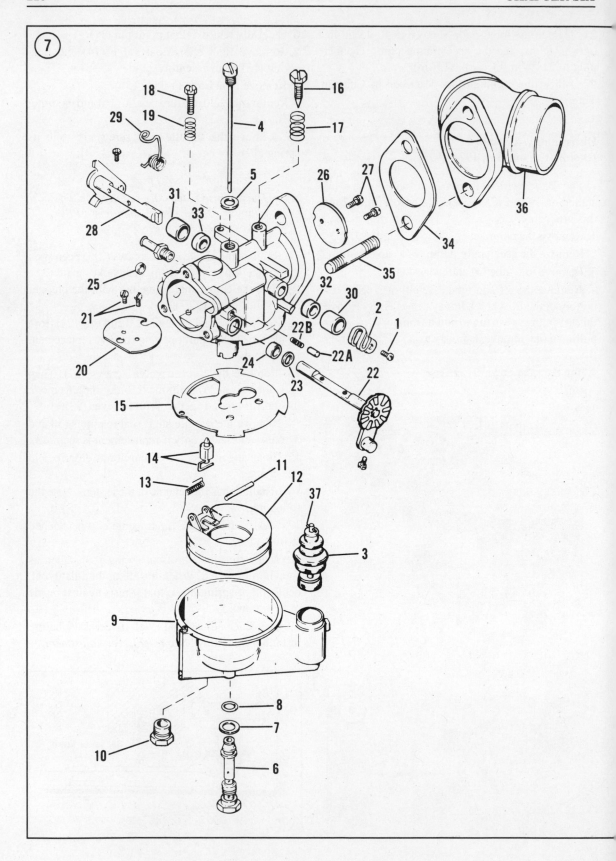

## BENDIX MODEL
## 16P12 CARBURETOR
## (1971-1975)

1. Accelerating pump lever screw
2. Accelerating pump lever
3. Accelerating pump
4. Idle tube
5. Idle tube gasket
6. Main fuel jet and tube assembly
7. Fiber washer
8. O-ring
9. Bowl
10. Bowl drain plug
11. Float pin
12. Float assembly
13. Float spring
14. Float valve
15. Bowl gasket
16. Idle mixture needle
17. Idle mixture needle spring
18. Throttle stop screw
19. Throttle stop screw spring
20. Choke valve
21. Choke valve screw
22. Choke shaft and lever
22A. Plunger
22B. Spring
23. Choke shaft seal retainer
24. Choke shaft seal
25. Choke shaft cup plug
26. Throttle valve
27. Throttle valve screw
28. Throttle shaft and lever
29. Throttle shaft spring
30. Throttle shaft seal retainer
31. Throttle shaft seal retainer
32. Throttle shaft seal
33. Throttle shaft seal
34. Manifold gasket
35. Manifold stud
36. Intake manifold
37. Accelerating pump shaft pin

20. Adjust the carburetor as described in Chapter Three.

### Disassembly/Assembly
### (Keihin Carburetor, 1976-early 1978)

Refer to **Figure 9**.

1. Disassemble the accelerating pump as follows:
   a. Remove accelerating pump housing (33) at the bottom of the float bowl (29).
   b. Remove the spring (32), diaphragm (31) and 2 O-rings (30).

2. Remove the float bowl (29).

3. Remove the float pin screw (6) and withdraw the float pin (5) and float (23).

4. Detach the fuel valve (21) from the float.

5. Remove the O-ring (28) from slot in float chamber wall.

6. Remove the rod (7) and boot (8) from the float bowl.

7. Remove the slow jet plug (27) and remove the slow jet (25).

8. Remove the main jet (26) and the main nozzle (24).

9. Remove the O-ring (20) from the slot in the intake mounting flange.

10. Remove the throttle lever nut (19) and washer (18). Then remove the throttle lever (17) and its spring (16).

11. If necessary, remove the 2 brackets (2 and 15) at the top of the carburetor housing.

12. The throttle and choke valve assemblies (40) are matched to the individual carburetor during manufacturing. If these parts are damaged, the carburetor must be replaced.

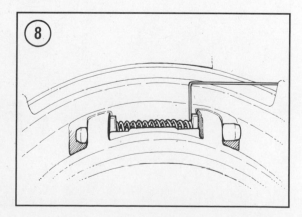

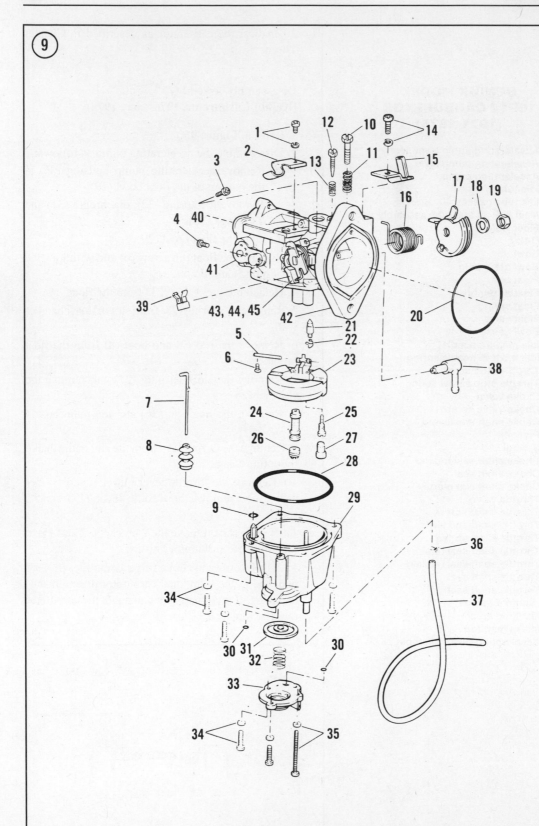

## KEIHIN CARBURETOR
## (1976-EARLY 1978)

1. Screw and washer
2. Bracket
3. Screw
4. Screw
5. Float pin
6. Screw
7. Rod
8. Boot
9. O-ring
10. Throttle stop screw
11. Spring
12. Low speed mixture screw
13. Spring
14. Screw and washer
15. Bracket
16. Spring
17. Throttle lever
18. Washer
19. Nut
20. O-ring
21. Fuel valve
22. Clip
23. Float assembly
24. Main nozzle
25. Slow jet
26. Main jet
27. Plug
28. O-ring
29. Float bowl
30. O-ring
31. Diaphragm
32. Spring
33. Housing
34. Screw and washer
35. Screw and washer
36. Clip
37. Hose
38. Fitting
39. Spacer (not standard)
40. Choke plate (not shown)
41. Choke lever
42. Mounting flange
43. Accelerator pump lever
44. Rocker arm
45. Rocker arm spring

13. Installation is the reverse of these steps. Note the following.

14. Check and adjust the float level before installing the float bowl. See *Float Level Measurement* in this chapter.

15. Adjust the carburetor as described in Chapter Three.

### Disassembly/Assembly
### (Keihin Carburetor, Late 1978-on)

Refer to **Figure 10** (late 1978-1980) or **Figure 11** (1981-on) for this procedure.

1. Remove the accelerating pump housing (28) at the bottom of the float bowl. Then remove the spring (27) and diaphragm (26).

2. Remove the O-ring (25A) from the accelerating pump housing.

3. Remove the float bowl (24).

4. Remove the float pin screw (6) and withdraw the float pin (5) and float (19).

5. Detach the fuel valve (17) from the float.

6. Remove the rubber boot (8) and accelerator pump rod (7) from the float bowl.

7. Remove the low speed plug (22) and remove the low speed jet (20).

8. Remove the main jet (21).

9. Remove the nut and washer from the throttle shaft. Then remove the throttle lever or fast idle cam (41) and its spring.

10. If necessary, remove the 2 brackets at the top of the carburetor housing.

11. The throttle and choke valve assemblies are matched to the individual carburetor during manufacturing. If these parts are damaged, the carburetor must be replaced. Do not remove them.

12. Installation is the reverse of these steps. Note the following.

13. Check and adjust the float level before installing the float bowl. See *Float Level Measurement* in this chapter.

14. Adjust the carburetor as described in Chapter Three.

### Inspection/Cleaning (All Models)

1. Clean all metal parts in a good grade of carburetor cleaner. This solution is available at most automotive or motorcycle supply stores, in a small, resealable tank with a dip basket (**Figure 12**). If it is tightly

6

# KEIHIN CARBURETOR (LATE 1978-1980)

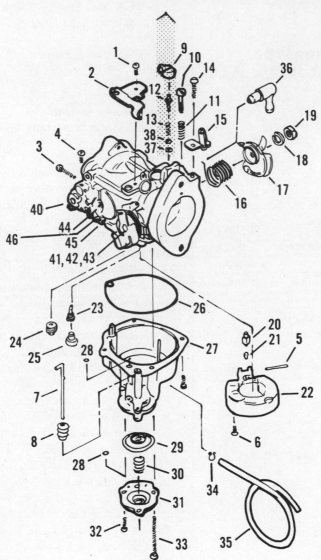

1. Screw and washer
2. Bracket
3. Fast idle adjusting screw
4. Screw
5. Float pin
6. Screw
7. Rod
8. Boot
9. Limiter cap
10. Throttle stop screw
11. Spring
12. Idle mixture screw
13. Spring
14. Screw and washer
15. Bracket
16. Spring
17. Throttle lever
18. Washer
19. Nut
20. Inlet valve
21. Clip
22. Float assembly
23. Low speed jet
24. Main jet
25. Plug
26. Gasket
27. Float bowl
28. O-ring (2)
29. Diaphragm
30. Spring
31. Housing
32. Screw and washer (5)
33. Screw and washer
34. Clip
35. Hose
36. Fitting
37. O-ring
38. Washer
39. Choke plate (not shown)
40. Choke lever shaft
41. Mounting flange
42. Accelerating pump lever
43. Rocker arm
44. Rocker arm spring
45. Intermediate lever
46. Fast idle cam

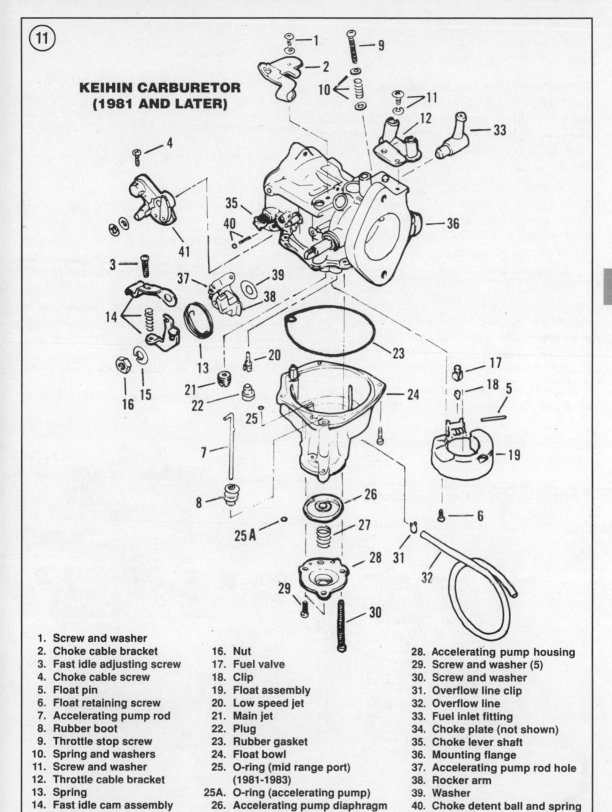

**KEIHIN CARBURETOR
(1981 AND LATER)**

1. Screw and washer
2. Choke cable bracket
3. Fast idle adjusting screw
4. Choke cable screw
5. Float pin
6. Float retaining screw
7. Accelerating pump rod
8. Rubber boot
9. Throttle stop screw
10. Spring and washers
11. Screw and washer
12. Throttle cable bracket
13. Spring
14. Fast idle cam assembly
15. Lockwasher

16. Nut
17. Fuel valve
18. Clip
19. Float assembly
20. Low speed jet
21. Main jet
22. Plug
23. Rubber gasket
24. Float bowl
25. O-ring (mid range port)
    (1981-1983)
25A. O-ring (accelerating pump)
26. Accelerating pump diaphragm
27. Accelerating pump spring

28. Accelerating pump housing
29. Screw and washer (5)
30. Screw and washer
31. Overflow line clip
32. Overflow line
33. Fuel inlet fitting
34. Choke plate (not shown)
35. Choke lever shaft
36. Mounting flange
37. Accelerating pump rod hole
38. Rocker arm
39. Washer
40. Choke detent ball and spring
41. Fast idle cam

sealed when not in use, the solution will last for several cleanings. Follow the manufacturer's instructions for correct soaking time.

2. Remove all parts from the cleaner and blow dry with compressed air. Blow out the jets with compressed air. Unless specified in the procedure, *do not* use a drill or piece of wire to clean them as minor gouges in a jet can alter the flow rate and upset the air/fuel mixture.

3. If the floats are suspected of leaking, put them in a small container of water and push them down. If the floats sink or if bubbles appear indicating a leak, the floats must be replaced.

4. Check the float needle and seat contact areas closely. Both contact surfaces should appear smooth without any gouging or other apparent damage. Replace both needle and seat as a set if any one part is worn or damaged.

5. Inspect the accelerating pump diaphragm for holes and cracks. Replace if necessary.

6. Replace the accelerating pump rod on Keihin carburetors if bent or worn.

7. Replace all worn O-rings and gaskets.

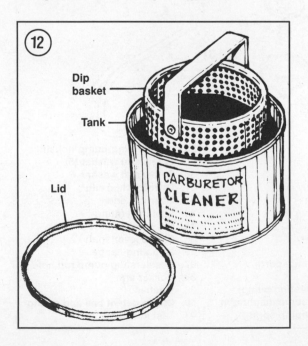

**12**

Dip
basket

Tank

Lid

CARBURETOR
CLEANER

## Fuel Level Measurement

### *Model DC Carburetor*

Refer to *Disassembly/Assembly, Model DC* in this chapter.

### *Bendix Carburetor, 1971-1975*

1. Remove the carburetor as described in this chapter.
2. Remove the float bowl.
3. Turn the carburetor over so that the float is at the top.
4. In this position, the bottom float surface must be 3/16 in. (4.76 mm) from the carburetor gasket surface (**Figure 13**). A 3/16 in. (4.76 mm) drill bit may be used as a gauge.
5. To adjust, bend the float tang (**Figure 14**) as required, using needlenose pliers.
6. Reinstall the float bowl and install the carburetor.

### *Keihin Carburetor, 1976-early 1978*

1. Remove the carburetor as described in this chapter.

Float level
3/16 inch

**13**

**14**

2. Remove the float bowl.

3. *Valve fully closed position:* Turn the carburetor to position the float bowl as indicated in **Figure 15**. Measure the float height from the float bowl gasket surface to the bottom float surface. Bend the float tang to adjust.

4. *Valve fully open position:* Repeat Step 3 to adjust the float to this position. Bend the float tang as required.

5. Reinstall the float bowl and install the carburetor.

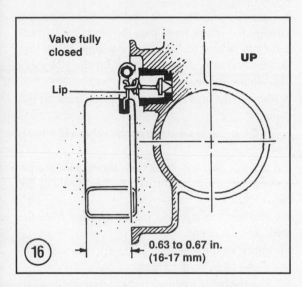

### Keihin Carburetor, late 1978-on

1. Remove the carburetor as described in this chapter.

2. Remove the float bowl.

3. Turn the carburetor to position the float bowl as shown in **Figure 16**. Measure the float height from the float bowl gasket surface to the bottom float surface. Bend the float tang to adjust to the dimensions shown in **Figure 16**.

4. Reinstall the float bowl and install the carburetor.

### Rejetting Carburetors

Do not try to solve a poor running engine problem by rejetting the carburetors if all of the following conditions hold true.

1. The engine has held a good tune in the past with the standard jetting.

2. The engine has not been modified (this includes the addition of accessory exhaust systems).

3. The motorcycle is being operated in the same geographical region under the same general climatic conditions as in the past.

4. The motorcycle was and is being ridden at average highway speeds.

If those conditions all hold true, the chances are that the problem is due to a malfunction in the

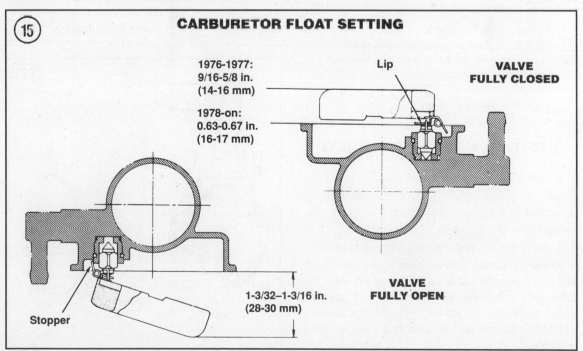

carburetor or in another component that needs to be adjusted or repaired. Changing carburetor jet size probably won't solve the problem. Rejetting the carburetors may be necessary if any of the following conditions hold true.

1. A non-standard type of air filter element is being used.

2. A non-standard exhaust system is installed on the motorcycle.

3. Any of the following engine components have been modified: pistons, cam, valves, compression ratio, etc.

*NOTE*
*When installing accessory engine equipment, manufacturers often enclose guidelines on rejetting the carburetor.*

4. The motorcycle is in use at considerably higher or lower altitudes or in a considerably hotter or colder climate than in the past.

5. The motorcycle is being operated at considerably higher speeds than before and changing to colder spark plugs does not solve the problem.

6. Someone has previously changed the carburetor jetting.

7. The motorcycle has never held a satisfactory engine tune.

*NOTE*
*If it is necessary to rejet the carburetors, check with a dealer or motorcycle performance tuner for recommendations as to the size of jets to install for your specific situation.*

## THROTTLE CABLE REPLACEMENT

1. Remove the seat and fuel tank.

2. Disconnect the throttle cable at the handlebar as described in Chapter Eight under *Handlebar Removal/Installation.*

3. Remove any throttle cable holding bolt or bracket.

4. At the carburetor, hold the lever up with one hand and disengage the cable end (**Figure 17**). Slip the cable out through the carburetor bracket. **Figure 17** shows the throttle cables for 1981 and later models. 1980 and earlier models are equipped with one throttle cable.

*NOTE*
*The piece of string attached in the next step will be used to pull the new throttle cable back through the frame so it will be routed in its original position.*

5. Tie a piece of heavy string or cord to the end of the throttle cable at the carburetor. Wrap this end with masking or duct tape. Do not use an excessive amount of tape as it will be pulled through the frame loop during removal. Tie the other end of the string to the frame.

6. Carefully pull the cable (and attached string) out through the frame loop, past the electrical harness and from behind the headlight housing. Make sure the attached string follows the same path as the cable through the frame.

7. Remove the tape and untie the string from the old cable.

8. Tie the string to the new throttle cable and wrap it with tape.

9. Carefully pull the string back through the frame, routing the new cable through the same path as the old cable.

10. Remove the tape and untie the string from the cable and the frame.

11. Lubricate the new cable as described under *Control Cables* in Chapter Three.

12. Slip the cable in through the carburetor bracket. Then, while holding the lever up with one hand, engage the cable(s) at the carburetor lever.

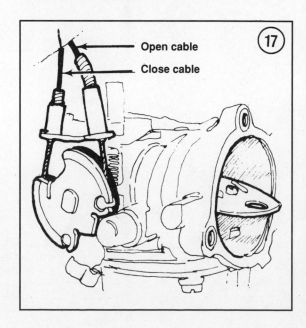

13. Attach the throttle cable to the throttle/switch housing and install the housing as described in Chapter Eight.

14. Operate the throttle grip and make sure the carburetor throttle linkage is operating correctly and with no binding. If operation is incorrect or there is binding, carefully check that the cables are attached correctly and there are no tight bends in the cables.

15. Adjust the throttle cable as described in Chapter Three.

16. Install all throttle cables attaching bolts or brackets.

17. Install the fuel tank and seat.

*WARNING*
*If engine speed increases in Step 18, correct the cause before riding the motorcycle. Do not operate the motorcycle in this unsafe condition.*

18. Start the engine and turn the handlebar from side to side. Do not operate the throttle. If the engine speed increases as the handlebar assembly is turned, the throttle cable is routed incorrectly. Remove the seat and fuel tank and recheck the cable routing.

## FUEL TANK

### Removal/Installation

Refer to **Figure 18** or **Figure 19** for this procedure.

1. Disconnect the battery negative cable.

2. Disconnect the fuel line at the fuel supply valve. Drain the fuel into an approved gasoline tank or can.

3. On dual tank models, disconnect one of the crossover hoses at the front of the tanks. Plug the hoses with a golf tee to prevent fuel leaks.

6

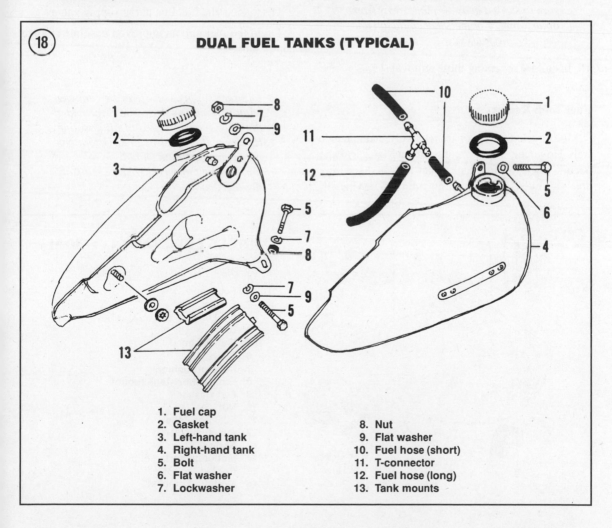

**(18) DUAL FUEL TANKS (TYPICAL)**

1. Fuel cap
2. Gasket
3. Left-hand tank
4. Right-hand tank
5. Bolt
6. Flat washer
7. Lockwasher
8. Nut
9. Flat washer
10. Fuel hose (short)
11. T-connector
12. Fuel hose (long)
13. Tank mounts

4A. *Single tanks:* Remove the front and rear mounting bolts and remove the fuel tank.

4B. *Dual tanks:*

  a. Remove the instrument cover at the top of the fuel tanks.

  b. If equipped with an odometer knob, unscrew it from the right-hand side.

  c. Remove the panel mounting screws and raise the panel away from the tanks.

*NOTE*
*Keep track of all washers and bolts during fuel tank removal. They must be returned to their original positions.*

  d. Remove the front and rear mounting bolts and nuts and remove the fuel tank(s).

*WARNING*
*Whenever removing the tank(s), make sure to store it in a safe place away from open flame or objects that could fall and damage the tank(s).*

15. Install by reversing these removal steps.

## Fuel Tank Repairs

Motorcycle fuel tanks are relatively maintenance free. However, a major cause of fuel tank leakage occurs when the fuel tank is not mounted securely and is allowed to vibrate during riding. When install-

ing the tank, make sure that the rubber dampers (**Figure 18** or **Figure 19**) at the front and rear of the tank are in position and that the tank is mounted securely at the front and back with the proper fastener(s).

If your fuel tank leaks, refer service to a qualified dealer or welding shop equipped to repair fuel tanks.

## FUEL SUPPLY VALVE

### Removal/Installation and Strainer Cleaning

The fuel supply valve removes particles which might otherwise enter the carburetor and possibly contaminate it.

Refer to **Figure 20** (1966-1974) or **Figure 21** (1975-on) for this procedure.

1. Turn the valve to the OFF position.

2. Disconnect the fuel line at the supply valve.

3. Drain the fuel into an approved gasoline storage can or tank.

*NOTE*
*The fuel supply valve can be removed without removing the gas tank, provided the tank(s) are drained of all gasoline. However, if the fuel strainer is contaminated, the fuel tank should be removed and flushed thoroughly before reinstalling the strainer.*

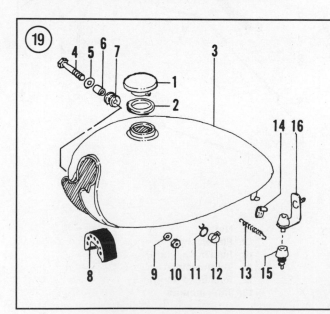

**SINGLE FUEL TANK (TYPICAL)**

1. Fuel cap
2. Gasket
3. Fuel tank
4. Bolt
5. Flat washer
6. Spacer
7. Rubber bushing
8. Front rubber tank mount
9. Flat washer
10. Nut
11. Fuel hose clip
12. Fuel hose clamp
13. Spring
14. Mounting spring clip
15. Rubber mount
16. Mounting bracket

4. Remove the fuel tank, if required, as described in this chapter.

5. Loosen the nut and remove the fuel supply valve.

6. Gently clean the strainer with a medium soft toothbrush and carefully blow out with compressed air. If the strainer is damaged or contaminated, replace it with a new one. On 1975 and later model supply valves, check the O-ring and replace if required.

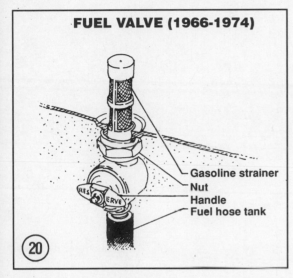

**FUEL VALVE (1966-1974)**

— Gasoline strainer
— Nut
— Handle
— Fuel hose tank

20

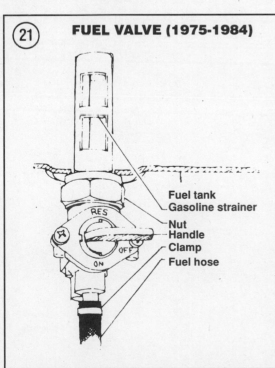

21 **FUEL VALVE (1975-1984)**

Fuel tank
Gasoline strainer
Nut
Handle
Clamp
Fuel hose

RES / OFF / ON

7. Coat the supply valve threads with Harley-Davidson Pipe Sealant with Teflon.

8. Install the supply valve. Tighten the nut securely.

9. Install the fuel tank.

10. Pour a small amount of fuel into the fuel tank and check the supply valve for fuel leaks. Repair any leaks before riding the motorcycle.

## EXHAUST SYSTEM

### Removal/Installation

A number of exhaust systems have been used on Harley-Davidson models covered in this manual. This procedure presents a general guideline for exhaust system removal and installation.

1. Place the bike on the sidestand.

2. Remove all exhaust pipe shields.

3. Loosen and remove the rear muffler clamps and brackets.

4. On models with exhaust pipe cross-over connections, loosen the clamp.

5. Remove the mufflers.

6. Loosen the exhaust pipes at the cylinder head and remove them.

7. Remove the exhaust port gaskets.

8. Examine all clamps and fasteners for damage. Replace parts as required.

9. Install by reversing these removal steps. Note the following.

10. Install new exhaust port-to-pipe gaskets.

11. Install all parts and secure fasteners finger tight only. Then tighten the exhaust flange bolts securely and work back to the mufflers. This will minimize exhaust leaks at the cylinder heads.

> *NOTE*
> *Any exhaust gas leak will cause an increase in exhaust noise. Check all connections from the cylinder head to the muffler with the engine running for looseness or improper fitting. Repair as required.*

### Exhaust System Care

The exhaust system greatly enhances the appearance of any motorcycle. And more importantly, the exhaust system is a vital key to the motorcycle's operation and performance. As the owner, you should periodically inspect, clean and polish the

6

exhaust system. Special chemical cleaners and preservatives compounded for exhaust systems are available at most motorcycle shops.

Severe dents which cause gas flow restrictions require the replacement of the damaged part.

Problems occurring within the exhaust pipes are normally cause by rust from the collection of water in the pipe. Periodically, or whenever the exhaust pipes are removed, turn the pipes to remove any trapped water.

# CHAPTER SEVEN

# ELECTRICAL SYSTEM

This chapter covers the following systems:

a. Charging.

b. Ignition.

c. Starting.

d. Lighting.

e. Directional signals.

f. Horn.

Refer to Chapter Three for routine ignition system maintenance. Electrical system specifications are found in **Table 1** (1966-early 1978) or **Table 2** (late 1978-1984). **Tables 1-6** are at the end of the chapter.

## CHARGING SYSTEM
### (1966-1969)

The charging system consists of the battery, generator and a voltage regulator.

Two styles of generators were used on 1966-1969 models. These were the standard (**Figure 1**) and fan-cooled (**Figure 2**) generators. The standard generator is a two-pole direct current two-brush unit. A voltage regulator controls the generator charging rate. When the battery charge is low or when current is used, the voltage regulator increases the generator charging rate. The charging rate decreases when the battery is near or at full charge and no current is being used.

The fan-cooled generator operates basically the same as the standard generator. The major differ-

ences are a larger housing and a higher current generating capacity. In addition, a fan is used to dissipate heat in the generator (7, **Figure 2**).

### Testing

Whenever the charging system is suspected of trouble, make sure the battery is fully charged before going any further. Clean and test the battery as described in Chapter Three. If the battery is in good condition, test the charging system as follows.

*NOTE*
*This section describes test procedures for both the standard and fan-cooled generators. Where differences occur, they will be pointed out.*

*CAUTION*
*It is important to read this procedure thoroughly before attempting to check the charging system to prevent component damage.*

### Preliminary Tests

The following tests and inspection should be performed before checking the generator charging. The following check can be performed with the generator installed in the frame.

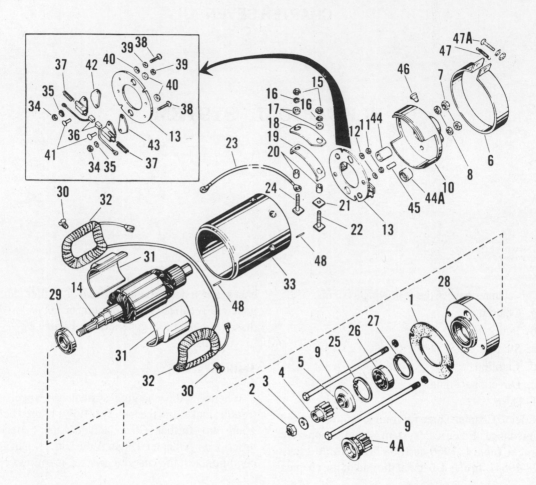

**STANDARD GENERATOR (1966-1969)**

1. Mounting gasket
2. Gear shaft nut
3. Gear shaft washer
4. Drive gear
4A. Drive gear with oil slinger
5. Drive end oil deflector
6. Brush cover strap
7. Commutator end cover nut
8. Commutator end cover washer
9. Frame screw
10. Commutator end cover
11. Brush cable nut
12. Brush cable washer
13. Brush holder mounting plate
14. Armature
15. Terminal screw nut
16. Terminal screw lockwasher
17. Insulating washer
18. Terminal insulator
19. Terminal bolt clip
20. Terminal screw bushing
21. Bracket insulator
22. Terminal screw
23. Positive brush cable
24. Terminal screw
25. Bearing retainer
26. Armature bearing
27. Bearing retainer
28. Drive end plate
29. Armature oil seal
30. Pole shoe seal
31. Pole shoe
32. Field coil
33. Frame
34. Terminal screw nut
35. Terminal screw lockwasher
36. Brush
37. Brush spring
38. Brush holder plate screw
39. Brush holder plate screw washer
40. Brush holder plate screw washer
41. Brush holder plate rivet
42. Brush holder insulation
43. Brush holder spacer
44. End cover bushing
44A. End cover bearing
45. Generator oil wick
46. Commutator end cover oil cup
47. Brush cover strap spring
47A. Brush cover screw, lockwasher and nut (1966 model)
48. End locating pin

## FAN COOLED GENERATOR (1966-1969)

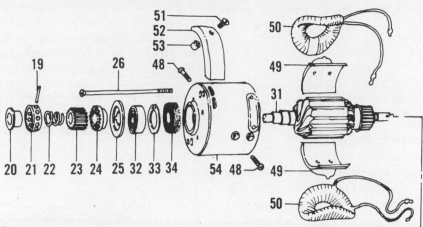

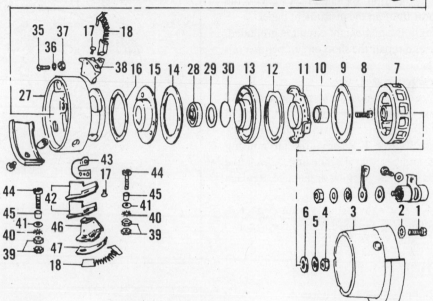

| | | |
|---|---|---|
| 1. Fan housing screw | 19. Clutch spring collar pin | 37. Brush holder screw nut |
| 2. Internal lock washer | 20. Clutch spring collar | 38. Brush holder (negative) |
| 3. Fan housing | 21. Oil slinger | 39. Terminal screw nut |
| 4. Armature shaft nut | 22. Clutch spring | 40. Terminal screw lockwasher |
| 5. Armature shaft lockwasher | 23. Drive gear | 41. Terminal screw insulating washer |
| 6. Armature shaft plain washer | 24. Clutch | 42. Field coil terminal insulator |
| 7. Fan | 25. Drive end oil deflector | 43. Field coil terminal |
| 8. Fan baffle plate screw | 26. Frame screw | 44. Terminal screw |
| 9. Fan baffle plate | 27. Frame end | 45. Terminal screw bushing |
| 10. Fan spacer | 28. Armature bearing | 46. Brush holder (positive) |
| 11. Fan housing spider | 29. Armature spacer shim | 47. Brush holder insulation |
| 12. End plate | (.020 in., 0.51 mm) | 48. Pole shoe screw |
| 13. Brush end bearing housing | 30. Bearing plate spring ring | 49. Pole shoe |
| 14. Drive end cover gasket | 31. Armature | 50. Field coil |
| 15. Inner oil retainer | 32. Armature bearing | 51. Air intake shield screw |
| 16. Commutator end bearing | 33. Drive end spring ring | 52. Air intake shield |
| shim (0-3) | 34. Felt retainer | 53. Spacing bushing |
| 17. Terminal screw | 35. Negative brush holder screw | 54. Generator frame |
| 18. Brush and spring | 36. Lockwasher | |

1. If your model is equipped with a fuse in the generator field circuit, check it first. The fuse is located near the regulator. Replace the fuse if blown. Make sure the fuse insulating sleeve is in good condition.

*NOTE*
*Refer to the wiring diagram for your model at the end of the book to see if a fuse is used.*

2. Check the generator signal light circuit for grounding as follows:
   a. Disconnect the wire or wires from the generator "A" terminal (**Figure 3**). Place the wires aside so that they do not contact any ground.
   b. Turn the ignition switch on and check the generator light on the instrument panel.
   c. If the light is on, the light circuit is grounded. Locate and repair the short circuit before continuing.

3. *Fan-cooled generators:*
   a. Disconnect the condenser at the generator "A" terminal. A damaged or shorted condenser will prevent the generator from charging correctly.
   b. Refer to **Figure 2**. Remove the fan housing screws and remove the housing. Check the brushes for wear, sticking or damage. Replace the brushes and recheck.

4. If Steps 1-3 test okay or have been corrected, perform the *Charging System Test* in this chapter.

**Charging System Test**

The test circuit is shown in **Figure 4**.
1. Attach a tachometer to the engine following the manufacturer's instructions.
2. Disconnect the generator "F" terminal wire.
3. Connect a jumper wire from the generator "F" terminal to a good ground on the engine.
4. Remove the wire(s) from the generator "A" terminal.
5. Connect the positive lead of a DC ammeter to the generator "A" terminal as shown in **Figure 4**.

*CAUTION*
*Do not run the engine with the generator field grounded any longer than necessary. Do not run the engine with no load on the generator (ammeter negative lead not to connected to battery) any longer than necessary.*

6. Start the engine and run it at 2,000 rpm.

7. Momentarily connect the ammeter negative terminal to the positive battery terminal. This is done by touching the ammeter negative lead to the regulator BAT terminal as shown in **Figure 4**. The charging amperage should be interpreted as follows:

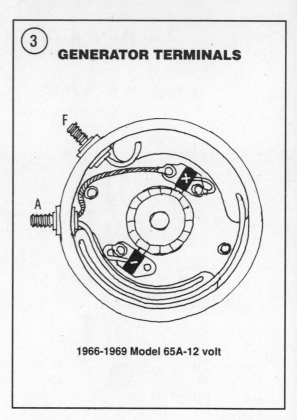

③ **GENERATOR TERMINALS**

**1966-1969 Model 65A-12 volt**

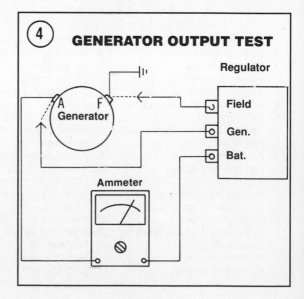

④ **GENERATOR OUTPUT TEST**

a. *Standard generator:* 10 amps or more.

b. *All air-cooled generators:* 20 amps or more.

7. If the charging amperage is not within specifications in Step 6, check the generator as described in this chapter. If the charging amperage is within specifications, the problem is in the regulator or wiring. Test the separate charging system components as described in this chapter.

> *CAUTION*
> *Disconnect the ammeter from the battery before turning the engine off.*

8. Polarize the generator after servicing it or the regulators, or after disconnecting any charging system wiring. See *Polarizing Generator* in this chapter.

## Polarizing Generator

After any charging system wiring or component has been disconnected, the generator must be polarized before starting the engine. Failure to polarize the generator may result in burned or stuck voltage regulator contacts and possible damage to the wiring and the generator windings.

> *NOTE*
> *Make sure the battery is in good condition before performing the following procedure.*

After all wiring is connected (do not start engine), *momentarily* connect a heavy jumper wire between BAT and GEN terminals on the regulator (**Figure 5**).

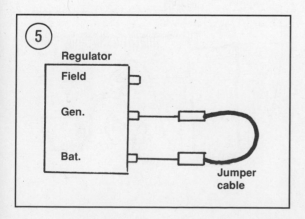

## Generator Removal/Installation

1. Disconnect and tag the generator "A" and "F" terminal wires.
2. Remove the 2 screws securing the generator to the gearcase.
3. *Fan-cooled generators:* Remove the footshifter and sidestand on footshift models or the clutch assembly and sidestand on handshift models.
4. Remove the generator from the left side of the motorcycle.
5. Installation is the reverse of these steps. Make sure all wiring is correctly attached.

## Brush Inspection

The overhaul of a generator is best left to an expert. This procedure shows how to replace defective brushes.

### Standard generator

Refer to **Figure 1** for this procedure.
1. Remove the commutator end cover nuts, washers and screws.
2. Gently tap the commutator end cover (10, **Figure 1**) off of the frame and armature shaft.
3. Remove the brush holder mounting plate (13) from the frame.
4. Disconnect the black brush wires and the generator positive brush cable (23) from the brush holder terminals.
5. Remove the brushes (36) from the brush holders.
6. Clean the brush holders with contact cleaner.
7. Measure the longest side of each brush with a caliper (**Figure 6**). Replace the brushes if they measure 1/2 in. or less.
8. Reverse Steps 1-5 to install the brushes.

### Air-cooled generators

Refer to **Figure 2** for this procedure.
1. Remove the fan housing screws and remove the housing (3, **Figure 2**).
2. Remove the armature shaft nut (4), lockwasher (5) and plain washer (6).
3. Using a claw-type puller, pull the fan (7) off of the armature shaft (32).
4. Remove the Woodruff key (8) from the armature (if so equipped).

7

5. Remove the baffle plate screws (9). Then remove the following parts in order:

  a. Baffle plate (10).

  b. Fan spacer (11).

  c. Fan housing spider (12).

  d. End plate (13).

*NOTE*
*When performing Step 6, the armature bearing should come off with the bearing housing. However, if the bearing should stay on the shaft, do not remove it but proceed with Step 7.*

6. Attach a claw-type puller to the brush end bearing housing (14) and remove it.

7. Remove the terminal screws from one of the brush holders (39 or 47). Then lift the brush and spring assembly (19) out of the brush holder. Repeat for the opposite holder.

8. Clean the brush holders with contact cleaner.

9. Check each brush for wear or damage. When installed, the brushes should contact the commutator fully. Brushes should be replaced when they are worn to half their original length. Replace the brushes if necessary.

10. Reverse Steps 1-7 to install the brushes.

## Generator Testing

Generator testing is best left to a Harley-Davidson dealer or to an electronic repair shop specializing in motorcycles because of the specialized tools required to check this component accurately.

## VOLTAGE REGULATOR
## (1966-1969)

### Testing

There are 4 basic regulator tests to be performed:

  a. Generating system test, which determines whether the generator or regulator is faulty.

  b. Cutout relay closing voltage.

  c. Voltage regulator setting.

  d. Current regulator setting.

Before testing the regulator, operate the motorcycle for at least 15 minutes so that the regulator is at normal operating temperature. Regulator cover and gasket must be in place.

*NOTE*
*Before making any voltage regulator test, be sure that the battery is in good condition and is at or near full charge. See Chapter Three for battery testing.*

### Delco-Remy Voltage Regulators

*Generating system test*

Refer to **Figure 7** for this procedure.

1. Turn the engine OFF.

2. Disconnect the regulator BAT terminal battery wire. Connect this wire to the negative terminal of a 0-30 amp DC ammeter. Connect the positive ammeter wire to the BAT regulator terminal.

3. Connect the positive terminal of a 0-20 volt DC voltmeter to the regulator GEN terminal. Connect the negative voltmeter terminal to a good ground.

4. Disconnect the regulator "F" terminal wire. Connect this to one terminal of a field control rheostat. Connect the other rheostat terminal to a good ground. Set the field control rheostat to the OPEN position.

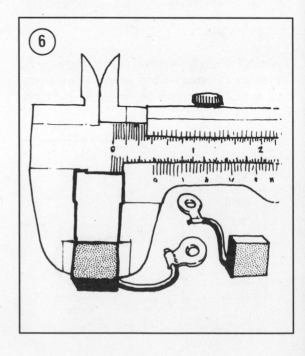

*NOTE*
*A rheostat is a variable electrical resistor used to regulate current. These can be purchased at most electronics stores.*

5. Start the engine and idle it at 2,000 rpm.

6. Slowly turn the field control rheostat toward the DIRECT position until the ammeter indicates or 10 amps. Interpret results as follows:

   a. If generator output is correct as listed above, the regulator is faulty.

   b. If there is no ammeter reading, or if the reading is low—below 10 amps—observe the voltmeter reading. If the voltmeter reads 12 volts or less, the generator is faulty and requires repair.

   c. If the ammeter reading is low, but the voltmeter reads 15 volts or more, the cut-out relay in the voltage regulator is defective.

7. If the voltage regulator is defective, perform the following test procedures.

### Cutout relay closing voltage

With the engine turned off, make the same testing connections as specified in *Generating system test*. See **Figure 7**.

1. Turn the field control rheostat to the OPEN position.

2. Start the engine and idle it at approximately 1,500 rpm.

3. Slowly turn the field control rheostat toward the DIRECT position and observe the voltmeter reading. It should increase slowly and then "kick" suddenly. The cutout relay closing voltage is that indicated on the voltmeter just before it "kicks." Cutout relay closing voltage should be 11.8-13.0 volts.

4. If the cutout relay closing voltage is not within the specifications in Step 3, have a Harley-Davidson dealer adjust the cutout relay. If the relay still does not test correctly, replace it with a new unit.

### Voltage regulator setting

Refer to **Figure 8** for this procedure.

1. Turn the engine off.

2. Write down the motorcycle's voltage regulator part number.

3. Disconnect wire to battery at voltage regulator BAT terminal and connect it to one terminal of a 1/4-ohm, 100-watt resistor. Connect the other resistor terminal to the voltage regulator BAT terminal.

4. Connect the positive terminal of a 0-20 volt DC voltmeter to the voltage regulator BAT terminal. Connect the negative voltmeter terminal to a good ground.

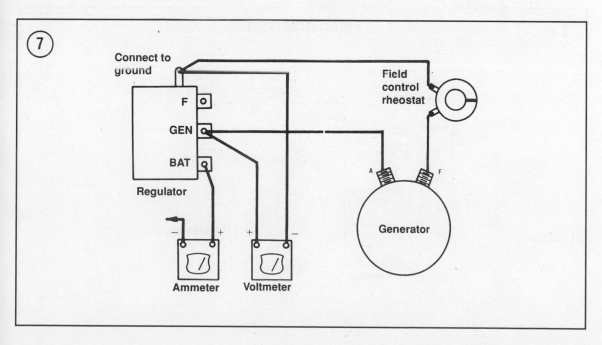

5. Disconnect the generator wire from the voltage regulator "F" terminal and connect it to one terminal of a field control rheostat. Connect the other field control rheostat terminal to the voltage regulator "F" terminal. Turn the field control rheostat control to DIRECT.

6. Start the engine and idle it at 2,000 rpm. Turn the field control rheostat to OPEN, then to DIRECT position. Observe the voltmeter reading. The voltmeter reading is the regulated voltage of the upper regulator contacts. Maintain engine speed and slowly turn the field control rheostat toward the OPEN position until the voltmeter reading drops slightly and stabilizes. The voltmeter reading at this point is the regulated voltage of the lower regulator contacts. The lower regulator contact voltage must be 0.1-0.3 volts less than that of the upper contacts. If the regulated voltage is incorrect, have a Harley-Davidson dealer adjust the voltage regulator. If adjustment does not correct the voltage reading, replace the voltage regulator.

### *Correct regulator setting*

This test is performed only on motorcycles with 3-unit voltage regulators. See **Figure 9** for this procedure.

1. Disconnect the voltage regulator BAT terminal wire. Connect this wire to the negative lead of a 0-30

amp DC ammeter. Connect the positive ammeter lead to the voltage regulator BAT terminal.

2. Connect the positive lead of a 0-20 volt DC voltmeter to the voltage regulator BAT terminal. Connect the negative voltmeter lead to a good ground.

3. Start the engine and idle it at 2,000 rpm. Turn on the headlight and connect an additional load, such as an adjustable carbon pile, across the battery until the voltmeter indicates one volt less than the regulated voltage.

4. Observe the ammeter reading. If it is not within the specifications listed in **Table 4**, have a Harley-Davidson dealer adjust the voltage regulator. If adjustment does not correct the voltage reading, replace the voltage regulator.

### CHARGING SYSTEM (1970-1984)

The charging system consists of the battery, alternator and a solid state rectifier/voltage regulator. See **Figure 10** (1970-1975) or **Figure 11** (1976-1984).

The alternator generates an alternating current (AC) which the rectifier converts to direct current (DC). The regulator maintains the voltage to the battery and load (lights, ignition, etc.) at a constant voltage regardless of variations in engine speed and load.

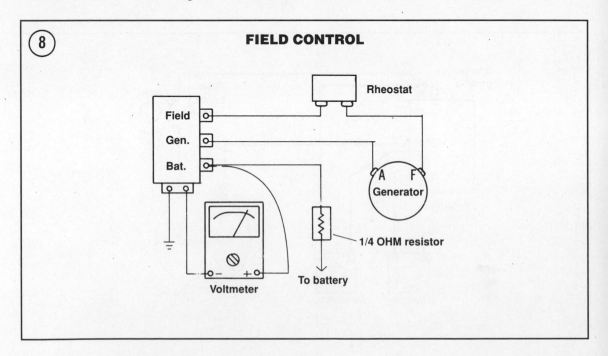

## Service Precautions

Before servicing the charging system, observe the following precautions to prevent damage to any charging system component.

1. Never reverse battery connections. Instantaneous damage may occur.

2. If it becomes necessary to use a booster battery, make sure to connect cables positive-to-positive and negative-to-negative.

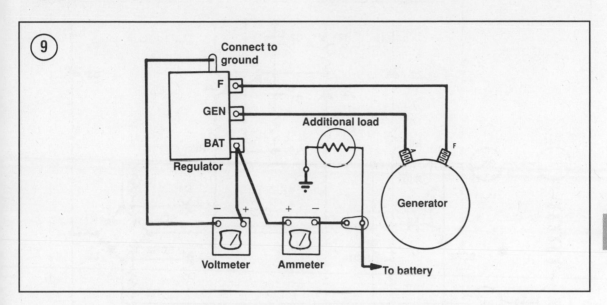

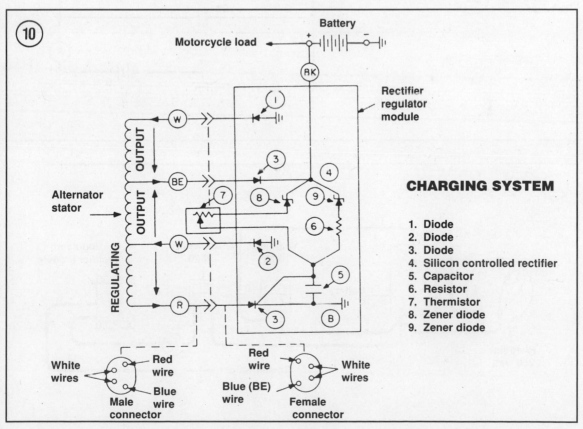

**CHARGING SYSTEM**

1. Diode
2. Diode
3. Diode
4. Silicon controlled rectifier
5. Capacitor
6. Resistor
7. Thermistor
8. Zener diode
9. Zener diode

(11)

## CHARGING SYSTEM (1976-1984)

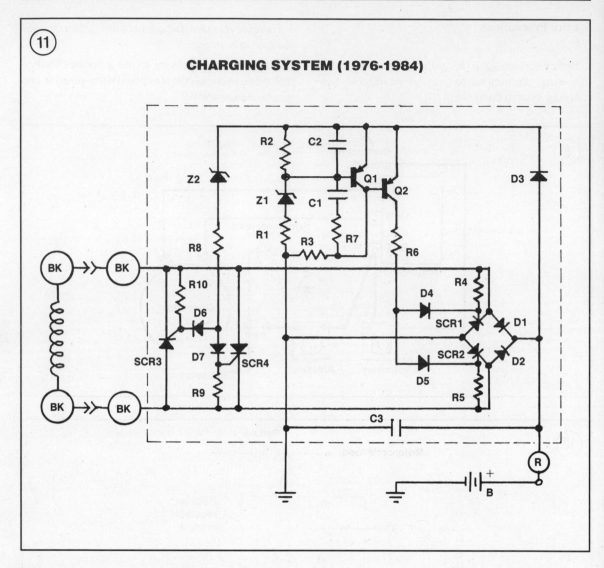

(12)

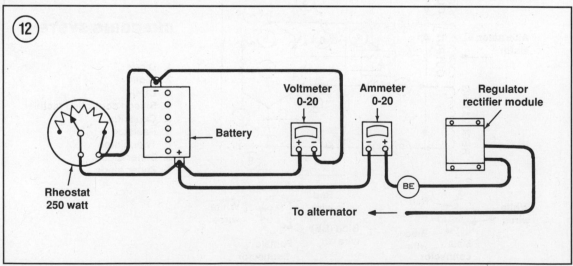

3. Do not short across any connections.

4. Never attempt to polarize an alternator.

5. Never start the engine with the alternator disconnected from the voltage regulator/rectifier, unless instructed to do so in testing.

6. Never start or run the engine with the battery disconnected.

7. Never attempt to use a high-output battery charger to assist in engine starting.

8. Never disconnect any engine-to-voltage regulator terminal with engine running.

9. Before charging battery, disconnect the negative battery lead.

10. Do not mount the voltage regulator/rectifier unit at another location.

11. Make sure that the battery negative terminal is connected to both engine and frame.

### Preliminary Testing

Whenever the charging system is suspected of trouble, make sure the battery is fully charged before going any further. Clean and test the battery as described in Chapter Three. If the battery is in good condition, test the charging system as follows.

### Testing (1970-1975)

Refer to **Figure 12** for this procedure.

1. Attach a tachometer to the engine following the manufacturer's instructions.

2. Start engine and let it idle to bring to normal operating temperature. Shut the engine off.

3. Connect a voltmeter, ammeter and rheostat or carbon pile as shown in **Figure 12**. Record the ambient air temperature with a thermometer.

4. The rheostat or carbon pile must be turned OFF.

5. Start engine and idle at 3,600 rpm.

6. With a fully charged battery, voltmeter should indicate 13.8 to 15.0 volts, depending on ambient temperature. Ammeter should indicate approximately 3.5 amps.

*NOTE*
*The load voltage will vary with ambient temperature. The voltage should be within the upper and lower limits in the graph in* ***Figure 13***.

7. Slow engine idle to 2,000 rpm. Adjust rheostat or carbon pile so that voltmeter reads exactly 13.0 volts. Ammeter must indicate at least 10.5 amps.

8. If regulated voltage obtained in Step 6 was okay, but output current in Step 7 was less than 10.5 amps, continue with Step 10.

9. Stop the engine.

10. Disconnect the voltage regulator/rectifier module plug. Check voltage regulator/rectifier with an ohmmeter. **Table 5** lists test connections. Replace the voltage regulator/rectifier module if any of the resistance checks in **Table 5** are incorrect.

11. Disconnect the alternator stator connector. With an ohmmeter, check the stator connections listed in

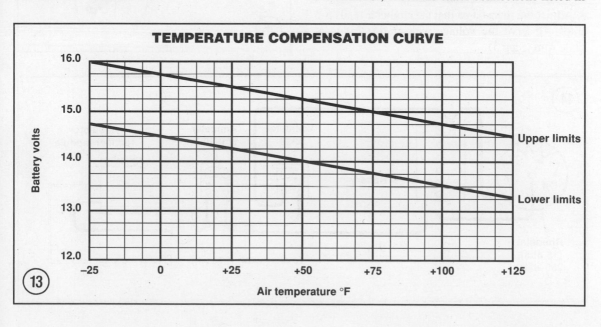

**TEMPERATURE COMPENSATION CURVE**

Figure 13 — Battery volts vs. Air temperature °F, showing Upper limits and Lower limits.

**Table 6**. Replace the stator if any of the resistance checks are incorrect.

12. If the resistance checks in Step 11 are okay, perform the following stator output voltage test. Note that during this test the engine is operated with the voltage regulator/rectifier module disconnected.

   a. Connect a 0-150 volt AC voltmeter between both white stator terminals. Start the engine and idle at 2,000 rpm. Voltmeter should indicate 50-100 volts.

   b. Stop the engine.

   c. Connect the 0-150 volt AC voltmeter between the stator red and blue terminals. Start the engine and idle at 2,000 rpm. Voltmeter should read 75-125 volts.

   d. Replace the stator if any test results in "a" or "c" are incorrect.

### Testing (1976-on)

Refer to **Figure 14** for this procedure.

1. Connect a tachometer to the engine following the manufacturer's instructions.

2. Connect a voltmeter, ammeter and rheostat or carbon pile as shown in **Figure 14**. The rheostat or carbon pile must be turned off. Record the ambient air temperature with a thermometer.

3. Start the engine and run it at 3,600 (1976-early 1978) or 2,000 (late 1978-on) rpm with no load on the rheostat. Observe the voltage.

4. Adjust the rheostat so that the ammeter reads 5-6 amps. Observe the voltage with engine idling at same rpm (Step 3).

5. Check that the voltmeter reads within the limits shown in **Figure 13** during the previous tests. If the voltage is higher or lower, the regulator is faulty and must be replaced. If the voltage readings are correct, proceed to Step 6.

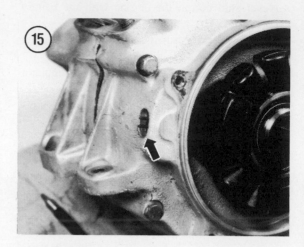

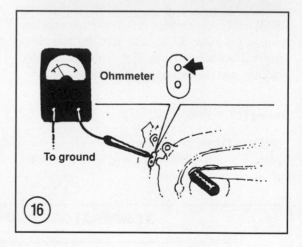

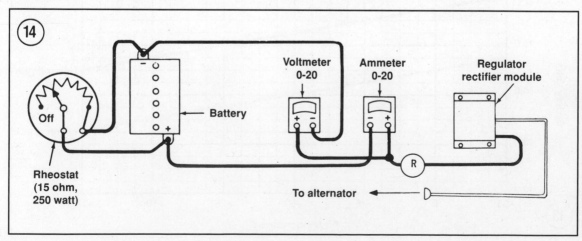

*NOTE*
*For example, if the ambient tempera-*
*ture is 75° F, the maximum voltage*
*would be 15 volts and the minimum*
*voltage would be 13.8 volts. See **Figure***
***13**.*

6. Run the engine at 2,000 rpm. Adjust the rheostat so that the voltmeter reads 13.0 volts. Check that the ammeter reads at least 14 amps.

7. If the amp reading in Step 6 is incorrect, perform the *Alternator Test* in this chapter.

## Alternator Test

1. Check for a grounded stator as follows:
   a. With the ignition switch turned off, disconnect the regulator/rectifier-to-stator terminal at the engine crankcase (**Figure 15**).
   b. Referring to **Figure 16**, connect an ohmmeter between either stator pin or socket and ground. Switch the ohmmeter to R × 1.
   c. The ohmmeter should indicate infinite resistance (no continuity) across either stator pin or socket.
   d. If any other reading is recorded, the stator is grounded and must be replaced.
   e. If the stator reading is correct, proceed to Step 2.

2. With the engine OFF and an ohmmeter set on R × 1, check the resistance between the stator contact pins. The resistance should be about 0.2-0.4 ohms; if the resistance is incorrect, replace the stator.

3. Disconnect the regulator/rectifier and connect an AC voltmeter (0-150 volts) across the stator contact pins in the front of the crankcase. Test as follows:

   a. *1976-early 1978:* Start the engine and check the voltage reading at various rpm readings. The voltmeter should read 19-26 volts per each 1,000 rpm.
   b. *Late 1978-on:* Start the engine and idle at 3,000 rpm. The voltage should read 60 volts or higher.

4. If the stator coil resistance in Step 1 and Step 2 is okay, but the alternator output (Step 3) is low, the rotor has probably been demagnetized. Replace the rotor as described in this chapter.

5. If the alternator is okay, but the system output was below normal, the regulator is faulty and must be replaced.

## Regulator/Rectifier Test

There is no test for the regulator alone. The regulator can be checked by substituting a known good regulator.

*NOTE*
*Because most repair shops and dealers*
*do not accept returns on electrical*
*parts, it's a good idea to refer regula-*
*tor/rectifier replacement testing to a*
*Harley-Davidson dealer.*

## Rotor Removal/Installation

Refer to **Figure 17** for this procedure.

1. Remove the clutch as described in Chapter Five.

2. Try to remove the rotor (**Figure 18**) by pulling it off of the crankshaft by hand. If the rotor is tight, attach a rotor puller to the rotor as shown in **Figure 19**. Use the Harley-Davidson rotor puller (part No. HD-95960-52B) or an equivalent.

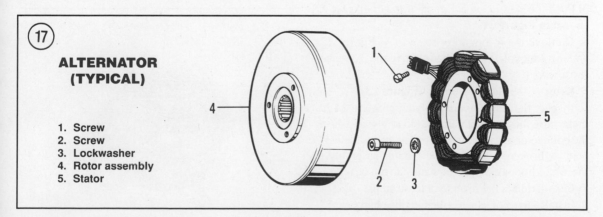

**(17)**

**ALTERNATOR (TYPICAL)**

1. Screw
2. Screw
3. Lockwasher
4. Rotor assembly
5. Stator

*CAUTION*
*If the rotor is on tight, don't try to re-move it without a puller; any attempt to do so will damage the engine and/or rotor. Many aftermarket pullers are available from motorcycle dealers or mail order houses. If you can't buy or borrow one, have a dealer remove the rotor.*

3. If a puller is being used to break the rotor loose, turn the rotor puller center bolt with a wrench until the rotor is free.

*CAUTION*
*If normal rotor removal attempts fail, do not force the puller as the threads may be stripped out of the rotor causing expensive damage. Take the bike to a dealer and have the rotor removed.*

4. Remove the rotor from the crankshaft. Remove the puller from the rotor.

*CAUTION*
*Carefully inspect the inside of the rotor (**Figure 20**) for small bolts, washers or other metal debris that may have been picked up by the magnets. These small metal bits can cause severe damage to the alternator stator assembly.*

5. Install by reversing these removal steps.
6. Install the clutch as described in Chapter Five.

## Stator Removal/Installation

Refer to **Figure 17** for this procedure.
1. Remove the rotor as described in this chapter.
2. Disconnect the stator electrical connector.
3. Push the stator connector out of the crankcase as shown in **Figure 21**.
4. Remove the 4 mounting screws (A, **Figure 22**) and the 2 lockplates (B). Then remove the 2 connector screws (C).
5. Remove the stator assembly (**Figure 23**).
6. Inspect the stator wires and plug (**Figure 24**) to make sure they are not damaged in any way. Check the connector electrical contacts. They should not be bent or broken off. Replace the stator assembly if these conditions exist, as the stator can not be rebuilt.
7. Installation is the reverse of these steps.
8. Install the rotor as described in this chapter.

## Voltage Regulator Removal/Installation

The regulator cannot be rebuilt; if damaged it must be replaced.

1. Disconnect the voltage regulator connector at the crankcase.

2. Remove the left-hand side cover.

3. Disconnect the regulator lead from the circuit breaker at the positive battery terminal.

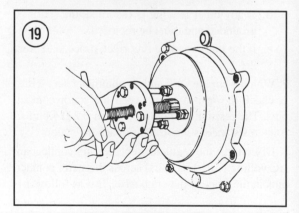

4. Remove the regulator mounting bolts and remove the regulator assembly.

5. Install a new regulator by reversing Steps 1-4.

## CIRCUIT BREAKER

Models from 1966-early 1978 are equipped with a single contact point automatic advance circuit breaker. On 1966-1969 models, the circuit breaker

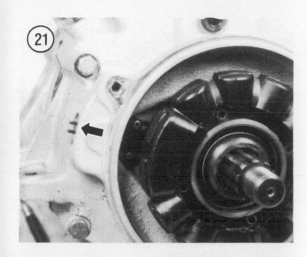

is a self-contained unit (**Figure 25**). On 1970-early 1978 models, the circuit breaker is incorporated into the right gearcase cover (**Figure 26**).

The breaker cam rotates at 1/2 crankshaft speed. There are 2 lobes on the cam. The narrow lobe times the front cylinder; the wide lobe times the rear cylinder. A single coil fires both spark plugs simultaneously, but one spark always occurs in a cylinder which is on its exhaust stroke.

When the breaker points are closed, current flows from the battery through the primary windings of the ignition coil, thereby building a magnetic field around the coil. As the points open, the magnetic field collapses. When the field collapses, a very high voltage (up to approximately 15,000 volts) is induced in the secondary windings of the ignition coil. This high voltage is sufficient to jump the gap at each spark plug.

The condenser assists the coil in producing high voltage, and also serves to protect the points. Inductance of the ignition coil primary current tends to keep a surge of current flowing through the circuit even after the points have started to open. The condenser stores this surge and thus prevents arcing at the points.

### Removal (1966-1969)

Refer to **Figure 25** for this procedure.

1. Thoroughly clean the circuit breaker area.

2. Remove the screw and lockwasher and remove the circuit breaker cover from the circuit breaker.

3. Disconnect the circuit wire connector.

4. Remove the circuit breaker stem clamp nut and clamp. Then lift the circuit breaker out of the crankcase.

7

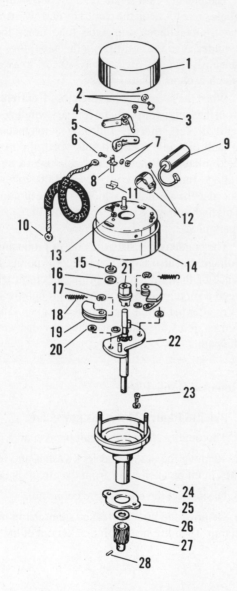

㉕ CIRCUIT BREAKER (1966-1969)

1. Cover
2. Screw and lockwasher
3. Lock screw
4. Contact point lever
5. Contact point and support
6. Wire stud screw
7. Wire stud nut and lockwasher
8. Wire stud
9. Condenser
10. Cable
11. Wire stud insulator
12. Condenser bracket and screw
13. Eccentric screw
14. Base
15. Fiber washer
16. Washer
17. E-clip
18. Flyweight spring
19. Flyweight
20. Washer
21. Cam
22. Camshaft
23. Screw and washer
24. Stem
25. Gasket
26. Shaft washer
27. Gear
28. Gear pin
29. Nut
30. Clamp

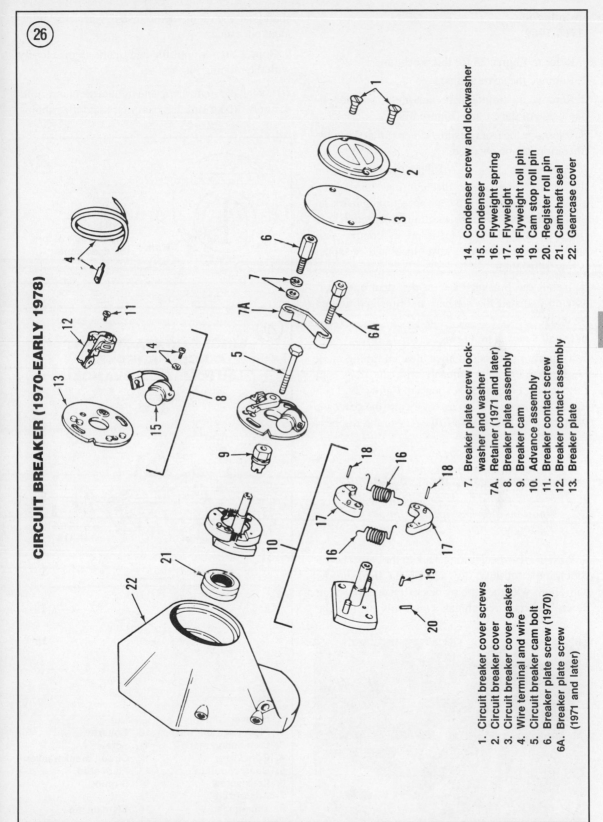

(26)

**CIRCUIT BREAKER (1970-EARLY 1978)**

1. Circuit breaker cover screws
2. Circuit breaker cover
3. Circuit breaker cover gasket
4. Wire terminal and wire
5. Circuit breaker cam bolt
6. Breaker plate screw (1970)
6A. Breaker plate screw
    (1971 and later)
7. Breaker plate screw lockwasher and washer
7A. Retainer (1971 and later)
8. Breaker plate assembly
9. Breaker cam
10. Advance assembly
11. Breaker contact screw
12. Breaker contact assembly
13. Breaker plate

14. Condenser screw and lockwasher
15. Condenser
16. Flyweight spring
17. Flyweight
18. Flyweight roll pin
19. Cam stop roll pin
20. Register roll pin
21. Camshaft seal
22. Gearcase cover

7

## Installation
## (1966-1969)

Refer to **Figure 25** for this procedure.

1. Remove the spark plugs.

2. Remove the timing inspection plug from the left-hand side of the engine (**Figure 27**).

3. Telescope the front pushrod cover so that the valve operation can be checked.

4. Turn engine clockwise until the front piston is on the compression stroke, then continue turning the engine clockwise until the advance timing mark on the flywheel is centered in the timing inspection hole (**Figure 28**). The cylinder is on its compression stroke just after its intake valve closes and the tappet is low in its guide.

5. Install the breaker base on the stem and shaft assembly. Install the washers and nuts and tighten securely.

6. Install a new stem O-ring.

7. Turn the shaft gear to align the cam timing mark (3, **Figure 29**) approximately with the fiber cam follower (2, **Figure 29**) as shown in **Figure 29**. Then install the circuit breaker assembly into the gear case with the wire facing toward the rear of the engine.

*NOTE*
*Positioning the wire toward the rear of the engine will position the circuit breaker points to the outside of the engine, thus allowing access to the adjusting screws when the cover is removed.*

8. With the flywheel timing mark in the center of the timing hole, observe how closely the mark on the cam aligns with the rubbing block. If it does not align closely, lift the circuit breaker assembly and turn the

shaft gear one tooth, then reinstall. Check the alignment once again.

9. Repeat Step 8 until the cam mark aligns closely with the rubbing block.

10. Install the stem clamp and tighten the clamp bolts securely. Make sure the cam mark and the rubbing

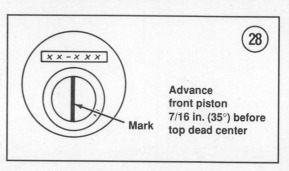

Advance
front piston
7/16 in. (35°) before
top dead center

Mark

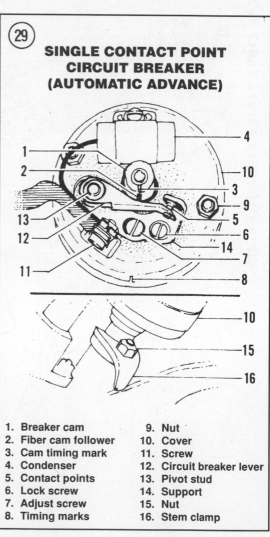

## SINGLE CONTACT POINT CIRCUIT BREAKER (AUTOMATIC ADVANCE)

1. Breaker cam
2. Fiber cam follower
3. Cam timing mark
4. Condenser
5. Contact points
6. Lock screw
7. Adjust screw
8. Timing marks
9. Nut
10. Cover
11. Screw
12. Circuit breaker lever
13. Pivot stud
14. Support
15. Nut
16. Stem clamp

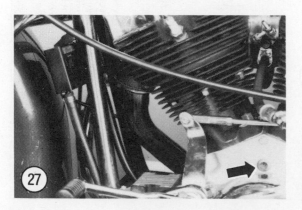

block are in correct alignment after tightening the clamp.

11. Adjust the ignition timing as described in Chapter Three.

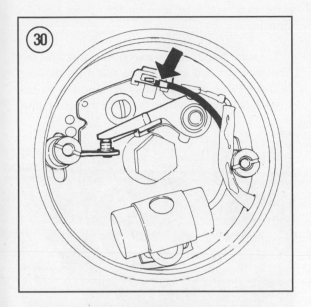

## Circuit Breaker Disassembly/Inspection/ Reassembly (1966-1969)

Refer to **Figure 25** to disassemble and reassemble the circuit breaker assemblies. Observe the following.

1. Clean all metal parts in solvent and allow to dry thoroughly.

2. Make sure the base plate is cleaned of all dirt or grease before installing the contact breakers.

3. Check the gear (27, **Figure 25**) for worn, chipped or broken teeth. Replace the gear if it shows these conditions.

4. Check the pin hole in the gear (27). If the hole is enlarged or distorted, check the mating hole in the end of the camshaft (22). Replace both parts if necessary. Replace the pin if it shows signs of wear.

5. Check the flyweight springs (18, **Figure 25**) for sagging or breakage. Check the flyweight pivot posts on the camshaft (22, **Figure 25**) for looseness. Check the flyweight (19, **Figure 25**) pivot holes. Replace worn or damaged parts.

6. Replace all O-rings and gaskets during reassembly.

## Removal/Installation (1970-early 1978)

Refer to **Figure 26** for this procedure.

1. Remove the circuit breaker cover (2, **Figure 26**) and gasket.

2. Remove the wire terminal from the circuit breaker contact terminal post (**Figure 30**).

3. Remove the circuit breaker plate screws (**Figure 31**) and remove the plate assembly.

4. Remove the circuit breaker cam bolt (A, **Figure 32**).

5. Pull the advance assembly out of the gearcase cover (B, **Figure 32**).

6. Installation is the reverse of these steps. Note the following.

7. The advance assembly must be installed so that the pin in the back of the plate (**Figure 33**) engages with the slot on the end of the crank (**Figure 34**).

*NOTE*
*If the breaker cam (9, **Figure 26**) was removed from the advance assembly, refer to **Inspection** in this chapter for correct installation. The breaker cam can be installed backwards.*

8. Adjust the ignition timing as described in Chapter Three.

### Disassembly/Inspection/Reassembly

1. Pull back on the flyweights (A, **Figure 35**) and pull the breaker cam (B, **Figure 35**) off of the advance assembly. See **Figure 36**.

2. Disconnect the flyweight springs from the flyweights (**Figure 37**).

3. Slip the flyweights off of the advance assembly pivots (A, **Figure 38**).

4. Inspect the condenser and contact breakers points as described in Chapter Three.

5. Inspect the flyweights and springs (**Figure 39**) for wear or damage. Check also that the springs are not stretched or distorted.

6. Check the flyweight pivots (B, **Figure 38**) on the advance plate for wear or looseness. Replace the advance plate if necessary.

7. Check the breaker cam for excessive wear at the slots at base of cam (A, **Figure 40**). Check the cam surface (B, **Figure 40**) for pitting, deep scoring or excessive wear. Replace the cam if necessary.

8. Assembly is the reverse of these steps. Note the following.

9. When installing the cam onto the advance assembly, align flat surface on cam (A, **Figure 41**) with the roll pin installed in the assembly plate (B, **Figure 41**).

### IGNITION SYSTEM (LATE 1978-1979)

A breakerless electronic ignition system is installed on late 1978-1979 models. Refer to **Figure 42** and **Figure 43** for diagrams of the ignition circuit.

This system differs from breaker point systems in that the mechanical points are replaced by an electronic ignition timer driven by the camshaft. The timer consists of a rotor and sensor. The sensor feeds signals to a control module mounted on the back of the timer inner cover. The timer rotor is mounted on a mechanical flywheel centrifugal advance, just as a point cam would be in a breaker point ignition system.

The rotor has 2 lobes—a small one that fires the front cylinder and a large lobe that fires the rear cylinder. When the ignition switch and kill switch are ON, current flows from the battery to the control

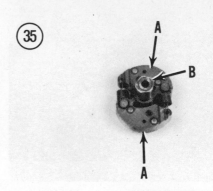

module where an oscillator produces a signal that flows through the sensor, creating a magnetic field around it. Current also flows through the ignition coil primary winding. When the leading edge of a rotor lobe enters the magnetic field, it slightly weakens the field. The control module detects the weakened field and stops current flow through the ignition coil primary winding.

From this point on, the system operates like a breaker point ignition system. The sudden stoppage of current flow in the ignition coil primary winding causes a rapid collapse of the coil's magnetic field, inducing a high voltage current into the ignition coil secondary winding and firing the spark plug.

The electronic ignition system eliminates breaker point maintenance, but regular inspection of the sensor air gap (the gap between the rotor lobes and the sensor) and of the mechanical advance operation is required.

### Sensor Air Gap

Inspect the sensor air gap every 2,500 miles. Refer to **Figure 44** for this procedure.

1. Remove the spark plug leads at the plugs and remove the spark plugs.

2. Remove the ignition timing cover.

3. Turn the crankshaft and center the wide rotor lobe in the sensor (**Figure 44**).

4. Measure the gap between the sensor and the rotor. It should be 0.004-0.006 in. (0.10-015 mm). If the gap is incorrect, loosen the 2 screws that attach the sensor (4, **Figure 45**) and move the sensor (5, **Figure 45**) until the gap is correct. Then, holding the sensor steady, tighten the screws.

> *NOTE*
> *If the engine doesn't run smoothly and spits back because of the standard lean fuel mixture, set the sensor air gap as*

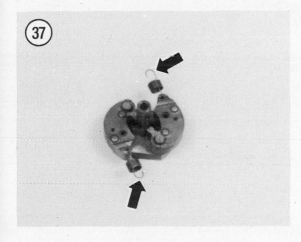

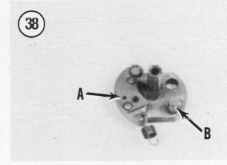

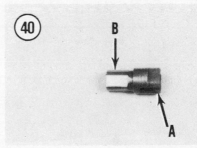

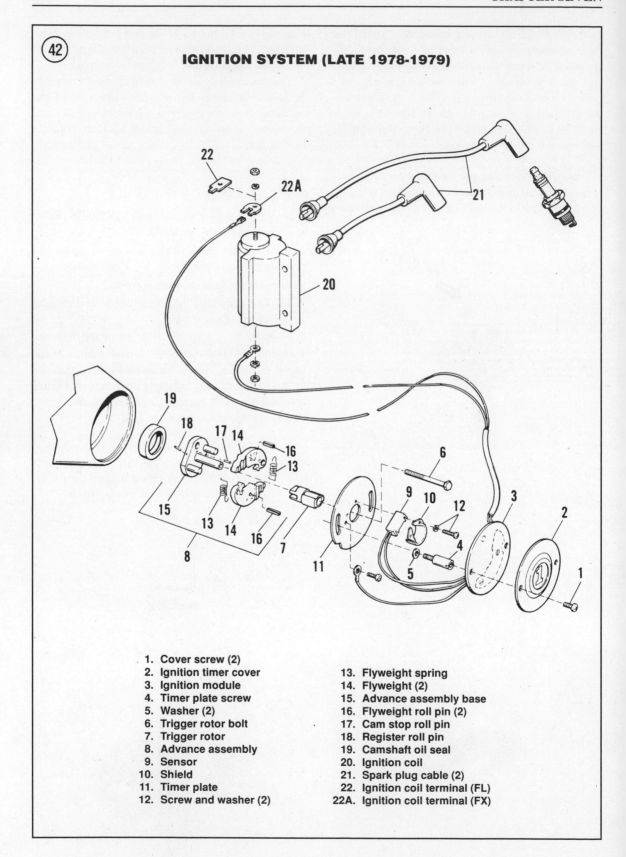

**42**

**IGNITION SYSTEM (LATE 1978-1979)**

1. Cover screw (2)
2. Ignition timer cover
3. Ignition module
4. Timer plate screw
5. Washer (2)
6. Trigger rotor bolt
7. Trigger rotor
8. Advance assembly
9. Sensor
10. Shield
11. Timer plate
12. Screw and washer (2)
13. Flyweight spring
14. Flyweight (2)
15. Advance assembly base
16. Flyweight roll pin (2)
17. Cam stop roll pin
18. Register roll pin
19. Camshaft oil seal
20. Ignition coil
21. Spark plug cable (2)
22. Ignition coil terminal (FL)
22A. Ignition coil terminal (FX)

*close to 0.004 in. as possible. This adjustment will strengthen the ignition signals for better combustion.*

5. Rotate the crankshaft to center the small lobe in the sensor. Measure the gap as before (Step 4) and correct it if necessary. Both gaps must be within the specified tolerance.

6. Install the spark plugs and check the ignition timing as described in Chapter Three.

### Ignition System Removal/Installation

Refer to **Figure 42** for this procedure.

1. Remove the ignition cover and the timing module (3, **Figure 42**).

2. Remove the 2 timer plate screws (4) and washers.

3. Unscrew the trigger rotor bolt (6) and remove the rotor (7).

4. Remove the advance assembly (8).

5. To disassemble the sensor (9) and shield (10) from the timing plate, remove the screws that hold them in place.

6. To disassemble the advance mechanism, perform the procedures as described under *Disassembly/Inspection/Reassembly.*

7. Installation is the reverse of these steps. Note the following.

8. Lubricate the advance mechanism pivots with WD-40 or Loctite Anti-Seize.

9. Make sure the timing rotor is seated squarely on the shaft.

10. Check the sensor air gap as described in this chapter.

11. Check the ignition timing as described in Chapter Three.

### IGNITION SYSTEM (1980-ON)

In 1980, a Magnavox electronic ignition was introduced. This system has a full electronic advance that replaced the mechanical advance on early 1978-1979 models. The inductive pickup unit is driven by the engine and generates pulses which are routed to the solid-state ignition control module. This control module computes the ignition timing advance and ignition coil dwell time, eliminating the need for mechanical advance and routine ignition service. Refer to **Figure 46** and **Figure 47** for diagrams of the ignition circuit.

*NOTE*
*On 1980-early 1981 models, the control module shuts off current to the ignition coil 4 seconds after the ignition switch*

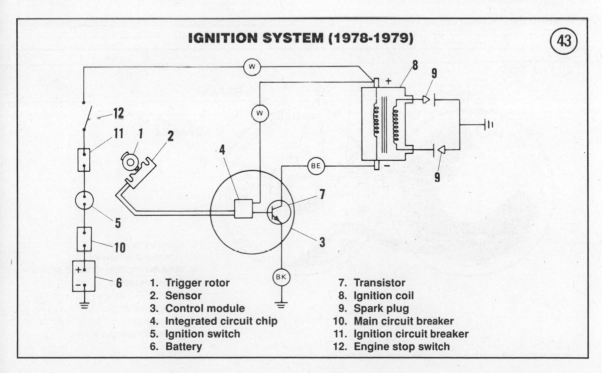

**IGNITION SYSTEM (1978-1979)** ④

1. Trigger rotor
2. Sensor
3. Control module
4. Integrated circuit chip
5. Ignition switch
6. Battery
7. Transistor
8. Ignition coil
9. Spark plug
10. Main circuit breaker
11. Ignition circuit breaker
12. Engine stop switch

*is turned ON. If the engine is not started within that time, the module must be reset by turning the ignition switch OFF and ON again. Later ignition modules allow 6 seconds for the engine to be started.*

The timing sensor is triggered by the leading and trailing edges of the 2 rotor slots shown in 7, **Figure 47**. As rpm increases, the control module "steps" the timing in 3 stages of advance.

### Ignition Component Removal

Refer to **Figure 46** for this procedure.

1. Disconnect the negative battery lead.

2. Drill out the outer cover rivets (12, **Figure 46**) with a 3/8 in. drill bit. Remove the outer cover.

3. Remove the inner cover (10) and its gasket.

4. Remove the screws securing the sensor plate (6) to the crankcase.

5. Disconnect the sensor wire connector at the crankcase. Then remove the connector from the wires. Withdraw the wires through the crankcase hole one wire at a time. Remove the sensor plate and wires.

6. Remove the rotor screw (15) and rotor (14).

7. Disconnect the ignition module wires at the ignition coil.

8. Remove the ignition module (4) as follows:

a. Loosen and remove the ground wire connection at the frame.

b. Remove the module mounting bolts and remove the module assembly.

### Inspection (All Models)

1. If necessary, follow the procedures in Chapter Two to troubleshoot the electrical system.

2. Check the ignition compartment for oil leakage. If present, remove the crankcase seal (**Figure 48**) by prying it out with a screwdriver or seal remover. Install a new seal by tapping it in place with a

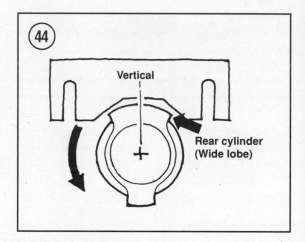

**Vertical**

**Rear cylinder (Wide lobe)**

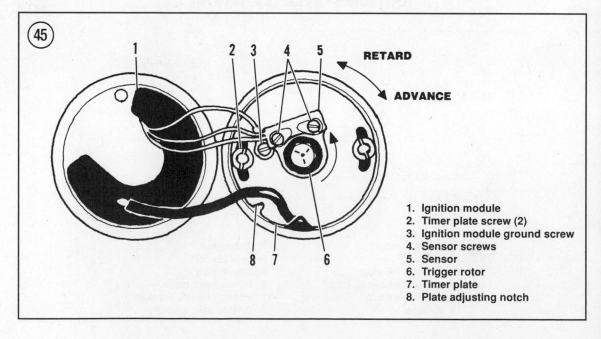

RETARD

ADVANCE

1. Ignition module
2. Timer plate screw (2)
3. Ignition module ground screw
4. Sensor screws
5. Sensor
6. Trigger rotor
7. Timer plate
8. Plate adjusting notch

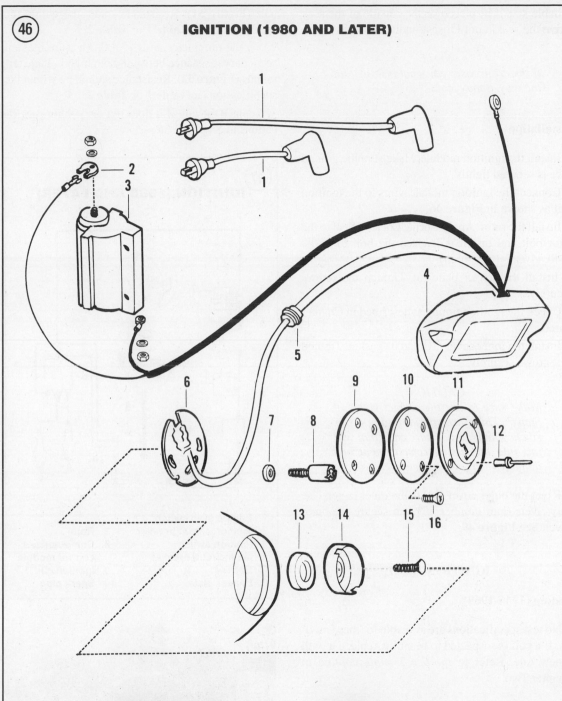

**IGNITION (1980 AND LATER)**

1. Spark plug cable (2)
2. Ignition coil terminal
3. Ignition coil
4. Ignition module
5. Connector
6. Sensor plate
7. Washer (2)
8. Sensor plate screw (2)
9. Gasket
10. Inner cover
11. Outer cover
12. Outer cover rivet (2)
13. Camshaft oil seal
14. Rotor
15. Rotor screw and star washer
16. Inner cover screws (2)

7

suitable size drift placed on the outside of the seal. Drive the seal in until it seats in the crankcase.

*NOTE*
*If the crankcase seal is not installed all the way, it may leak.*

### Installation

1. Install the ignition module. Make sure the ground wire is secured tightly.

2. Connect the ignition module wires to the ignition coil as shown in **Figure 46**.

3. Install the rotor. Apply Loctite Lock N' Seal to the rotor bolt and install it. Tighten the bolt to 75-80 ft.-lb. (103.5-108.8 N•m).

4. Install the sensor plate (6). Tighten the screws securely.

5. Check the ignition timing as described in Chapter Three.

6. Install the inner cover (10) and its gasket. Tighten the screws securely.

*CAUTION*
*Make sure to use the correct rivets in Step 7. These are special timing cover rivets which do not have ends that will fall into the timing compartment and damage the ignition components.*

7. Rivet the outer cover (11) to the inner cover. Use only rivets (part number 8699) to secure the outer cover. See **Figure 49**.

### IGNITION COIL

#### Testing (1966-1969)

No test specifications are available for these models. If a coil is suspected to be faulty, replace it with a new one. Refer to *Ignition Troubleshooting* in Chapter Two.

#### Testing (1970-On)

If the coil (**Figure 50**) condition is doubtful, there are several checks which can be made. Disconnect the coil secondary and primary wires before testing.

1. Set an ohmmeter on R × 1. Measure the coil primary resistance between both coil primary termi-

nals (**Figure 51**). Resistance should be within the specifications in **Table 1** or **Table 2**.

2. Set the ohmmeter on R × 100. Measure the coil secondary resistance between both high voltage terminals (**Figure 52**). Resistance should be within the specifications in **Table 1** or **Table 2**.

3. Replace the coil if it does not test within specifications in Step 1 or Step 2.

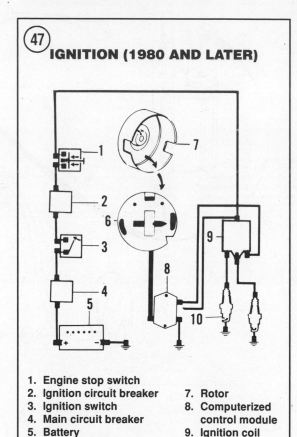

**47**

**IGNITION (1980 AND LATER)**

1. **Engine stop switch**
2. **Ignition circuit breaker**
3. **Ignition switch**
4. **Main circuit breaker**
5. **Battery**
6. **Sensor plate**
7. **Rotor**
8. **Computerized control module**
9. **Ignition coil**
10. **Spark plug**

**48**

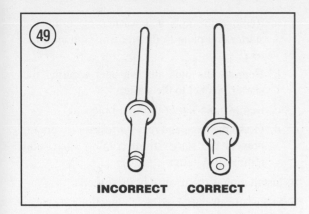

INCORRECT    CORRECT

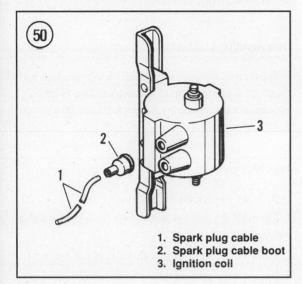

1. Spark plug cable
2. Spark plug cable boot
3. Ignition coil

### Removal/Installation

1. Remove the coil cover (if so equipped).

2. Disconnect all ignition coil wiring.

3. Remove the coil mounting bolts and remove the coil.

4. Installation is the reverse of these steps.

> *CAUTION*
> *When replacing an ignition coil on 1980 and later models, make sure the coil is marked ELECTRONIC ADVANCE. Installing an older type ignition coil could damage electronic ignition components. Early 1978-1979 models may use an old style coil.*

### Spark Plug Cable Replacement

To replace the coil wire, grasp the coil wire at the coil and pull it out (**Figure 50**). Reverse to install. Replace the cable boot (2, **Figure 50**) if worn or damaged. When reinstalling a coil wire, make sure the boot is secured to the coil to prevent moisture and dirt from entering.

7

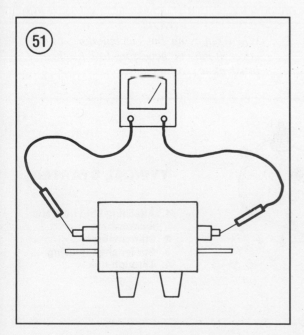

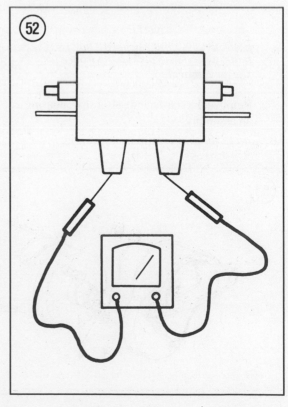

## STARTER

### Troubleshooting

Starter troubleshooting procedures are described in Chapter Two.

> *CAUTION*
> *Never attempt to operate the starter by pushing the starter button for more than 30 seconds at a time. If the engine fails to start, wait a minimum of 2 minutes to allow the starter to cool. The starter can be damaged by failing to observe this caution.*

### Removal/Installation

1. Disconnect the battery negative lead.

2. Disconnect the electric starter cable at the starter.

3A. *1966-early 1978:* Referring to **Figure 53**, perform the following:

  a. Remove the starter shaft housing-to-chain housing nuts and washers.

  b. Remove the starter-to-transmission end support plate.

> *NOTE*
> *It may be necessary to loosen and raise battery carrier to provide clearance for starter removal.*

  c. Remove the starter and starter shaft housing as an assembly.

3B. *Late 1978-on:* Perform the following:

  a. Remove the nuts and washers securing the starter end plate to the transmission and battery.

  b. Remove the nuts and washers securing the starter bracket to the starter.

  c. Remove the starter through-bolts.

  d. Grasp the front and rear starter end covers (to prevent the starter from coming apart) and remove the starter.

4. Installation is the reverse of these steps.

5. Make sure all starter wiring connections are clean and free of all dirt and road debris.

### Disassembly/Assembly

Starter motor repair is generally a job for electrical shops or a Harley-Davidson dealer. The following procedures describe how to check starter brush condition.

#### *Prestolite starters*

Refer to **Figure 54** for this procedure.

1. Clean all grease, dirt and carbon from the case and end covers.

2. Remove the starter through-bolts (11, **Figure 54**).

3. Remove the commutator end cover (12).

> *NOTE*
> *When removing the commutator end cover, it may be necessary to hold the brush plate (13) in place.*

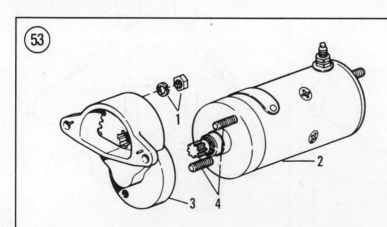

(53)

**TYPICAL STARTER**

1. Attaching stud nuts and lockwashers
2. Starter motor
3. Starter shaft housing
4. Through-bolt

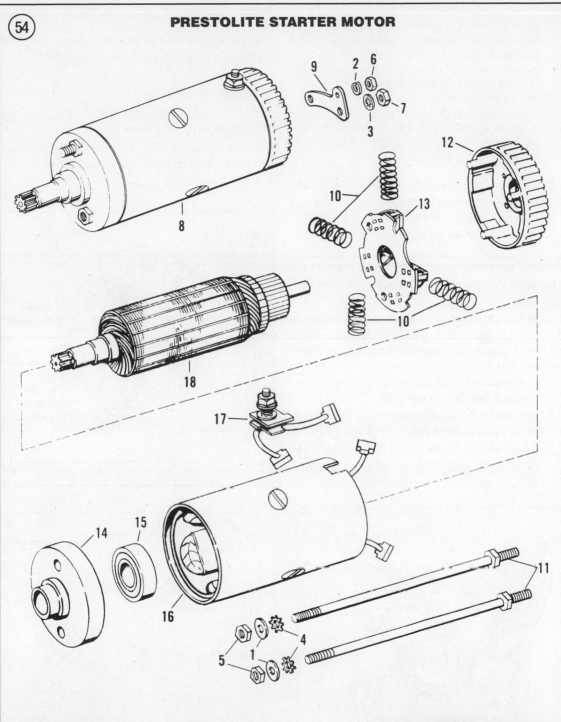

**PRESTOLITE STARTER MOTOR**

54

| | | |
|---|---|---|
| 1. Through-bolt washer | 7. Nut | 13. Brush plate and holder assembly |
| 2. Lockwasher | 8. Starter motor | 14. Drive end cover assembly |
| 3. Lockwasher | 9. Starter motor bracket | 15. Drive end ball bearing |
| 4. Through-bolt lockwasher | 10. Set of brush springs | 16. Assembled frame and field coils |
| 5. Through-bolt nut | 11. Set of through-bolts | 17. Terminal with set of brushes |
| 6. Nut | 12. Commutator end cover assembly | 18. Armature |

7

*NOTE*
*If removal of the brush plate and brushes is not required, pull the armature partway out to expose half of the 4 brushes' contact surfaces. Then install a spool (with a diameter approximately the same as that of the armature) into the starter from the opposite end to engage the brushes. That way the brushes will not fall out of their holders when the armature is removed.*

4. Remove the armature (18) and drive end cover (14) (with bearing) as an assembly.

5. If necessary, move the brushes out of the brush plate. Then remove the springs (10) and brush plate (13).

6. Inspect the starter as described in this chapter.

7. Installation is the reverse of these steps. Note the following:

  a. If the brushes and springs have been released from their holders, hold them in place with wire clips while installing the armature (**Figure 55**).

  b. Align the notch in the brush holder with the terminal insulator (**Figure 56**).

  c. The starter frame drive end cover is notched to fit with the drive end cover (**Figure 57**).

  d. The 2 ridges fixed on the commutator end cover must align with the terminal (**Figure 57**).

### Hitachi starters

Refer to **Figure 58** for this procedure.

1. Clean all grease, dirt and carbon from the case and end covers.

2. Remove the field wire.

3. Remove the commutator end cover screws and remove the cover (5, **Figure 58**).

4. Using a piece of wire, lift the brush springs and pull the brushes (6 and 7) out of the holder.

5. Remove the brush holder (9).

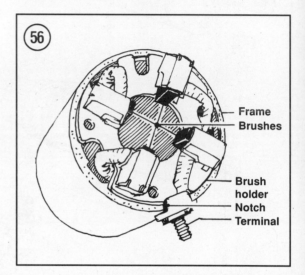

**56**

Frame
Brushes

Brush holder
Notch
Terminal

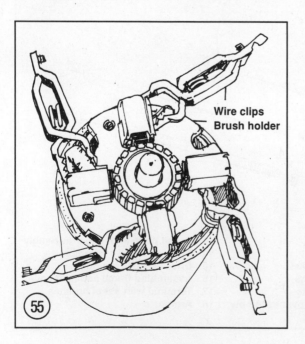

**55**

Wire clips
Brush holder

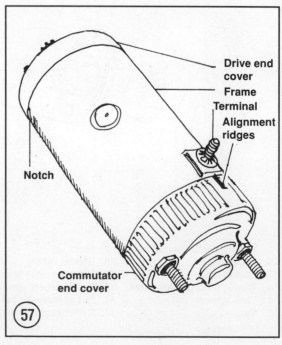

**57**

Drive end cover
Frame
Terminal
Alignment ridges

Notch

Commutator end cover

6. Remove the armature (11) and field frame (14).

7. Inspect the starter as described in this chapter.

8. Installation is the reverse of these steps. Note the following:

    a. Lubricate the armature bearings and the felt washer with high temperature grease.

    b. Install the armature into the frame.

    c. Install the front cover (10).

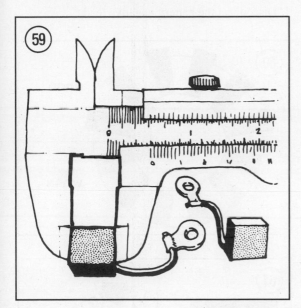

    d. Align the brush holder mounting screw holes with the frame and install the brush holder (9).

    e. Pull the brush springs up with a wire hook and install the brushes.

    f. Install the thrust washers (13) onto the armature shaft.

**Inspection (All Models)**

Refer to **Figure 54** or **Figure 58** for this procedure.

1. Measure the length of each brush with a vernier caliper (**Figure 59**). If the length is less than specified in **Table 1** or **Table 2**, it must be replaced. Replace the brushes as a set even though only one may be worn to this dimension.

2. Inspect the condition of the commutator. The mica in a good commutator is below the surface of the copper bars. On a worn commutator the mica and copper bars may be worn to the same level. See **Figure 60**. If necessary, have the commutator serviced by a dealer or motorcycle or automotive electrical repair shop.

3. Inspect the commutator copper bars for discoloration. If a pair of bars are discolored, grounded armature coils are indicated.

7

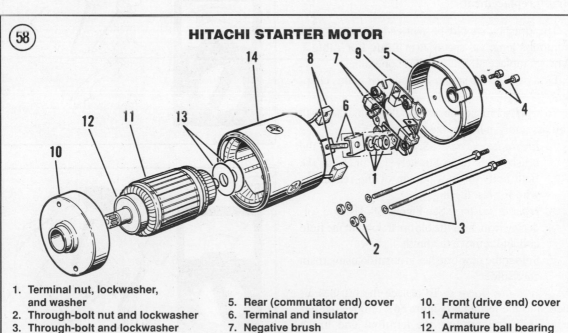

**HITACHI STARTER MOTOR**

1. Terminal nut, lockwasher, and washer
2. Through-bolt nut and lockwasher
3. Through-bolt and lockwasher
4. Rear cover screws and lockwashers
5. Rear (commutator end) cover
6. Terminal and insulator
7. Negative brush
8. Positive brush
9. Brush holder assembly
10. Front (drive end) cover
11. Armature
12. Armature ball bearing
13. Thrust washer
14. Frame

*NOTE*
*Step 4 and Step 5 apply to 1981 and later models only. Because of the internal wiring connections to the earlier model starters, there is no satisfactory field test to determine shorted or grounded field coils.*

4. Use an ohmmeter and check for continuity between the commutator bars (**Figure 61**); there should be continuity between pairs of bars. Also check continuity between the commutator bars and the shaft (**Figure 62**); there should be no continuity. If the unit fails either of these tests the armature is faulty and must be replaced.

5. Use an ohmmeter and inspect the field coil by checking continuity between the starter cable terminal and the starter case; there should be no continuity. Also check continuity between the starter cable terminal and each brush wire terminal; there should be continuity. If the unit fails either of these tests, the case/field coil assembly must be replaced.

6. Connect one probe of an ohmmeter to the brush holder plate and the other probe to each of the positive (insulated) brush holders; there should be no continuity. If the unit fails at either brush holder, the brush holder assembly should be replaced.

### Brush replacement

The brushes should be replaced if worn to the minimum length as specified in **Table 1** or **Table 2**. Always replace all brushes (4) at one time.

1. Disassemble the starter as described in this chapter.

2. To replace brushes attached to the frame and field coil assembly, perform the following:

    a. *Prestolite starters:* See **Figure 54**. Cut the brush leads approximately 1/2 in. from the field coils. Strip a portion of the remaining wire for resoldering.

    b. *Hitachi starters:* See **Figure 58**. Using a soldering iron, heat the old brush joint in the field coil and remove the brush.

    c. Solder the new brushes in position using rosin core solder.

3A. *Prestolite starters:* To replace the brushes attached to the terminal (**Figure 56**), remove the terminal nut and remove the terminal and brush assembly from inside the frame assembly. Installation is the reverse of these steps.

3B. *Hitachi starters:* Remove the brushes from the brush holder and replace.

## STARTER SOLENOID

### Removal/Installation (1966-early 1978)

Refer to **Figure 63** for this procedure.

1. Disconnect the battery negative lead.

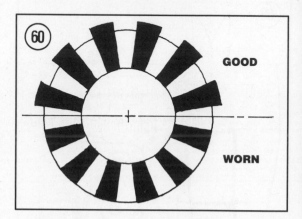

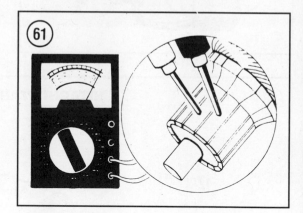

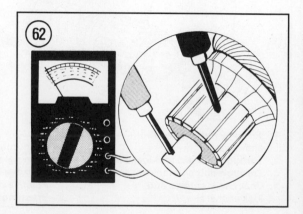

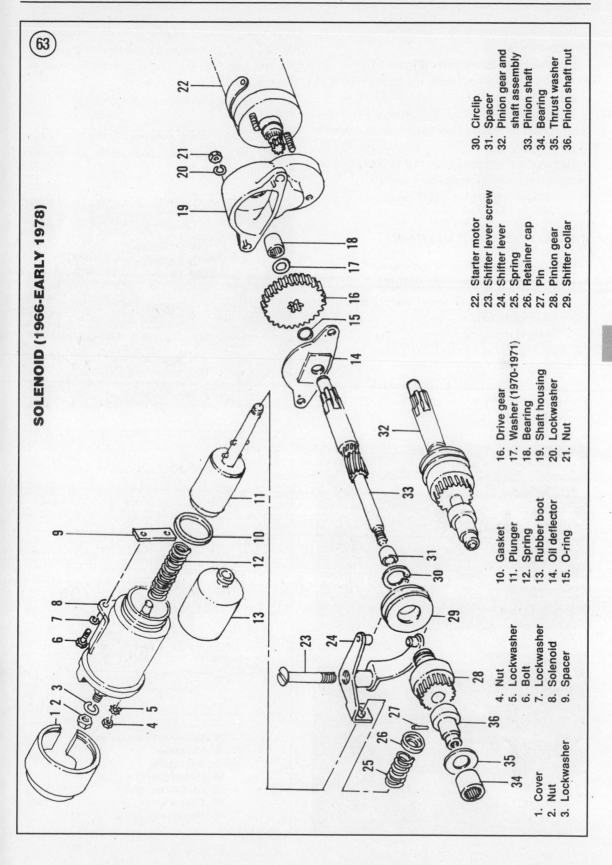

**SOLENOID (1966-EARLY 1978)**

1. Cover
2. Nut
3. Lockwasher
4. Nut
5. Lockwasher
6. Bolt
7. Lockwasher
8. Solenoid
9. Spacer
10. Gasket
11. Plunger
12. Spring
13. Rubber boot
14. Oil deflector
15. O-ring
16. Drive gear
17. Washer (1970-1971)
18. Bearing
19. Shaft housing
20. Lockwasher
21. Nut
22. Starter motor
23. Shifter lever screw
24. Shifter lever
25. Spring
26. Retainer cap
27. Pin
28. Pinion gear
29. Shifter collar
30. Circlip
31. Spacer
32. Pinion gear and shaft assembly
33. Pinion shaft
34. Bearing
35. Thrust washer
36. Pinion shaft nut

7

2. Remove the primary chain cover as described in Chapter Five.

3. Remove the solenoid cover (**Figure 64**).

4. Disconnect the wires from the starter solenoid (**Figure 65**).

5. Depress the retainer cap (26, **Figure 63**) at the front of the starter shaft housing and remove the pin (27) and spring (25).

6. Remove the solenoid (**Figure 66**) boot, gasket, plunger and plunger spring. See **Figure 67**.

8. Installation is the reverse of these steps.

### Removal/Installation (Late 1978-on)

Refer to **Figure 68** for this procedure.

1. Disconnect the negative battery cable.

2. Disconnect the small starter relay wire at the solenoid (**Figure 69**).

3. Disconnect the starter motor cable and the battery positive cable at the solenoid. See **Figure 69**.

4. Remove the solenoid mounting bolts. Then remove the solenoid, spacer, spring and felt gasket.

5. Installation is the reverse of these steps. Note the following.

6. Replace the felt gasket if worn or damaged.

7. Install the felt gasket on the solenoid. Then install the solenoid with the spacer inserted between the solenoid and the primary cover.

8. Tighten the solenoid mounting bolts to 12 ft.-lb. (16.6 N•m).

9. Attach the solenoid wires as follows:

   a. Attach the starter motor cable to the *short* solenoid stud.

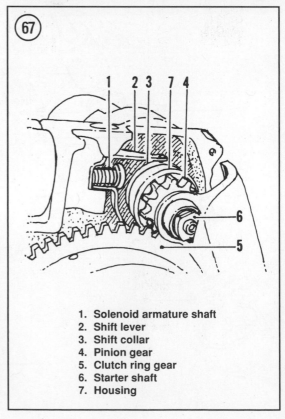

1. Solenoid armature shaft
2. Shift lever
3. Shift collar
4. Pinion gear
5. Clutch ring gear
6. Starter shaft
7. Housing

## SOLENOID (LATE 1978-ON)

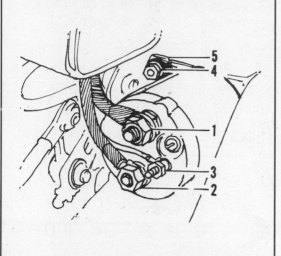

1. Battery positive terminal (long)
2. Starter terminal (short)
3. Starter relay wire (small terminal)
4. Bolt
5. Spacer

(69)

b. Attach the positive battery cable to the *long* solenoid stud.

c. Attach the starter relay wire (small diameter) to the small solenoid stud.

d. Place a lockwasher under each wire connector nut.

10. Reattach the negative battery cable when the solenoid is bolted tight and all solenoid wires are attached correctly.

## LIGHTING SYSTEM

### Headlight Replacement

These models are equipped with a sealed beam unit. Refer to **Figure 70**, **Figure 71** or **Figure 72**.

> *CAUTION*
> *Harley-Davidson warns against the use of an automotive type sealed beam headlight on these models. Because the current requirement for a motorcycle headlight is less than that of an automo-*

(68) **SOLENOID (LATE 1978-ON)**

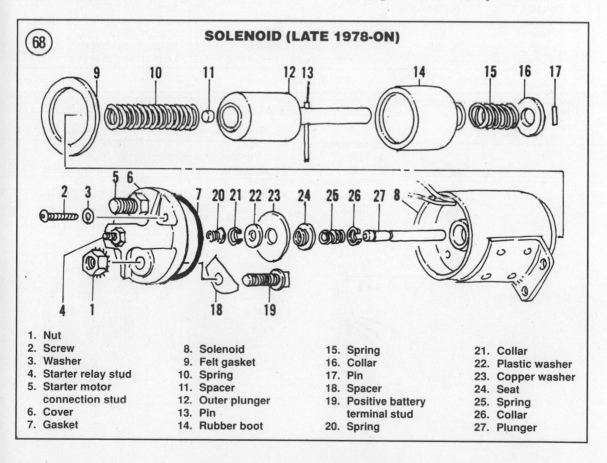

1. Nut
2. Screw
3. Washer
4. Starter relay stud
5. Starter motor connection stud
6. Cover
7. Gasket
8. Solenoid
9. Felt gasket
10. Spring
11. Spacer
12. Outer plunger
13. Pin
14. Rubber boot
15. Spring
16. Collar
17. Pin
18. Spacer
19. Positive battery terminal stud
20. Spring
21. Collar
22. Plastic washer
23. Copper washer
24. Seat
25. Spring
26. Collar
27. Plunger

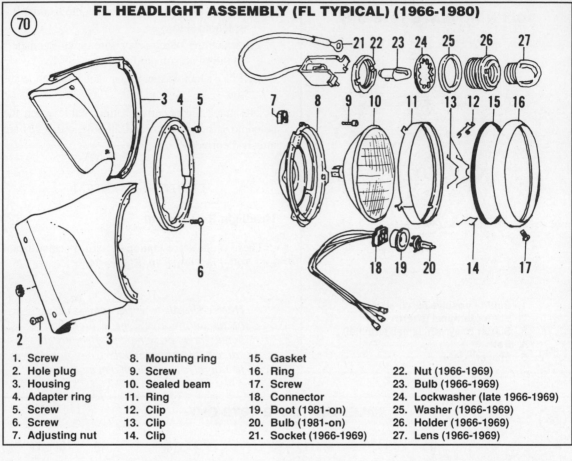

**FL HEADLIGHT ASSEMBLY (FL TYPICAL) (1966-1980)**

| | | |
|---|---|---|
| 1. Screw | 8. Mounting ring | 15. Gasket |
| 2. Hole plug | 9. Screw | 16. Ring |
| 3. Housing | 10. Sealed beam | 17. Screw |
| 4. Adapter ring | 11. Ring | 18. Connector |
| 5. Screw | 12. Clip | 19. Boot (1981-on) |
| 6. Screw | 13. Clip | 20. Bulb (1981-on) |
| 7. Adjusting nut | 14. Clip | 21. Socket (1966-1969) |

22. Nut (1966-1969)
23. Bulb (1966-1969)
24. Lockwasher (late 1966-1969)
25. Washer (1966-1969)
26. Holder (1966-1969)
27. Lens (1966-1969)

**HEADLIGHT ASSEMBLY (FX TYPICAL)**

| | | |
|---|---|---|
| 1. Bolt | 8. Cap | 15. Terminal boot |
| 2. Lockwasher | 9. Clip | 16. Connector |
| 3. Plug | 10. Terminal plate assembly | 17. Bulb |
| 4. Nut | 11. Housing | 18. Molding ring |
| 5. Lockwasher | 12. Clip | 19. Screw |
| 6. Washer | 13. Lockwasher | 20. Mounting ring |
| 7. Bracket | 14. Screw | 21. Molding ring |

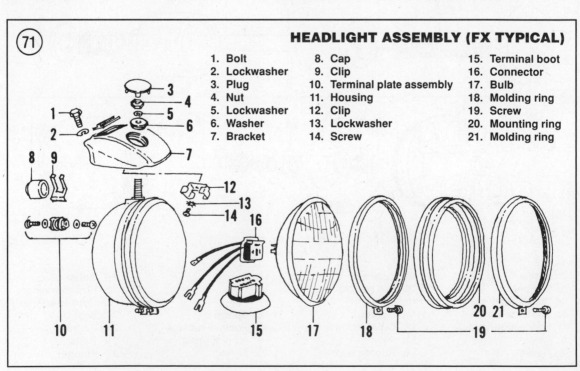

*bile, damage to the battery and/or alternator may occur.*

### 1966-early 1978 FX and 1966-1980 FL

Refer to **Figure 70** or **Figure 71** for this procedure.

1. Remove the headlight door retaining screw enough to allow removal of the headlight door.

2. Remove the retaining ring-to-headlight door screws and remove the retaining ring.

3. Disconnect the headlight connector block and remove the headlight.

4. Installation is the reverse of these steps. Clean the headlight connector with electrical contact cleaner if required.

5. Check and adjust the headlight beam as described in this chapter.

### Late 1978-on FX

Refer to **Figure 71** or **Figure 72** for this procedure.

1. Remove the outer headlight clamp screw and remove the clamp.

2. Carefully remove the headlight from the rubber mounting ring.

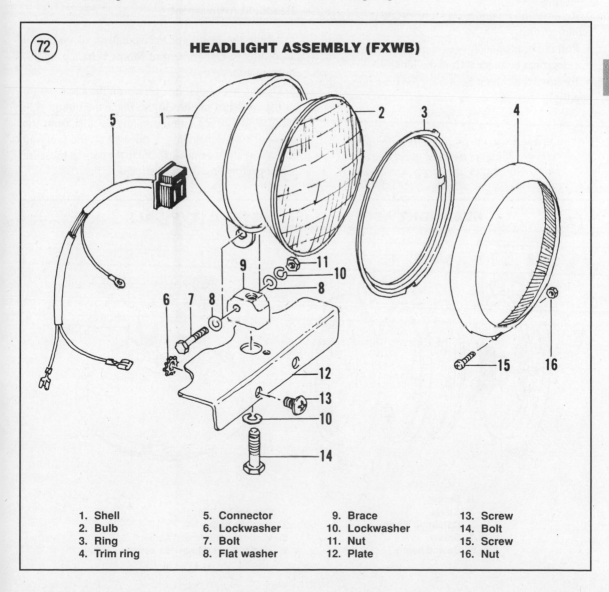

**(72) HEADLIGHT ASSEMBLY (FXWB)**

| | | | |
|---|---|---|---|
| 1. Shell | 5. Connector | 9. Brace | 13. Screw |
| 2. Bulb | 6. Lockwasher | 10. Lockwasher | 14. Bolt |
| 3. Ring | 7. Bolt | 11. Nut | 15. Screw |
| 4. Trim ring | 8. Flat washer | 12. Plate | 16. Nut |

3. Disconnect the connector from the headlight and remove the headlight.

4. Install by reversing these removal steps.

5. Check and adjust the headlight beam as described in this chapter.

### Headlight Replacement (1981-on FL)

These models use a sealed beam or quartz halogen bulb. Special handling of the quartz halogen bulb is required as described in this procedure.

Refer to **Figure 70** and **Figure 73** for this procedure.

1. Remove the headlight door screw and remove the headlight door.

2. Remove the retaining ring screws and remove the retaining ring.

3. Pull the headlight lens out slightly and disconnect the electrical connector from the lens.

4. Remove the rubber boot from the backside of the headlight lens.

*CAUTION*
*Do not touch the replaceable bulb glass with your fingers because of oil on your skin. Any traces of oil on the quartz halogen bulb will drastically reduce bulb life. Clean any traces of oil from the bulb with a cloth moistened in alcohol.*

*WARNING*
*Quartz halogen bulbs contain halogen gas. Wear eye safety glasses when handling this type of bulb.*

5. Compress the bulb wire clip together and remove the bulb.

6. Install by reversing these removal steps.

7. Adjust the headlight as described in this chapter.

### Headlight Adjustment

Adjust the headlight horizontally and vertically according to Department of Motor Vehicles regulations in your area.

1. Draw a horizontal line on a wall the same height as the center of the headlight. Place the motorcycle on a level surface approximately 25 feet from the wall (test pattern). Have a rider sit on the seat and make sure the tires are inflated to the correct pressure when performing this adjustment.

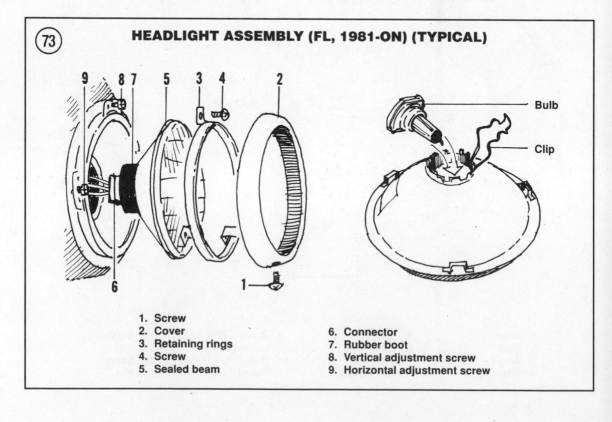

**⑦₃ HEADLIGHT ASSEMBLY (FL, 1981-ON) (TYPICAL)**

Bulb

Clip

1. Screw
2. Cover
3. Retaining rings
4. Screw
5. Sealed beam
6. Connector
7. Rubber boot
8. Vertical adjustment screw
9. Horizontal adjustment screw

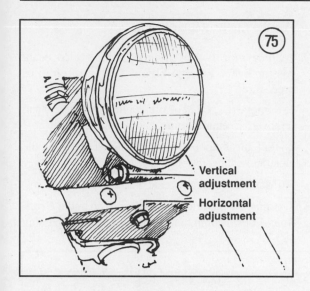

Vertical
adjustment

Horizontal
adjustment

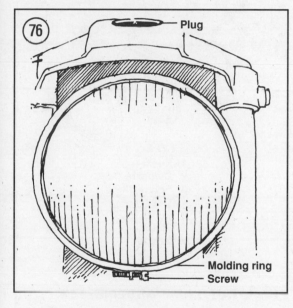

Plug

Molding ring
Screw

2. Aim the headlight at the wall and turn on the headlight. Switch the headlight to the high beam.

3. The top of the main beam should be even with, but not higher than the horizontal line. If the beam is incorrect, adjust as follows.

4A. *1966-on FL and 1971-early 1978 FX:*

    a. Remove the headlight door.

    b. Adjust the headlight by turning the horizontal or vertical adjustment screws in or out as required. **Figure 74** shows typical horizontal and vertical adjustment screws.

4B. *Late 1978-on FXWG and FXDG:* Referring to **Figure 75**, perform the following:

    a. Loosen the horizontal adjustment bolt underneath the front fork steering stem and turn the headlight housing side-to-side as required. Tighten the bolt.

    b. Loosen the vertical adjustment bolt and turn the headlight housing up and down as required. Tighten the bolt.

4C. *Late 1978-on FX (except FXWG and FXDF):* Referring to **Figure 76**, perform the following:

    a. Remove the snap plug on top of the headlight housing.

    b. Loosen the clamp nut behind the headlight bracket.

    c. As required, turn the headlight assembly up and down or side to side.

    d. Tighten the clamp nut and reinstall the snap plug.

5. Recheck the adjustment.

**7**

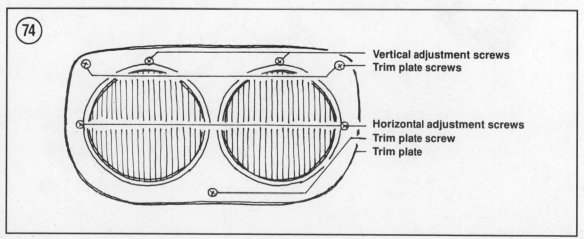

Vertical adjustment screws
Trim plate screws

Horizontal adjustment screws
Trim plate screw
Trim plate

## Taillight/Brake Light Replacement

Refer to **Figure 77** (typical) for this procedure.

1. Remove the rear lens.
2. Push in on the bulb and remove it.
3. Replace the bulb and install the lens.

## Turn Signal Light Replacement

1. Remove the turn signal lens.
2. Push in on the bulb and remove it.
3. Replace the bulb and install the lens.

## Passing Lights Replacement

Some models are equipped with sealed beam passing lights. Refer to **Figure 78** (typical) for this procedure.

1. Remove the sealed beam retaining ring screw and remove the retaining ring.

2. Pull the bulb partway out and disconnect the electrical connector.

3. Installation is the reverse of these steps.

4. Adjust the passing lamp as described under *Headlight Adjustment* in this chapter. Loosen the passing

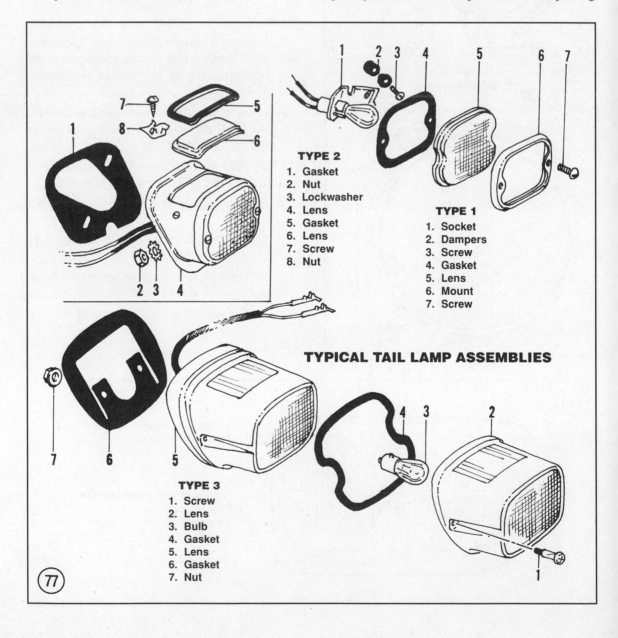

**TYPE 2**
1. Gasket
2. Nut
3. Lockwasher
4. Lens
5. Gasket
6. Lens
7. Screw
8. Nut

**TYPE 1**
1. Socket
2. Dampers
3. Screw
4. Gasket
5. Lens
6. Mount
7. Screw

**TYPICAL TAIL LAMP ASSEMBLIES**

**TYPE 3**
1. Screw
2. Lens
3. Bulb
4. Gasket
5. Lens
6. Gasket
7. Nut

⑦⑦

lamp mounting bolts or screws to adjust the light position.

## SWITCHES

Switches can be tested for continuity with an ohmmeter (see Chapter One) at the switch connector plug by operating the switch in each of its operating positions and comparing results with the switch operation. When testing switches, do the following:

a. First check the main fuse or circuit breaker.

b. Check the battery as described in Chapter Three and bring the battery to the correct state of charge, if required.

c. When separating 2 connectors, pull on the connector housings and not the wires.

d. After locating a defective circuit, check the connectors to make sure they are clean and properly connected. Check all wires going into a connector housing to make sure each wire is properly positioned and that the wire end is not loose.

e. To connect connectors properly, push them together until they click into place.

### Neutral Switch Removal/Installation

The neutral switch is installed in the top of the shifter cover (**Figure 79**). To replace the switch, disconnect the electrical connector at the switch. Unscrew the switch with a socket, then remove the switch (and shims, if used). Installation is the reverse of removal. If shims are used, be sure to install the same number as were removed.

### Handlebar Switch Replacement (1966-1971)

Refer to **Figure 80** or **Figure 81** for this procedure. This switch is used to operate the horn and starter.

*Early type*

1. Remove the switch housing screws and remove the housing.

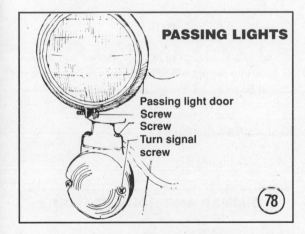

PASSING LIGHTS

Passing light door
Screw
Screw
Turn signal
screw

78

79

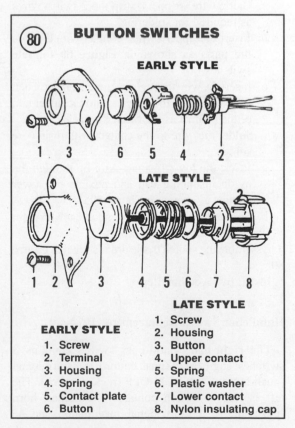

80    BUTTON SWITCHES

EARLY STYLE

1   3        6   5   4   2

LATE STYLE

1   2        3       4   5   6   7   8

**EARLY STYLE**
1. Screw
2. Terminal
3. Housing
4. Spring
5. Contact plate
6. Button

**LATE STYLE**
1. Screw
2. Housing
3. Button
4. Upper contact
5. Spring
6. Plastic washer
7. Lower contact
8. Nylon insulating cap

7

2. With a screwdriver, carefully remove the switch terminal from the housing.

3. Remove the spring, contact plate and button from the housing.

4. To replace the switch terminal, perform the following:

a. Cut the terminal wires at the terminal.

b. Using wire strippers, strip the 2 switch wires as required for resoldering.

c. Solder the 2 switch wires to the new switch terminal with rosin core solder.

5. Installation is the reverse of these steps.

### Late type

1. Remove the switch housing screws and separate the housing.

2. Disassemble the switch in the order shown in **Figure 80** for late switches.

3. To replace the switch wires or contacts, perform the following:

a. Cut the switch wires at the upper and lower switch contacts.

b. Remove the wires or contacts as required.

c. Using wire strippers, strip the 2 switch wires as required for soldering.

d. Insert the upper switch contact wire through the parts as shown in **Figure 80** for late switches.

e. Insert the wire through the upper contact. Then fray the end of the wire around the upper end of the contact as shown in **Figure 80**.

f. Solder the wire to the contact with rosin core solder.

g. Insert the lower contact wire up through the nylon insulating cap and next to the lower contact as shown in **Figure 80**.

h. Solder the wire to the lower contact with rosin core solder.

4. Assemble the switch in the order shown in **Figure 80**.

5. Install by reversing Step 1.

### Handlebar Switch Replacement (1972-on)

The right-hand handlebar switch contains 3 switches: engine start (push button), right turn signal (push button) and RUN-OFF (rocker switch). The left handlebar switch contains 3 switches: horn (push button), left turn signal (push button) and the headlight HI-LO beam (rocker switch). The individual switches (both push and rocker) can be replaced.

Refer to **Figure 82** (1972-1981) or **Figure 83** (1982-on).

1. Remove the screws securing the switch housing to the handlebar. Then carefully separate the switch housing assembly to gain access to the defective switch.

2. Disconnect the switch wire at the electrical connection.

3A. *1972-1981:*

a. To remove a rocker type switch, remove the switch screw and remove the switch.

b. To remove a push button switch, remove the screw and retainer. Then pull the switch out of the housing.

3B. *1982-on:* Replace the defective switch by removing the holding screw. Then pull the switch out of the housing.

4. Installation is the reverse of these steps. Make sure to route the wires to prevent damaging them when tightening the switch housing.

### HORN

The horn is an important safety device and should be kept in working order. If the horn is damaged, it should be replaced immediately.

### Removal/Installation

Refer to **Figure 84**.

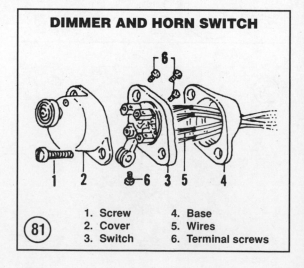

**DIMMER AND HORN SWITCH**

| | |
|---|---|
| 1. Screw | 4. Base |
| 2. Cover | 5. Wires |
| 3. Switch | 6. Terminal screws |

(81)

1. Remove all components as necessary to gain access to the horn.

2. Disconnect the horn electrical connector.

3. Remove the horn mounting attachments and remove the horn.

4. Installation is the reverse of these steps.

## Adjustment

The horn on these models cannot be disassembled. If the horn is damaged, it must be replaced. Refer to **Figure 84** for this procedure.

1. If the horn fails to blow properly, check for broken or frayed horn wires. If these are okay, perform the following.

2. While pushing the horn button, turn the contact point adjustment screw clockwise until the horn clicks one time. The adjustment screw is located in the back of the horn.

3. Then turn the adjustment screw counterclockwise while pushing the horn button until the best horn tone is obtained.

4. If the horn fails to operate after performing Steps 1-3, replace the horn.

## FUSES AND CIRCUIT BREAKERS

Each model uses either fuses or circuit breakers to protect the electrical circuit. Refer to the wiring diagram at the back of the book for your model for the type of protection used.

Whenever a failure occurs in any part of the electrical system, check the fuse box to see if a fuse has blown. On models equipped with circuit breakers, each circuit breaker is self-resetting and will automatically return power to the circuit when the electrical fault is found and corrected.

*CAUTION*
*If the electrical fault on circuit breaker equipped models is not found and corrected, the breakers will cycle on and off continuously. This will cause the motorcycle to run erratically and eventually the battery will lose its charge.*

If a fuse has blown, it will be evident by blackening of the fuse or by a break in the metal link in the fuse.

Usually the trouble can be traced to a short circuit in the wiring connected to the blown fuse or circuit

7

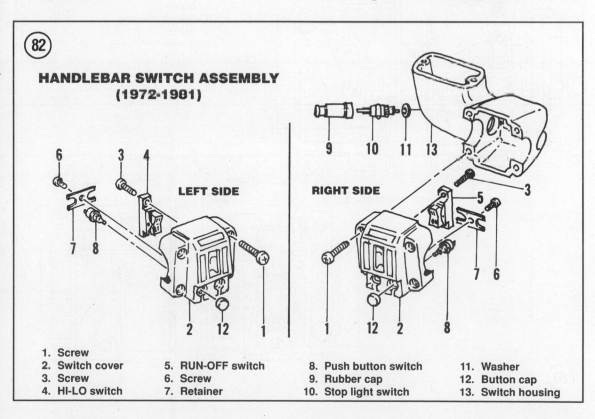

**(82)**

**HANDLEBAR SWITCH ASSEMBLY (1972-1981)**

LEFT SIDE

RIGHT SIDE

| | |
|---|---|
| 1. Screw | 8. Push button switch |
| 2. Switch cover | 9. Rubber cap |
| 3. Screw | 10. Stop light switch |
| 4. HI-LO switch | 11. Washer |
| 5. RUN-OFF switch | 12. Button cap |
| 6. Screw | 13. Switch housing |
| 7. Retainer | |

breaker. This may be caused by worn-through insulation or by a wire which has worked loose and shorted to ground. Occasionally, the electrical overload which causes the fuse to blow may occur in a switch or motor. By following the wiring diagrams at the end of the book, the circuits protected by each fuse can be determined.

A blown fuse or tripped circuit breaker should be treated as more than a minor annoyance; it should serve also as a warning that something is wrong in the electrical system. Before replacing a fuse, determine what caused it to blow and then correct the trouble.

> *WARNING*
> *Never replace a fuse with one of a higher amperage than that of the original fuse. Never use metal foil or other metallic material to bridge fuse terminals. Failure to follow these basic rules could result in heat or fire damage to major parts or loss of the entire vehicle.*

Replace a defective fuse by pulling it out of its holder and snapping a new one in place. To replace a circuit breaker, disconnect the wires and remove it. Reverse to install.

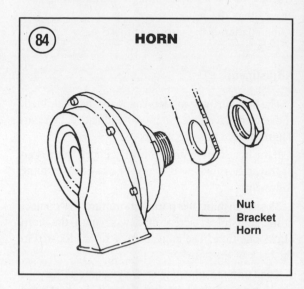

(84) **HORN**

Nut
Bracket
Horn

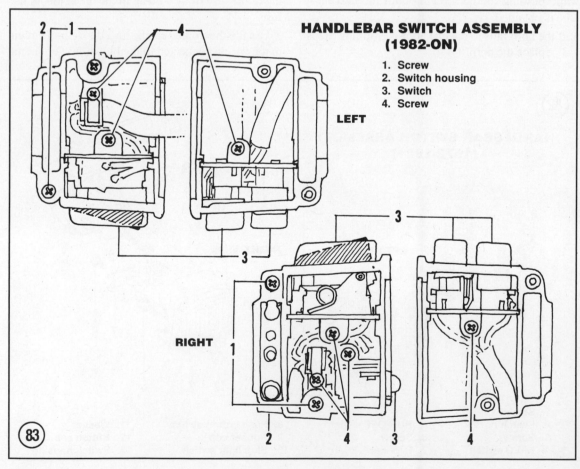

**HANDLEBAR SWITCH ASSEMBLY**
**(1982-ON)**

1. Screw
2. Switch housing
3. Switch
4. Screw

LEFT

RIGHT

(83)

## Table 1 ELECTRICAL SPECIFICATIONS (1970-EARLY 1978)

| | |
|---|---|
| **1970-1975** | |
| Stator output voltage @ 2,000 rpm | |
| White-to-white connection | 50-100 volts |
| Blue-to-red connection | 75-125 volts |
| Stator resistance check | |
| White-to-white connection | 0.3-1.0 ohms* |
| White-to-blue connection | See text |
| Blue-to-red connection | 1.5-2.0 ohms* |
| Any pin to module base | 100 K ohms (minimum)** |
| **1976-early 1978** | |
| Charging checks | See text |
| **Ignition coil resistance (all)** | |
| Primary | 4.7-5.7 ohms |
| Secondary | 16,000-20,000 ohms |
| **Starter brush length** | |
| Prestolite | 1/4 in. (6.35 mm) |
| Hitachi | 7/16 in. (11.11 mm) |

\* Ohm reading of 0 indicates short circuit.
\*\* Any reading less than 100 K ohms indicates short circuit.

## Table 2 ELECTRICAL SPECIFICATIONS (LATE 1978-1984)

| | |
|---|---|
| **Alternator** | |
| AC voltage output | 19-26 volt per 1,000 rpm |
| Stator coil resistance | 2-4 ohms |
| **Ignition coil resistance** | |
| Late 1978-1979 | |
| Primary | 4.7-5.7 ohms |
| Secondary | 16,500-20,000 ohms |
| 1980-on | |
| Primary | 3.3-3.7 ohms |
| Secondary | 16,500-19,500 ohms |
| **Regulator** | |
| Voltage output @ 3,600 rpm | 13.8-15 volts |
| Amps @ 3,600 rpm | |
| Late 1978-1980 | 16 amps |
| 1981 | 17.8 amps |
| 1982-on | 22 amps |
| **Starter brush length (minimum)** | |
| Prestolite | 1/4 in. (6.35 mm) |
| Hitachi | 7/16 in. (11.11 mm) |

## Table 3 VOLTAGE REGULATOR RANGE

| Delco-Remy Part No. | Voltage Regulator Setting |
|---|---|
| 1118 388 | 7.5 |
| 11119 187C | 7.2-7.5 |
| 11119 187D | 7.2-7.5 |
| 1118 307 | 7.0[1] |
| 1118 307 | 7.4 |
| 1118 794 | 7.0[1] |
| 1118 794 | 7.4[2] |
| 1118 995 | 7.0-7.3 |
| 1118 989 | 6.5-6.8 |

1. Models 125-165 generators.
2. Models 58 and 61 generators.

### Table 4 CURRENT REGULATOR SPECIFICATIONS

| Delco-Remy Part No. | Current Regulator Setting |
|---|---|
| 1118 388 | 18 amperes |
| 1119 187C | 13.5-16.5 amperes |
| 1119 187D | 13.5-16.5 amperes |
| 11119 614 | 9.0-11.0 amperes |

### Table 5 MODULE RESISTANCE CHECKS

| Positive Connection | Negative Connection | Indication |
|---|---|---|
| First white wire | Module base | Infinity |
| Module base | First white wire | 3-15 ohms |
| Second white wire | Module base | Infinity |
| Module base | Second white wire | 3-15 ohms |
| Blue | Black | 3-15 ohms |
| Black | Blue | Infinity |
| Red | Module base | Infinity |
| Module Base | Red | Infinity |

### Table 6 STATOR RESISTANCE CHECKS

| Meter Connections | Resistance |
|---|---|
| White-White | 0.3 to 1.0 ohm |
| 1st and 2nd White-Blue | Both meter indications must be the same |
| Red-Blue | 1.5-2.0 ohms |
| Any lead-Ground | Infinity |

# CHAPTER EIGHT

# FRONT SUSPENSION AND STEERING

This chapter discusses service operations on suspension components, steering, wheels and related items. Tightening torques for 1970-1984 models are listed in **Table 1** (1970-early 1978) and **Table 2** (late 1978-1984). **Tables 1** and **2** are at the end of the chapter. For 1966-1969 models, use the general torque specifications listed in Chapter One.

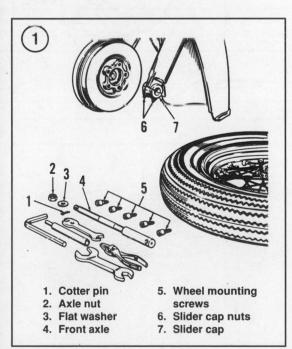

1. **Cotter pin**
2. **Axle nut**
3. **Flat washer**
4. **Front axle**
5. **Wheel mounting screws**
6. **Slider cap nuts**
7. **Slider cap**

## FRONT WHEEL

### Removal/Installation (1966-1971 FL Models With Drum Brake)

Refer to **Figure 1** (typical) for this procedure.

1. Support the bike so that the front wheel clears the ground.

2. Remove and discard the cotter pin.

3. Loosen the axle nut and remove it and the washer.

4. Remove the wheel mounting screws (5, **Figure 1**).

5. Loosen the fork slider cap nuts (6, **Figure 1**). Do not remove them.

*NOTE*
*Apply the front brake when performing Step 6 to hold the brake drum secure as the front wheel is removed.*

6. Withdraw the axle and remove the front wheel.

7. Installation is the reverse of these steps. Note the following.

8. Clean the brake drum and wheel mating surfaces of all grease and road debris. If the mating surfaces are not clean, the wheel may not run true.

9. Tighten the front wheel mounting fasteners in the following order:

a. Wheel mounting screws to 35 ft.-lb. (48.3 N•m). See 5, **Figure 1**. Tighten screws in a criss-cross pattern.

b. Axle nut to 50 ft.-lb. (69 N•m). See 2, **Figure 1**.

c. Fork slider cap nuts to 11 ft.-lb. (15.2 N•m). See 6, **Figure 1**.

10. Install a new cotter pin and bend it over completely. Never reuse an old cotter pin as it may break and fall out.

11. With the front wheel still clear of the ground, spin the front wheel to make sure the wheel spins true.

12. Adjust the front brake as described in Chapter Three.

### Removal/Installation (1972-early 1978 FL Models With Disc Brake)

1. Support the bike so that the front wheel clears the ground.

2. Remove and discard the cotter pin (if so equipped).

3. Remove the axle nut and lockwasher (if so equipped).

4. Loosen the fork slider cap nuts (**Figure 2**).

5. Push the axle out with a punch or drift.

6. Slide the wheel out of the brake caliper and remove it.

*NOTE*
*Insert a piece of wood in the caliper in place of the disc. That way, if the brake lever is inadvertently squeezed, the piston will not be forced out of the cylinder. If this does happen, the caliper might have to be disassembled to reseat the piston and the system will have to be bled. By using the wood, bleeding the brake is not necessary when installing the wheel.*

*CAUTION*
*Do not set the wheel down on the disc surface, as it may be scratched or warped. Either lean the wheel against a wall or set it on 2 blocks of wood (**Figure 3**).*

7. Installation is the reverse of these steps. Note the following.

8. *Carefully* insert the disc between the pads when installing the front wheel.

9. Tighten the slider cap nuts to 11 ft.-lb. (15.2 N•m). Tighten the axle nut to 50 ft.-lb. (69 N•m). Make sure gap between the slider cap and fork slider is equal on both sides.

### Removal/Installation (1970-1972 FX Models)

Refer to **Figure 4** (typical) for this procedure.

1. Support the bike so that the front wheel clears the ground.

2. Remove the brake clevis pin (1, **Figure 4**).

3. Loosen the front brake adjuster and disconnect the brake cable at the front hub.

4. Remove the axle nut and lockwasher.

5. Remove the brake anchor and brake shoe centering cap screw (4, **Figure 4**) and lockwasher.

6. Loosen the axle pinch bolt (6, **Figure 4**).

7. Remove the axle by tapping it out with an aluminum drift.

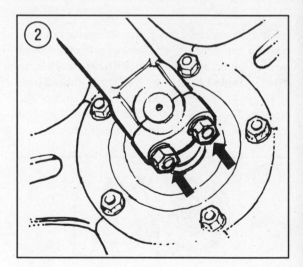

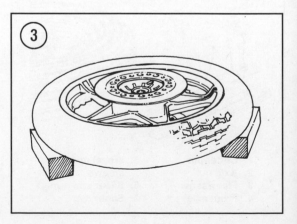

8. Remove the front wheel and brake drum as an assembly.

9. If necessary, pull the brake drum out of the wheel assembly.

10. Installation is the reverse of these steps. Note the following.

11. Grease the wheel bearing and pack one ounce of grease into the wheel hub.

12. Install the axle lockwasher and axle nut. Tighten the axle nut to 50 ft.-lb. (69 N•m). Tighten the axle pinch bolt securely.

13. Adjust the front brake as described in Chapter Three.

### Removal/Installation (1973-early 1978 FX)

1. Support the bike so that the front wheel clears the ground.

2. On 1974-1978 models, remove the brake caliper(s) mounting bolt, washers and locknut. Lift and support the caliper(s) with a bungee cord to prevent damage to the brake hose.

3. Remove the axle nut and lockwasher. On 1974-1978 models, remove the flat washer also.

4. Loosen the fork slider cap nuts (**Figure 2**).

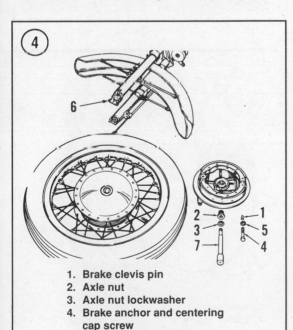

1. **Brake clevis pin**
2. **Axle nut**
3. **Axle nut lockwasher**
4. **Brake anchor and centering cap screw**
5. **Brake anchor and centering cap screw lockwasher**
6. **Front axle pinch bolt**
7. **Front wheel axle**

5. Push the axle out with an aluminum drift.

6. Slide the wheel out of the brake caliper (1973 models) and remove partway.

7. Pull the speedometer drive out of the front wheel.

8. Remove the front wheel.

*NOTE*
*Insert a piece of wood in the caliper(s) in place of the disc. That way, if the brake lever is inadvertently squeezed, the piston will not be forced out of the cylinder. If this does happen, the caliper might have to be disassembled to reseat the piston and the system will have to be bled. By using the wood, bleeding the brake is not necessary when installing the wheel.*

*CAUTION*
*Do not set the wheel down on the disc surface, as it may be scratched or warped. Either lean the wheel against a wall or set it on 2 wood blocks (**Figure 3**).*

9. Installation is the reverse of these steps. Note the following.

10. Engage the speedometer drive with the front wheel engagement dogs.

11A. *1973:* Carefully insert the disc between the pads when installing the front wheel.

11B. *1974-1978:* Install the front wheel and axle. Then install the brake caliper(s) over the brake disc. Install the caliper mounting bolt, washers and locknut. Tighten the bolts as described in Chapter Ten.

12. Install the axle washers.

13. Tighten the axle nut to 50 ft.-lb. (69 N•m). Tighten the slider cap nuts to 11 ft.-lb. (15.2 N•m). Make sure the gap between the slider cap and fork slider is equal on both sides.

### Removal/ Installation (Late 1978-1982 FL Models)

Refer to **Figure 5** (typical) for this procedure.

1. Support the bike so that the front wheel clears the ground.

2. Remove the axle nut and lockwasher.

3. Loosen the fork slider cap nuts.

4. Push the axle out with an aluminum drift. Remove the spacer, if so equipped.

5. Slide the wheel out of the brake caliper and remove partway. Then pull the speedometer drive out of the front wheel.

6. Remove the front wheel.

7. Remove the hub cap from the wheel.

*NOTE*
*Insert a piece of wood in the caliper(s) in place of the disc. That way, if the brake lever is inadvertently squeezed, the piston will not be forced out of the cylinder. If this does happen, the caliper might have to be disassembled to reseat the piston and the system will have to be bled. By using the wood, bleeding the brake is not necessary when installing the wheel.*

*CAUTION*
*Do not set the wheel down on the disc surface, as it may be scratched or warped. Either lean the wheel against a wall or set it on 2 wood blocks (Figure 3).*

8. Installation is the reverse of these steps. Note the following.

9. Install the hub cap onto the front wheel.

10. Engage the speedometer drive with the front wheel engagement dogs.

11. *Carefully* insert the disc between the pads when installing the front wheel.

12. Tighten the axle nut to 50 ft.-lb. (69 N•m). Tighten the slider cap nuts to 11 ft.-lb. (15.2 N•m). Make sure the gap between the slider cap and fork slider is equal on both sides.

13. Perform the *Front Wheel Bearing End Play Check* as described in this chapter.

**Removal/Installation (Late 1978-on FX Models Except 1984 FXE/FXWG)**

Refer to **Figure 6** (typical) for this procedure.

1. Support the bike so that the front wheel clears the ground.

2. Remove the brake caliper(s) mounting bolts, washers and locknut. Lift the caliper off and secure the brake caliper(s) with a bungee cord to prevent damage to the brake hose.

*NOTE*
*Insert a piece of wood in the caliper(s) in place of the disc. That way, if the*

*brake lever is inadvertently squeezed, the piston will not be forced out of the cylinder. If this does happen, the caliper might have to be disassembled to reseat the piston and the system will have to be bled. By using the wood, bleeding the brake is not necessary when installing the wheel.*

3. Remove the axle nut, lockwasher and washer.

4. Loosen the fork slider cap nuts.

5. Push the axle out with an aluminum drift.

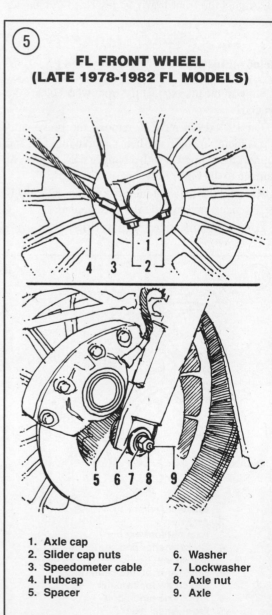

**⑤**

**FL FRONT WHEEL
(LATE 1978-1982 FL MODELS)**

1. **Axle cap**
2. **Slider cap nuts**
3. **Speedometer cable**
4. **Hubcap**
5. **Spacer**
6. **Washer**
7. **Lockwasher**
8. **Axle nut**
9. **Axle**

6. Pull the speedometer drive out of the front wheel and remove the front wheel.

*CAUTION*
*Do not set the wheel down on the disc surface, as it may be scratched or warped. Either lean the wheel against a wall or set it on 2 wood blocks (***Figure 3***).*

7. Remove the front axle spacers from the wheel.

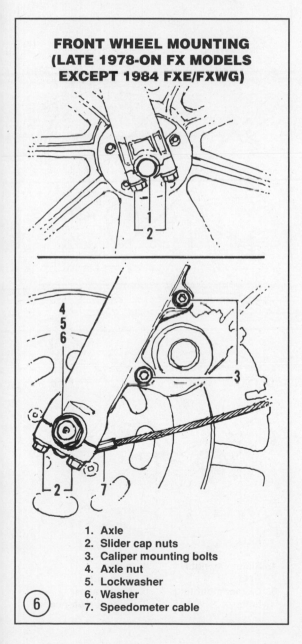

**FRONT WHEEL MOUNTING (LATE 1978-ON FX MODELS EXCEPT 1984 FXE/FXWG)**

1. **Axle**
2. **Slider cap nuts**
3. **Caliper mounting bolts**
4. **Axle nut**
5. **Lockwasher**
6. **Washer**
7. **Speedometer cable**

⑥

*NOTE*
*On late 1978-1983 FXWG models, a spacer is used between the front wheel and the left fork leg. On 1983 FXWG models, a spacer with a groove on the outside diameter is used between the front wheel and the right fork leg. The spacer with the groove on 1983 models must be installed on the right-hand side.*

8. Installation is the reverse of these steps. Note the following.
9. Engage the speedometer drive with the front wheel engagement dogs, if so equipped.
10. Install the front wheel and axle. Tighten the axle nut to 50 ft.-lb. (69 N•m). Tighten the fork slider cap nuts to 11 ft.-lb. (15.2 N•m).

*NOTE*
*New caliper mounting bolts must be installed when reinstalling the brake caliper(s) as described in Step 11.*

11. Install the brake caliper(s) over the brake disc. Install the caliper mounting bolts, washers and nuts. Tighten the bolts to 25-30 ft.-lb. (34.5-41.4 N•m).
12. Perform the *Front Wheel Bearing End Play Check* as described in this chapter.

**Removal/Installation (1984 FXWG and FXE)**

Refer to **Figure 7** (typical) for this procedure.
1. Support the bike so that the front wheel clears the ground.
2. Remove the brake caliper mounting bolts and lift the caliper(s) away from the brake disc. Support the caliper(s) with a bungee cord so that the weight of the caliper is not supported by the brake line.

*NOTE*
*Insert a piece of wood in the caliper(s) in place of the disc. That way, if the brake lever is inadvertently squeezed, the piston will not be forced out of the cylinder. If this does happen, the caliper might have to be disassembled to reseat the piston and the system will have to be bled. By using the wood, bleeding the brake is not necessary when installing the wheel.*

3. Remove the axle nut and lockwasher.
4. Loosen the fork slider cap nuts.

**8**

5. Push the axle out with a punch or drift.

6. Pull the wheel away from the fork sliders slightly and remove the speedometer drive gear from the wheel.

7. Remove the axle spacer (if so equipped). Remove the wheel.

> *CAUTION*
> *Do not set the wheel down on the disc surface, as it may be scratched or warped. Either lean the wheel against a wall or set it on 2 wood blocks (**Figure 3**).*

8. Installation is the reverse of these steps. Note the following.

9. Align the notches in the gearcase with the speedometer drive dogs.

10. Tighten the slider cap nut on the axle nut side to 11 ft.-lb. (15.2 N•m). Tighten the axle nut to 50 ft.-lb. (69 N•m). Then loosen the slider cap nuts and tighten both slider cap nuts to 11 ft.-lb. (15.2 N•m). Make sure gap between the slider cap and fork slider is equal on both sides.

11. Perform the *Front Wheel Bearing End Play Check* as described in this chapter.

12. *Carefully* insert the disc between the pads when installing the brake caliper(s). Tighten the brake caliper bolts to specifications in Chapter Ten.

## Inspection (All Models)

1. Remove any corrosion on the front axle with a piece of fine emery cloth.

2. Visually check the rims for cracks, fractures, dents or bends. On spoked wheels, the rims can be replaced by a Harley-Davidson dealer. On cast wheels, the wheel must be replaced if damaged.

> *WARNING*
> *Do not try to repair any damage to cast wheels as it will result in an unsafe riding condition.*

3. Measure the axial runout (side play) and radial runout (up-and-down play) of the wheel with a dial indicator as shown in **Figure 8**. The maximum allowable axial and radial runout is 0.031 in. (0.78 mm). If the runout exceeds this dimension, check the wheel bearings. Some of this condition can be corrected on spoke wheels as described in this chapter. If the wheel bearings are in good condition on cast

wheels and no other cause can be found, the wheel will have to be replaced as it cannot be serviced.

**Front Wheel Bearing End Play Check**

After tightening the axle nut, check the wheel bearing end play as follows.

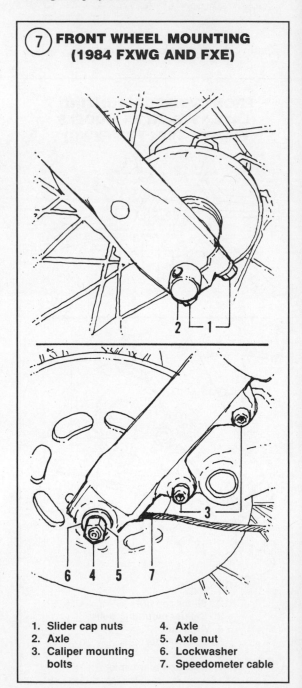

**(7) FRONT WHEEL MOUNTING (1984 FXWG AND FXE)**

| | |
|---|---|
| 1. Slider cap nuts | 4. Axle |
| 2. Axle | 5. Axle nut |
| 3. Caliper mounting bolts | 6. Lockwasher |
| | 7. Speedometer cable |

1. Mount a dial indicator so that the plunger contacts the end of the axle.

2. Grasp the front wheel and try to move it back and forth and measure the axle end play.

3. The standard end play is 0.004-0.018 in. (0.10-0.46 mm). If the end play does not fall within this range, replace the axle spacer with a longer or shorter one. Spacers of various lengths are available from Harley-Davidson dealers.

## FRONT HUB

### Disassembly/Inspection/Assembly (1966 Roller Bearing Hubs)

Refer to **Figure 9** for this procedure.

*NOTE*
*Label all parts during removal so that they can be reinstalled in their original positions. Plastic bags or small boxes work well.*

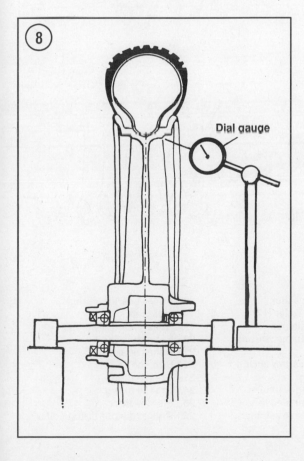

Dial gauge

1. Grasp the wheel and check for excessive side play and for excessive radial (up-and-down) play.

2. Remove the front wheel as described in this chapter.

3. Remove the thrust bearing cover screws and lock-washers.

4. Remove the following parts in order:
   a. Thrust cover (3, **Figure 9**).
   b. Cork grease retainer (4).
   c. Thrust bearing housing (5).
   d. Gasket (6).
   e. Thrust bearing adjusting shims (7).
   f. Thrust washer (8).
   g. Thrust bearing sleeve (9).
   h. Thrust washer (10).

5. Hold the hub assembly over a large pan or box and remove the 12 bearing rollers (11) and the roller retainer (12).

6. Remove the roller retainer thrust washer (13).

7. Working from the opposite side of the hub, remove the following parts in order:
   a. Spring lock ring (14).
   b. Retaining washer (15).
   c. Hub inner sleeve (16).
   d. Cork grease retainer (17).
   e. Spring lock ring (18).
   f. Roller bearing washer (19).

8. Remove the opposite side roller retainer (21) and bearing rollers (20) (14 total) as described in Step 5.

9. Clean all parts in solvent and allow to dry thoroughly.

10. Inspect all parts for wear or damage.

11. If excessive sideplay was detected at the hub (Step 1), one or more thrust bearing adjusting shim(s) can be added. Perform the following:
   a. Assemble the thrust bearing outer cover (3), thrust bearing housing (5) and thrust bearing housing gasket (6).
   b. Install all of the thrust bearing adjusting shims (7) onto the thrust bearing sleeve (9) and assemble onto the thrust bearing outer cover (3).
   c. Check clearance with a feeler gauge. Clearance should be 0.005-0.007 in. (0.13-0.18 mm). Note that the cork grease retainer (4) is not installed when adjusting side play.
   d. Add shims as required to obtain correct clearance. Adjustment shims are available from Harley-Davidson dealers. Each shim is 0.002 in. (0.05 mm) thick.

12. If excessive radial (up and down) play was detected at the hub (Step 1), oversize bearing rollers (11 and 20, **Figure 9**) can be installed. Rollers are available from 0.001 in. (0.03 mm) undersize in steps of 0.0002 in. (0.005 mm). Select rollers to provide 0.0010 to 0.0015 in. (0.025-0.038 mm) radial clearance.

13. Reverse these steps to assemble the hub. Note the following.

14. Closed side of each bearing retainer (12 and 21, **Figure 9**) faces toward the center of the hub.

15. If the hub grease fitting (24) was removed, make sure to reinstall it with a new washer (25). Failure to install a washer will cause the grease fitting to crimp the adjusting shims.

16. Make sure all lock rings seat completely in the hub grooves.

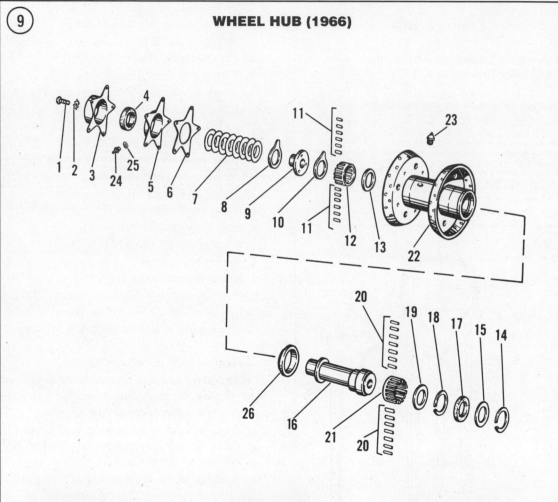

**(9) WHEEL HUB (1966)**

1. Thrust bearing cover screw
2. Thrust bearing cover screw lockwasher
3. Thrust bearing outer cover
4. Cork grease retainer
5. Thrust bearing housing
6. Thrust bearing housing gasket
7. Thrust bearing adjusting shim
8. Thrust washer
9. Thrust bearing sleeve
10. Thrust washer
11. Bearing roller (12)
12. Roller retainer
13. Roller retainer thrust washer
14. Roller bearing spring lock ring
15. Retaining washer
16. Hub inner sleeve
17. Cork grease retainer
18. Roller bearing spring lock ring
19. Roller bearing washer
20. Bearing roller
21. Roller retainer
22. Hub shell
23. Grease fitting
24. Grease fitting
25. Plain washer
26. Roller retainer thrust collar

17. Inject one ounce of brake grease into the hub after installation.

*WARNING*
*Do not overlubricate or the brakes may become contaminated.*

### Disassembly/Inspection/Assembly (1967-1972 FL Models)

The front hub on these models uses permanently lubricated and sealed ball bearings. These bearings require no periodic lubrication. The bearings require replacement when excessive hub play and/or bearing noise is detected. Refer to **Figure 10**. A press is required to remove the bearings.

1. Check the front wheel as described in this chapter under *Front Wheel*.

2. Remove the front wheel as described in this chapter.

3. Remove brake drum-to-hub screws and remove the brake drum (2, **Figure 10**).

4. Slide the bearing spacer (3) out of the hub.

5. Remove the wheel hub bearing locknut (4) using Harley-Davidson tool 94630-67 or equivalent (**Figure 11**). This nut has *left-hand* threads.

6. Carefully pry the seal (**Figure 12**) from the hub and remove the spacer.

7. Clean the hub and all spacers thoroughly in solvent. Do not clean the sealed bearings in solvent. Instead wipe them off with a clean shop rag.

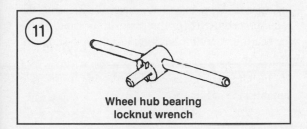

(11)

**Wheel hub bearing locknut wrench**

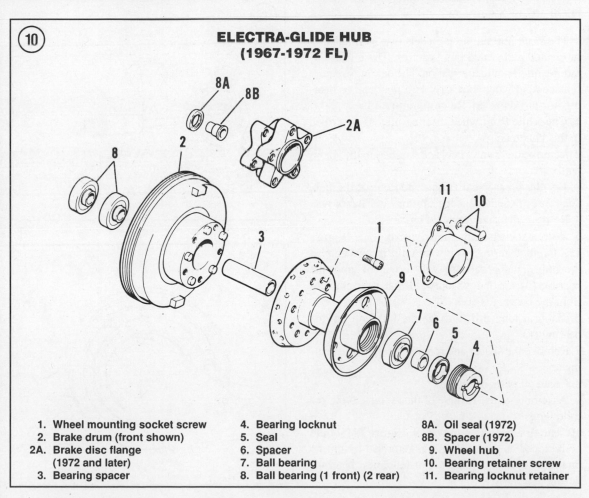

(10)

**ELECTRA-GLIDE HUB (1967-1972 FL)**

| | | |
|---|---|---|
| 1. Wheel mounting socket screw | 4. Bearing locknut | 8A. Oil seal (1972) |
| 2. Brake drum (front shown) | 5. Seal | 8B. Spacer (1972) |
| 2A. Brake disc flange (1972 and later) | 6. Spacer | 9. Wheel hub |
| 3. Bearing spacer | 7. Ball bearing | 10. Bearing retainer screw |
| | 8. Ball bearing (1 front) (2 rear) | 11. Bearing locknut retainer |

8

8. Turn the inner bearing race by hand (**Figure 13**) and check for excessive noise or play. If necessary, have a Harley-Davidson dealer or machine shop remove and install new bearings (7 and 8), as a press is required.

9. Assembly is the reverse of these steps. Note the following.

10. Pack the cavity between the bearings with grease.

11. Tighten the locknut (4, **Figure 10**) with the special tool. On all except 1970-1972 models, stake it to the hub in 4 places so that it cannot loosen. On 1970-1972 models, staking the locknut is not necessary; drive the retainer into the locknut slots with a chisel to lock the nut in position.

12. Clean the brake drum to hub mating surfaces thoroughly, then assemble. Tighten the screws in a criss-cross pattern to prevent wheel runout.

## Disassembly/Inspection/Assembly (1970-1972 FX)

The front hub on these models uses permanently lubricated and sealed ball bearings. These bearings require no periodic lubrication. The bearings require replacement when excessive hub play and/or bearing noise is detected. Refer to **Figure 14**.

1. Check the front wheel bearing play as described in this chapter.

2. Remove the front wheel as described in this chapter.

3. Pry the grease seal (**Figure 12**) out of the hub. Take care not to damage the hub during seal removal.

4. Remove the snap ring (2, **Figure 14**).

5. With a socket on the outer bearing race (**Figure 15**), tap the bearing (3) inward until it seats against the hub. This procedure will move the opposite bearing (4) far enough outward that it disengages from the bearing spacer (5).

6. Using a long drift through the hub, tap out the bearing (4).

7. Remove the spacer and remove the opposite bearing.

8. Clean all parts in solvent.

9. Assembly is the reverse of these steps. Note the following.

10. The first bearing removed (4, **Figure 14**) should be replaced as the bearing was removed by hitting on the inner race. It is best to install 2 new bearings during assembly.

11. Pack both new bearings with new brake grease.

12. Align the bearing (4) with the hub and install it with a large socket on the outside bearing race (**Figure 15**). Drive the bearing in until it seats against the hub shoulder.

13. Install the snap ring (2) with its flat side against the bearing. Make sure the snap ring seats in the groove completely.

14. Working from the opposite side, install the bearing spacer. Then install the bearing (3) using the same procedure as for the opposite bearing. Drive the bearing in until it seats against the hub shoulder.

15. Install a new grease seal by driving it in with a suitable size socket.

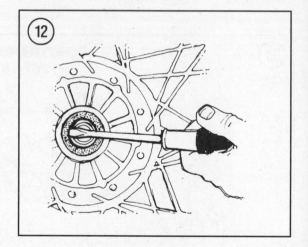

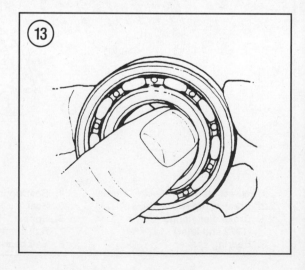

### Disassembly/Inspection/Assembly (1973-early 1978 Interchangeable 16 in. Cast and Spoked Hubs)

The front hub on these models uses tapered roller bearings that require removal and repacking every 10,000 miles. The bearings require replacement when excessive hub play and/or bearing noise is detected. Refer to **Figure 16**.

The bearing assembly is the same for both front and rear spoked and cast 16 in. wheels.

1. Check the front wheel bearing play as described under *Front Wheel* in this chapter.

2. Remove the front wheel as described in this chapter.

*NOTE*
*Label all parts so that they may be reinstalled in their original positions.*

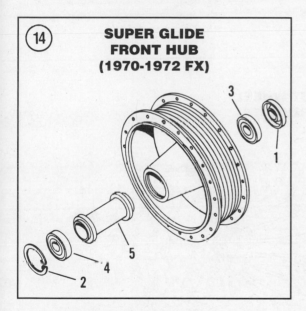

**SUPER GLIDE FRONT HUB (1970-1972 FX)**

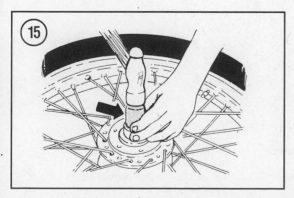

3. Remove the circlip (1, **Figure 16**) and washer.

4. Pry the oil seal out of the hub (**Figure 12**). Then remove the spacer and bearing.

5. Repeat Step 3 and Step 4 for the opposite side.

6. Clean all parts in solvent. Clean bearing cones in hub with solvent also.

7. Check the roller bearing cones and cups for wear, pitting or excessive heat (bluish tint). Replace the bearing cones and cups as a complete set. Replace the bearing cups as described in Step 8.

8. If the bearing cups require replacement, remove them using a suitable puller.

9. Installation is the reverse of these steps. Note the following.

10. If bearing end play was incorrect as described in this chapter, install a longer or shorter spacer (7) as required.

11. Apply new bearing grease to both bearing cones before assembly.

12. Install a new oil seal into the hub so that its lip is 3/16 to 1/4 in. (4.76-6.35 mm) below the hub edge.

13. Make sure the snap rings seat completely in their hub grooves.

### Disassembly/Inspection/Assembly (1973-on FX Laced Wheels)

*NOTE*
*The bearing cones and cups on these models are matched pairs. Label all parts so that they may be returned to their original positions.*

Refer to **Figure 17**, **Figure 18**, **Figure 19** or **Figure 20** for this procedure.

1. Remove the front wheel as described in this chapter.

2. Remove both oil seals (**Figure 12**).

3. Remove the spacer and both bearing cones.

4. Check the roller bearing cones and cups for wear, pitting or excessive heat (bluish tint). Replace the bearing cones and cups as a complete set. Replace the bearing cups as described in Step 5. If the bearing cones and cups do not require replacement, proceed to Step 7.

5. If the bearing cups require replacement, remove them using a suitable puller.

6. Remove the center spacer.

7. Clean the inside and outside of the hub with solvent. Dry with compressed air.

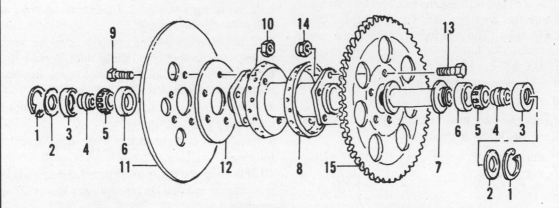

## 16 IN. SPOKE WHEEL
## (1973-EARLY 1978)

| | | |
|---|---|---|
| 1. Circlip | 6. Bearing race | 11. Brake disc |
| 2. Washer | 7. Spacer | 12. Spacer |
| 3. Oil seal | 8. Hub | 13. Bolt |
| 4. Spacer | 9. Bolt | 14. Nut |
| 5. Bearing | 10. Nut | 15. Driven sprocket |

## 16 IN. CAST WHEEL
## (1973-EARLY 1978)

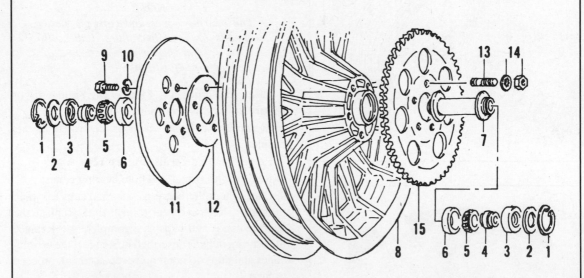

| | | |
|---|---|---|
| 1. Circlip | 6. Bearing race | 11. Brake disc |
| 2. Washer | 7. Spacer | 12. Spacer |
| 3. Oil seal | 8. Wheel assembly | 13. Stud |
| 4. Spacer | 9. Bolt | 14. Washer and nut |
| 5. Bearing | 10. Lockwasher | 15. Driven sprocket |

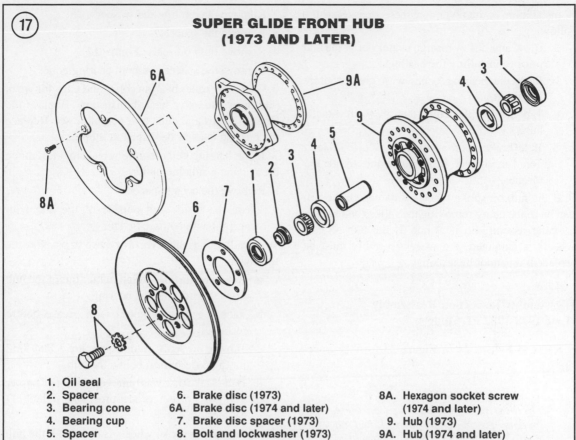

**(17)**

## SUPER GLIDE FRONT HUB
## (1973 AND LATER)

1. Oil seal
2. Spacer
3. Bearing cone
4. Bearing cup
5. Spacer
6. Brake disc (1973)
6A. Brake disc (1974 and later)
7. Brake disc spacer (1973)
8. Bolt and lockwasher (1973)
8A. Hexagon socket screw (1974 and later)
9. Hub (1973)
9A. Hub (1974 and later)

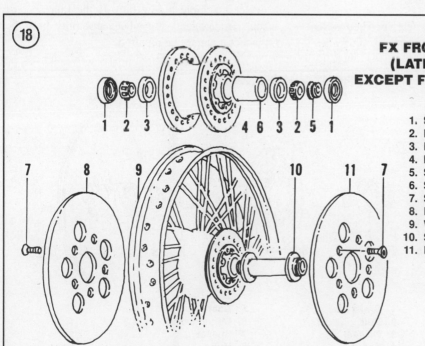

**(18)**

## FX FRONT WHEEL
## (LATE 1978-ON
## EXCEPT FXE AND FXWG)

1. Seal
2. Bearing
3. Bearing race
4. Hub
5. Spacer
6. Spacer
7. Screw
8. Brake disc
9. Wheel assembly
10. Spacer
11. Brake disc

**8**

8. Installation is the reverse of these steps. Note the following:

    a. Blow any dirt or foreign matter out of the hub prior to installing the bearings.

    b. Pack the bearing cones with grease before installation.

    c. Tap the oil seals in until they are flush with the hub.

    d. If the brake disc was removed, refer to Chapter Ten for correct procedures and tightening torques.

9. If the hub on spoke wheels is damaged, the hub can be replaced by removing the spokes and having a dealer assemble a new hub. If the hub on cast wheels is damaged, the wheel assembly must be replaced; it cannot be repaired.

## Disassembly/Inspection/Reassembly (Late 1978-1982 FL Models)

Refer to **Figure 21** or **Figure 22** for this procedure.

*NOTE*
*The bearing cones and cups on these models are matched pairs. Label all parts so that they may be returned to their original positions.*

1. Remove the front wheel as described in this chapter.

2. Remove the circlips and washers.

3. Remove the spacers.

4. Remove both oil seals (**Figure 12**).

5. Remove the spacer and both bearing cones.

6. Check the roller bearing cones and cups for wear, pitting or excessive heat (bluish tint). Replace the bearing cones and cups as a complete set. Replace the bearing cups as described in Step 7.

7. If the bearing cups require replacement, remove them using a suitable puller.

8. Remove the center spacer.

9. Clean the inside and outside of the hub with solvent. Dry with compressed air.

10. Installation is the reverse of these steps. Note the following:

    a. Blow any dirt or foreign matter out of the hub prior to installing the bearings.

    b. Pack the bearing cones with grease before installation.

    c. Tap the oil seals in until they are 13/64-7/32 in. (5.16-5.56 mm) below the hub.

    d. If the brake disc was removed, refer to Chapter Ten for correct procedures and tightening torques.

11. If the hub on spoke wheels is damaged, the hub can be replaced by removing the spokes and having a dealer assemble a new hub. If the hub on cast wheels is damaged, the wheel assembly must be replaced; it cannot be repaired.

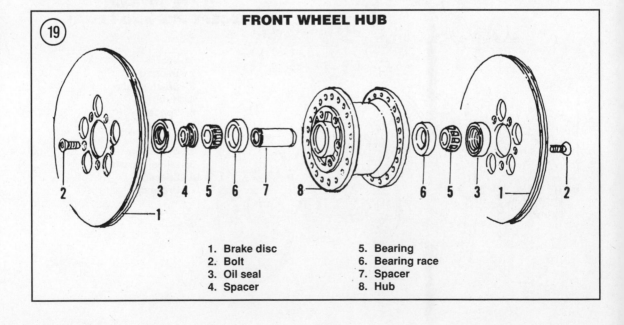

**FRONT WHEEL HUB**

1. Brake disc
2. Bolt
3. Oil seal
4. Spacer
5. Bearing
6. Bearing race
7. Spacer
8. Hub

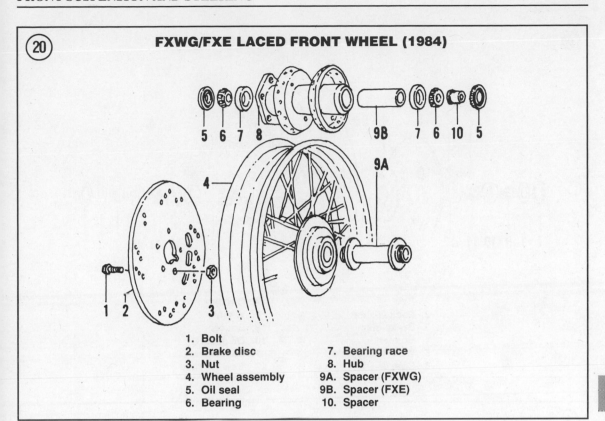

**FXWG/FXE LACED FRONT WHEEL (1984)**

1. Bolt
2. Brake disc
3. Nut
4. Wheel assembly
5. Oil seal
6. Bearing
7. Bearing race
8. Hub
9A. Spacer (FXWG)
9B. Spacer (FXE)
10. Spacer

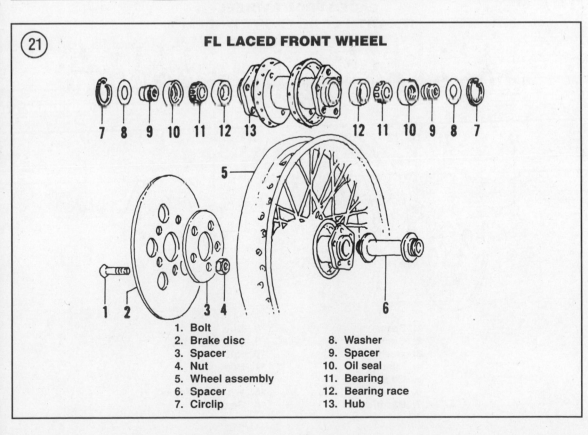

**FL LACED FRONT WHEEL**

1. Bolt
2. Brake disc
3. Spacer
4. Nut
5. Wheel assembly
6. Spacer
7. Circlip
8. Washer
9. Spacer
10. Oil seal
11. Bearing
12. Bearing race
13. Hub

8

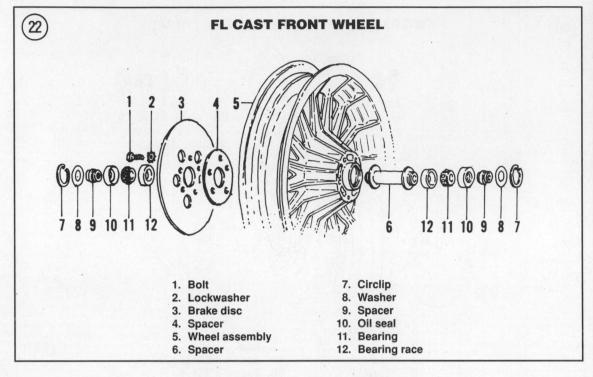

**FL CAST FRONT WHEEL**

1. Bolt
2. Lockwasher
3. Brake disc
4. Spacer
5. Wheel assembly
6. Spacer
7. Circlip
8. Washer
9. Spacer
10. Oil seal
11. Bearing
12. Bearing race

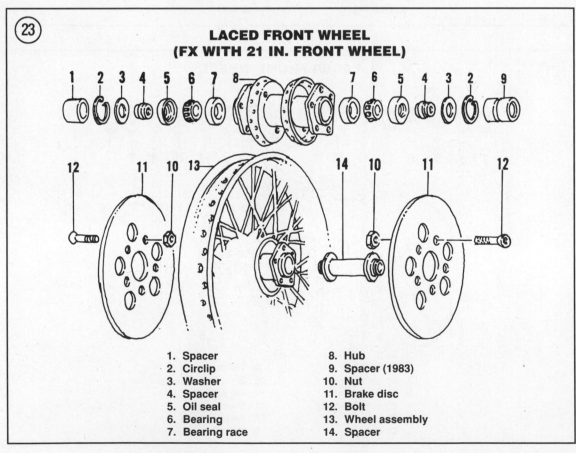

**LACED FRONT WHEEL
(FX WITH 21 IN. FRONT WHEEL)**

1. Spacer
2. Circlip
3. Washer
4. Spacer
5. Oil seal
6. Bearing
7. Bearing race
8. Hub
9. Spacer (1983)
10. Nut
11. Brake disc
12. Bolt
13. Wheel assembly
14. Spacer

**Disassembly/Inspection/Reassembly (Late 1978-on FX 21 in. Laced Front Wheel)**

Refer to **Figure 23** for this procedure.

*NOTE*
*The bearing cones and cups on these models are matched pairs. Label all parts so that they may be returned to their original positions.*

1. Remove the front wheel as described in this chapter.

2. Remove the left-hand spacer. On 1983 models, remove the right-hand spacer.

3. Remove the circlips and washers.

4. Remove the spacers.

5. Remove both oil seals.

6. Remove the spacer and both bearing cones.

7. Check the roller bearing cones and cups for wear, pitting or excessive heat (bluish tint). Replace the bearing cones and cups as a complete set. Replace the bearing cups as described in Step 8.

8. If the bearing cups require replacement, remove them using a suitable puller.

9. Remove the center spacer.

10. Clean the inside and outside of the hub with solvent. Dry with compressed air.

11. Installation is the reverse of these steps. Note the following:

    a. Blow any dirt or foreign matter out of the hub prior to installing the bearings.

    b. Pack the bearing cones with grease before installation.

    c. Tap the oil seals in until they are flush with the hub.

    d. If the brake disc was removed, refer to Chapter Ten for correct procedures and tightening torques.

    e. After the wheel is installed, the bearing end play should be checked as described in this chapter under *Front Wheel*.

12. If the hub is damaged, the hub can be replaced by removing the spokes and having a dealer assemble a new hub.

## WHEEL BALANCE

An unbalanced wheel results in unsafe riding conditions. Depending on the degree of imbalance and the speed of the bike, the rider may experience anything from a mild vibration to a violent shimmy and loss of control.

*NOTE*
*Be sure to balance the wheel with the brake disc(s) attached as they also affect the balance.*

Before attempting to balance the wheels, check to be sure that the wheel bearings are in good condition and properly lubricated. The wheel must rotate freely.

1. Remove the wheel to be balanced.

2. Mount the wheel on a fixture such as the one in **Figure 24** so it can rotate freely.

3. Give the wheel a spin and let it coast to a stop. Mark the tire at the lowest point.

4. Spin the wheel several more times. If the wheel keeps coming to rest at the same point, it is out of balance.

5. Tape a test weight to the upper (or light) side of the wheel. See **Figure 25**.

6. Experiment with different weights until the wheel, when spun, comes to rest at a different position each time.

7. Remove the test weight and install the correct size weight.

*NOTE*
*On spoked wheels, weights are attached to the spokes (**Figure 26**). On cast wheels, weights are attached to the rim (**Figure 27**).*

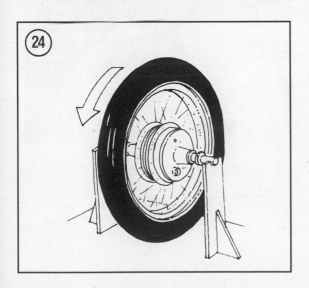

(24)

## Spoke Inspection and Replacement

Spokes loosen with use and should be checked periodically. The "tuning fork" method for checking spoke tightness is simple and works well. Tap the center of each spoke with a spoke wrench (**Figure 28**) or the shank of a screwdriver and listen for a tone. A tightened spoke will emit a clear, ringing tone and a loose spoke will sound flat or dull. All the spokes in a correctly tightened wheel will emit tones of similar pitch but not necessarily the same precise tone. The tension of the spokes does not determine wheel balance.

Bent, stripped or broken spokes should be replaced as soon as they are detected, as they can destroy an expensive hub.

Unscrew the nipple from the spoke and depress the nipple into the rim far enough to free the end of the spoke; take care not to push the nipple all the way in. Remove the damaged spoke from the hub and use it to match a new spoke of identical length. If necessary, trim the new spoke to match the original and dress the end of the thread with a thread die. Install the new spoke in the hub and screw on the nipple; tighten it until the spoke's tone is similar to the tone of the other spokes in the wheel. Periodically check the new spoke; it will stretch and must be retightened several times before it takes a final seat.

## Spoke Adjustment

If all spokes appear loose, tighten all on one side of the hub, then tighten all on the other side. One-half to one turn should be sufficient; do not overtighten.

After tightening the spokes, check rim runout to be sure you haven't pulled the rim out of shape.

One way to check rim runout is to mount a dial indicator on the front fork or swing arm, so that it bears against the rim.

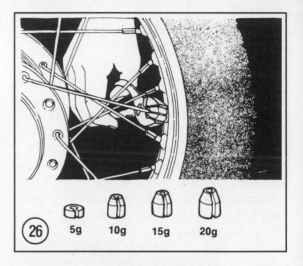

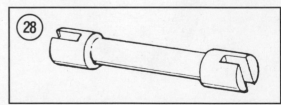

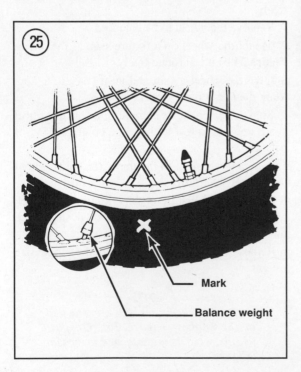

If you don't have a dial indicator, improvise one as shown in **Figure 29**. Adjust the position of the bolt until it just clears the rim. Rotate the rim and note whether the clearance increases or decreases. Mark the tire with chalk or light crayon at areas that produce significantly large or small clearances. Clearance must not change by more than 0.08 in. (2.03 mm).

To pull the rim out, tighten spokes which terminate on the same side of the hub and loosen spokes which terminate on the opposite side of the hub (**Figure 30**). In most cases, only a slight amount of adjustment is necessary to true a rim. After adjustment, rotate the rim and make sure another area has not been pulled out of true. Continue adjustment and checking until runout is less than 0.08 in. (2.03 mm).

### Rim Replacement

If the rim becomes bent or damaged, it should be replaced. A bent or dented wheel can cause serious handling problems.

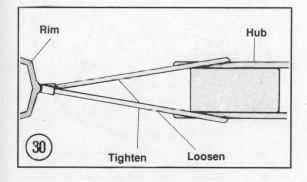

If the spokes are not bent or damaged also, they may be reused. This procedure describes how to replace the rim without removing the spokes.

1. Remove the tire as described in this chapter.

2. Securely fasten the spokes together with wire, string or tape at each point where they cross.

3. Place the replacement rim on top of the old rim and align the nipple holes of both rims. This is to make sure the replacement rim is the correct one. When the rims are aligned correctly, mark one spoke and its corresponding nipple hole on the new rim.

4. Remove the nipples from the spokes using a spoke wrench. If they are coated with dirt or rust, clean them in solvent and allow to dry. Then check the nipples for signs of cracking or other damage. Spoke nipples in this condition can strip when the wheel is later trued. Replace all nipples as necessary.

5. Lift the hub and spokes out of the old rim, making sure not to knock the spokes out of alignment.

6. Position the hub and spokes into the new rim, making sure to align the marks made in Step 3. Then insert the spokes into the rim until they are all in place.

7. Place a drop of oil onto the threaded end of each spoke and install the nipples. Thread the nipples halfway onto the spokes and stop them before they make contact with the rim.

8. Lift the wheel and stand it up on the workbench. Check the hub to make sure it is centered in the rim. If not, reposition it by hand.

9. With the hub centered in the rim, thread the nipples until they just seat against the rim. True the

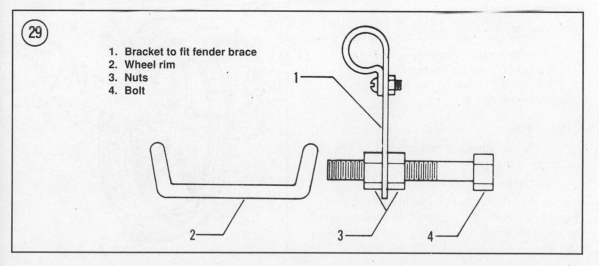

1. Bracket to fit fender brace
2. Wheel rim
3. Nuts
4. Bolt

wheel as described under *Spoke Adjustment* in this chapter.

## Seating Spokes

When spokes loosen or when installing new spokes, the head of the spoke should be checked for proper seating in the hub. If it is not seated correctly, it can loosen further and may cause severe damage to the hub. If one or more spokes require reseating, hit the head of each spoke with a punch. True the wheel as described under *Spoke Adjustment* in this chapter.

## TIRE CHANGING

The stock cast wheel is aluminum and the exterior appearance can easily be damaged. Special care must be taken with tire irons when changing a tire to avoid scratches and gouges to the outer rim surface. Insert scraps of leather between the tire irons and the rim to protect the rim from damage.

The cast wheels on all models are designed for use with either tubeless or tube-type tires. Tire removal and installation are basically the same for tube and tubeless tires; where differences occur they are noted. Tire repair is different and is covered in separate procedures.

When removing a tubeless tire, take care not to damage the tire beads, inner liner of the tire or the wheel rim flange. Use tire levers or flat-handled tire irons with rounded edges.

## Removal

*NOTE*
*A bead breaker machine will be required to break the tire bead on 16 inch wheels. Refer tire service to a Harley-Davidson dealer.*

1. Remove the valve core to deflate the tire.

2. Press the entire bead on both sides of the tire into the center of the rim.

3. Lubricate the beads with soapy water.

4. Insert the tire iron under the bead next to the valve (**Figure 31**). Force the bead on the opposite side of the tire into the center of the rim and pry the bead over the rim with the tire iron.

*NOTE*
*Insert scraps of leather between the tire irons and the rim to protect the rim from damage.*

5. Insert a second tire iron next to the first to hold the bead over the rim. Then work around the tire with the first tool prying the bead over the rim (**Figure**

32). On tube-type tires, be careful not to pinch the inner tube with the tools.

6. On tube-type tires, use your thumb and push the valve from its hole in the rim to the inside of the tire. Carefully pull the tube out of the tire and lay it aside.

*NOTE*
*Step 7 is required only if it is necessary to remove the tire completely from the rim, such as for tire replacement or tubeless tire repair.*

7. Stand the wheel upright. Insert a tire tool between the second bead and the same side of the rim that the

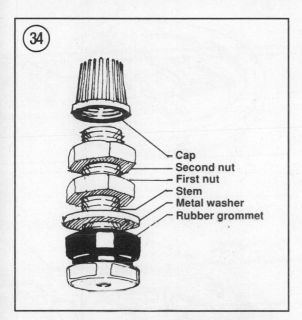

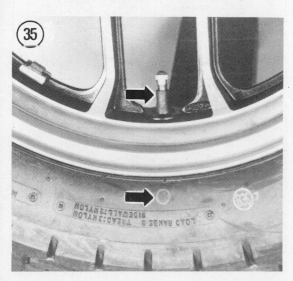

first bead was pried over (**Figure 33**). Force the bead on the side opposite the tool into the center of the rim. Pry the second bead off the rim, working around the wheel with 2 tire irons as with the first bead.

8. On tubeless tires, inspect the rubber grommet (**Figure 34**) where the valve stem seats against the inner surface of the wheel. Replace it if it's starting to deteriorate or has lost its resiliency. This is a common location of air loss.

**Installation**

1. Carefully inspect the tire for any damage, especially inside.

2. A new tire may have balancing rubbers inside. These are not patches and should not be disturbed. A colored spot near the bead indicates a lighter point on the tire. This spot should be placed next to the valve stem (**Figure 35**). In addition, most tires have directional arrows labeled on the side of the tire to indicate which direction the tire should rotate. Make sure to install the tire accordingly.

3. On tube-type tires, inflate the tube just enough to round it out. Too much air will make installation difficult. Place the tube inside the tire.

4. Lubricate both beads of the tire with soapy water.

5. Place the backside of the tire into the center of the rim and insert the valve stem through the stem hole in the wheel. The lower bead should go into the center of the rim and the upper bead outside. Work around the tire in both directions (**Figure 36**). Use a tire iron for the last few inches of bead (**Figure 37**).

6. Press the upper bead into the rim opposite the valve. Pry the bead into the rim on both sides of the initial point with a tire tool, working around the rim to the valve (**Figure 38**).

7. On tube-type tires, wiggle the valve to be sure the tube is not trapped under the bead. Set the valve stem squarely in its hole before screwing on the valve nut to hold it against the rim.

8. Check the bead on both sides of the tire for an even fit around the rim.

9. On tube-type tires, inflate the tire slowly to seat the beads in the rim. It may be necessary to bounce the tire to complete the seating.

10. On tubeless tires, place an inflatable band around the circumference of the tire. Slowly inflate the band until the tire beads are pressed against the rim. Inflate the tire enough to seat it, deflate the band and remove it.

8

*WARNING*
*Never exceed 40 psi (2.81 kg/cm$^2$)in-*
*flation pressure as the tire could burst.*
*Never stand directly over the tire while*
*inflating it.*

11. Inflate the tire to the required pressure (Chapter
Three). Tighten the valve stem locks and screw on
the cover cap.

12. Balance the wheel assembly as described in this
chapter.

## TIRE REPAIRS
## (TUBE-TYPE TIRES)

Patching a motorcycle tube is only a temporary
fix. A motorcycle tire flexes too much and the patch
could rub right off. However, a patched tube should
get you far enough to buy a new tube.

### Tire Repair Kits

The repair kits can be purchased from motorcycle
dealers and some auto supply stores. When buying,
specify that the kit you want is for motorcycles.

There are 2 types of tire repair kits:

  a. Hot patch.
  b. Cold patch.

Hot patches are stronger because they actually
vulcanize to the tube, becoming part of it. However,
they are far too bulky to carry for roadside repairs
and the strength is unnecessary for a temporary
repair.

Cold patches are not vulcanized to the tube; they
are simply glued to it. Though not as strong as hot
patches, cold patches are still very durable. Cold
patch kits are less bulky than hot patches and more
easily applied under adverse conditions. A cold
patch kit contains everything necessary and tucks
easily in with your emergency tool kit.

### Tube Inspection

1. Remove the inner tube as described under *Tire
Changing* in this chapter.

2. Install the valve core into the valve stem (**Figure
39**) and inflate the tube slightly. Do not overinflate.

3. Immerse the tube in water a section at a time. Look
carefully for bubbles indicating a hole. Mark each
hole and continue checking until you are certain that

all holes are discovered and marked. Also make sure
that the valve core is not leaking; tighten it if necessary.

*NOTE*
*If you do not have enough water to*
*immerse sections of the tube, try run-*
*ning your hand over the tube slowly and*
*very close to the surface. If your hand*
*is damp, it works even better. If you*
*suspect a hole anywhere, apply some*
*saliva to the area to verify it.*

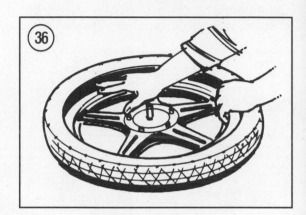

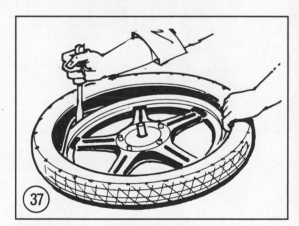

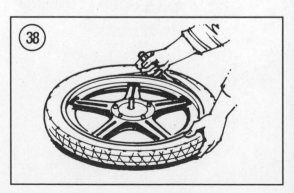

4. Apply a cold patch according to the manufacturer's instructions.

5. Dust the patch area with talcum powder to prevent it from sticking to the tire.

6. Carefully check the inside of the tire casing for small rocks or sand which may have damaged the tube. If the inside of the tire is split, apply a patch to the area to prevent it from pinching and damaging the tube again.

7. Check the inside of the rim.

8. Deflate the tube prior to installation in the tire.

## TIRE REPAIRS
### (TUBELESS TYPE)

Patching a tubeless tire on the road is very difficult. If both beads are still in place against the rim, a can of pressurized tire sealant may inflate the tire and seal the hole. The beads must be against the wheel for this method to work. Another solution is to carry a spare tube that could be temporarily installed and inflated. This may enable you to get to a service station where the tire can be correctly repaired. Be sure that the tube is designed for use with a tubeless tire. The tire industry recommends that tubeless tires be patched from the inside. Therefore, do not patch the tire with an external type plug. If you find an external patch on a tire, it is recommended that it be patch-reinforced from the inside.

Due to the variations of material supplied with different tubeless tire repair kits, follow the instructions and recommendations supplied with the repair kit.

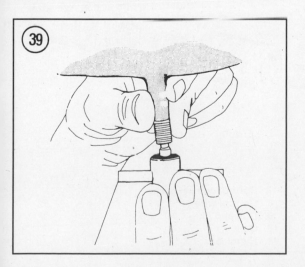

## HANDLEBAR AND
## THROTTLE CONTROL

Because of the vast number of handlebar and cable configurations for the models covered in this manual, this procedure presents a general guideline to handlebar removal and installation. Take Polaroid pictures or make sketches of the handlebar assembly if you are unsure about the routing of control cables. Bent or damaged handlebars should be replaced immediately.

### Removal

1. Support the bike on the sidestand.

> *CAUTION*
> *Cover the fuel tank with a heavy cloth or plastic tarp to protect it from accidental spilling of brake fluid. Wash any spilled brake fluid off any painted or plated surface immediately, as it will destroy the finish. Use soapy water and rinse thoroughly.*

2. Remove the fairing or windshield, if so equipped.

3. Remove the mirrors.

4. Remove the throttle control assembly as described in this chapter.

5. On models equipped with disc brake, remove the bolts securing the master cylinder and lay it on the fuel tank. It is not necessary to disconnect the hydraulic brake line.

6. If so equipped, slacken the clutch cable and disconnect it at the hand lever.

7. Remove the clamps securing the electrical cables to the handlebar.

8. Check the handlebar for any remaining electrical wiring or equipment and either disconnect or remove it.

9. Remove the handlebar clamp bolts and remove the handlebar.

10. Install by reversing these steps.

### Throttle Control Assembly

#### *Spiral grip*

The spiral grip was used on early models. Refer to **Figure 40** (typical) for this procedure.

1. Disconnect the throttle control wire at the carburetor and circuit breaker.

2. Remove the handlebar end screw (1, **Figure 40**). This screw may be difficult to remove; an impact driver will be helpful.

3. Remove the handlebar grip sleeve (3). Remove the screw and spring (2) from inside the grip.

4. Remove the 2 control coil plunger rollers (5) and the roller pin (4).

5. Pull the control coil plunger (6) and the control wire (8) from the handlebar. If the control wire is broken, pull the other piece from the carburetor or circuit breaker. The control wire is secured in the plunger by the control coil set screw (7).

6. Remove the control coils (10) and the end plug (9).

7. Clean all parts in solvent. Replace any worn or damaged parts.

8. Installation is the reverse of these steps. Note the following:

  a. Make sure that the screw (7) registers in the control coil end plug groove.

  b. Lubricate the control wire with graphite grease or oil as it is inserted into the control coil.

  c. Install both rollers over the roller pin, with the round side of roller up as it is positioned on the motorcycle.

  d. The easiest way to install the handlebar end screw (1) is to grasp the grip sleeve assembly (3) and apply slight pressure against the screw as it is started. Make sure to tighten the screw securely.

  e. After completing assembly at handlebar, connect control wire at carburetor and circuit breaker. Be sure that the carburetor lever opens and closes fully as the grip is turned. There should be 1/4 in. (6.35 mm) between the control clip on carburetor and the throttle control coil when the carburetor lever is fully closed against its stop.

  f. The end of the spark control wire must point directly at the adjuster stud hole on the circuit breaker when the circuit breaker is in its fully advanced position. End of control coil should extend approximately 3/8 in. (9.52 mm) beyond the clamp.

### *Drum throttle grip—1980 and earlier*

Refer to **Figure 41** (typical) for this procedure.

1. Remove both throttle control coil clamp screws and separate the upper and lower clamps.

2. Disconnect the control wire at the throttle grip. On late 1978 and later models, the ferrule disconnects from the control wire.

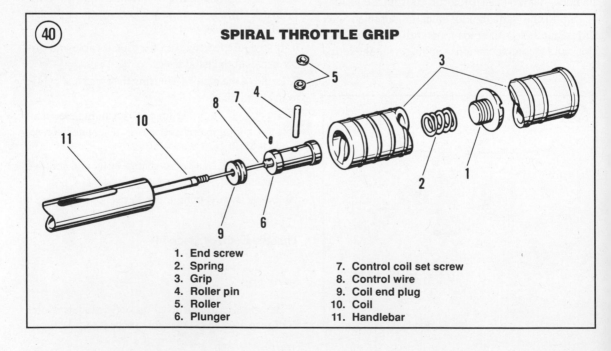

**SPIRAL THROTTLE GRIP**

1. End screw
2. Spring
3. Grip
4. Roller pin
5. Roller
6. Plunger
7. Control coil set screw
8. Control wire
9. Coil end plug
10. Coil
11. Handlebar

3. Remove worn or broken control wire by sliding it through the control adjuster locknut. To replace the control wire on early 1978 and earlier models, unsolder or cut off the control wire ferrule (16, **Figure 41**) before removing the wire.

4. Lubricate replacement wire with graphite grease and slide it into the control wire casing.

5. If necessary, solder ferrule onto new wire flush with wire end.

6. Attach the control wire onto the throttle grip and assemble the throttle assembly.

7. Adjust the throttle assembly as described in Chapter Three.

### *Drum throttle grip—1981 and later*

On 1981 and later models, a throttle closing cable was added and the throttle grip travel stop screw was eliminated. Refer to **Figure 42**.

1. Remove the throttle clamp screws and the upper clamp (2, **Figure 42**).

2. Slide the cylindrical ferrules out of their throttle grip seats and remove the ferrules from their cable ball ends.

3. Disconnect the lower ends of the cables from the carburetor. Note the routing of the cables, then pull the cable and lower clamp assembly up away from the motorcycle.

4. Loosen the cable elbow locknuts at the lower grip clamp and unscrew the elbows from the clamp.

5. Inspect the throttle spring and replace if necessary. Make sure the friction spring is securely seated on the friction adjuster screw (7).

6. Lubricate the insides of the grip clamps, the end of the handlebar and the cable ends and ferrules lightly with graphite grease.

7. Screw the throttle cable elbows into the lower grip clamp. The "open" cable elbow (9) has a larger thread than the "close" cable (12) and it goes in the left side of the lower clamp. The "close" cable has a spring at its lower end.

8

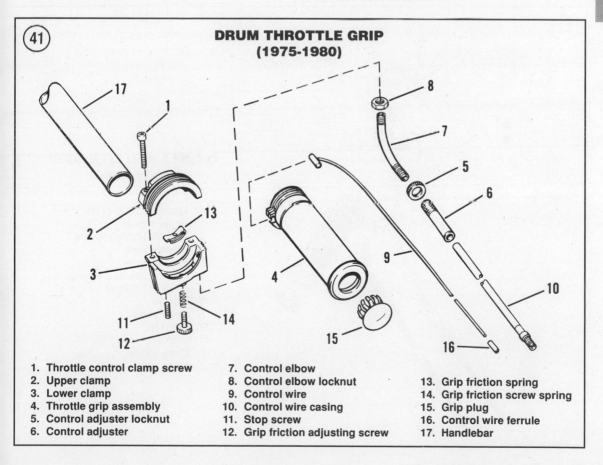

**DRUM THROTTLE GRIP (1975-1980)**

| | | |
|---|---|---|
| 1. Throttle control clamp screw | 7. Control elbow | |
| 2. Upper clamp | 8. Control elbow locknut | 13. Grip friction spring |
| 3. Lower clamp | 9. Control wire | 14. Grip friction screw spring |
| 4. Throttle grip assembly | 10. Control wire casing | 15. Grip plug |
| 5. Control adjuster locknut | 11. Stop screw | 16. Control wire ferrule |
| 6. Control adjuster | 12. Grip friction adjusting screw | 17. Handlebar |

8. Slide the throttle grip over handlebar, install the cylindrical ferrules on the cable ends and seat the ferrules in the grip notches.

9. Route the cables through the throttle cable clamp and connect them to the carburetor. The open cable goes through the inside fitting and the close cable (with spring) goes through the outside fitting (**Figure 43**).

10. Install the upper grip clamp and tighten the screws securely. Turn the cable elbows so there is no stress on the cables and tighten the elbow locknuts at the lower grip clamp.

11. Adjust the throttle cables as described in Chapter Three.

### STEERING HEAD

**Disassembly/Reassembly (FL Models)**

Refer to **Figure 44, Figure 45** and **Figure 46** (typical) for this procedure.

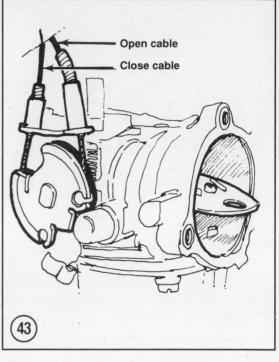

43

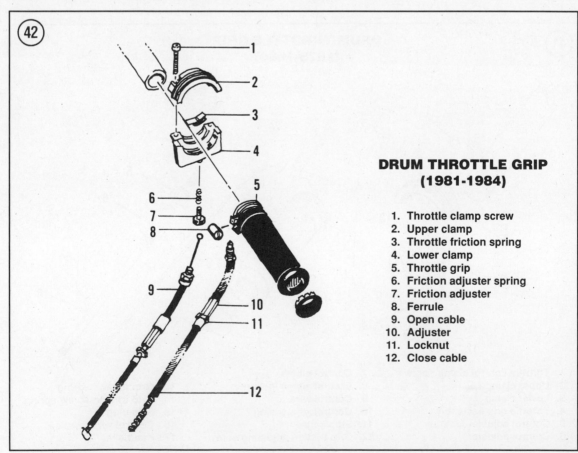

42

**DRUM THROTTLE GRIP**
**(1981-1984)**

1. Throttle clamp screw
2. Upper clamp
3. Throttle friction spring
4. Lower clamp
5. Throttle grip
6. Friction adjuster spring
7. Friction adjuster
8. Ferrule
9. Open cable
10. Adjuster
11. Locknut
12. Close cable

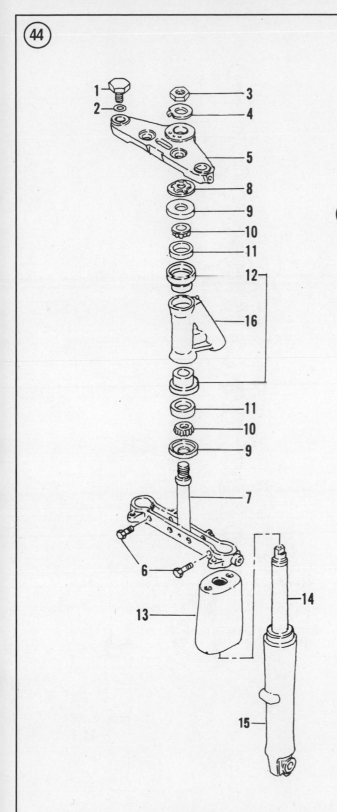

**STEERING ASSEMBLY
(FL NON-ADJUSTABLE FORK)**

1. Fork cap
2. Oil seal
3. Fork stem nut
4. Lockwasher
5. Upper bracket
6. Pinch bolt
7. Steering stem and
   lower bracket
8. Bearing seat
9. Dust shield
10. Bearing
11. Bearing race
12. Steering head cup
13. Cover
14. Fork tube
15. Slider
16. Frame

8

## STEERING ASSEMBLY (FL ADJUSTABLE FORK)

1. Steering damper adjusting screw
2. Spring
3. Spider spring cover
4. Spider spring
5. Pressure disc
6. Friction washer
7. Anchor plate
8. Friction washer
9. Pressure disc
10. Fork stem nut
11. Upper bracket
12. Bolt and washer
13. Head bearing nut
14. Dust shield
15. Bearing cone
16. Bearing race
17. Bearing cup
18. Frame
19. Bearing cup
20. Bearing race
21. Bearing cone
22. Bracket bolt
23. Bracket bolt washer
24. Steering stem and lower bracket
25. Bracket bolt washer
26. Fork cap
27. Filler screw
28. O-ring
29. Cotter pin
30. Nut
31. Bracket clamp bolt
32. Lower bracket
33. Fork tube
34. Slider

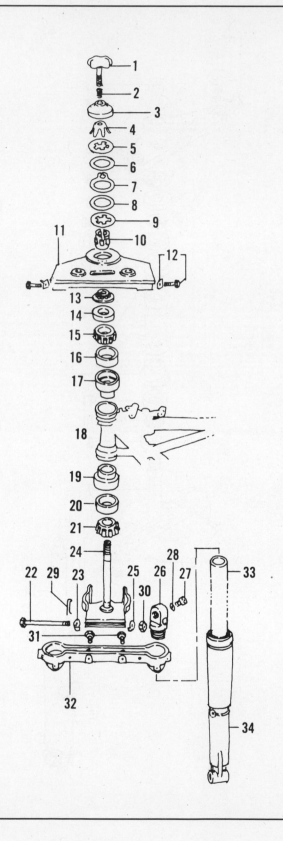

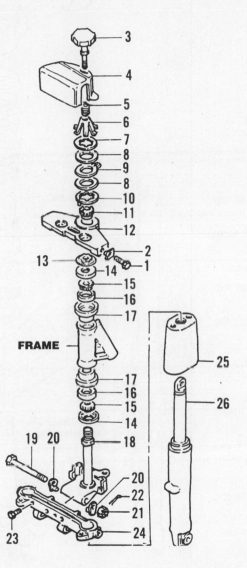

FRAME

## STEERING ASSEMBLY (FL ADJUSTABLE FORK)

1. Bolt
2. Washer
3. Damper adjusting screw
4. Cover
5. Spring
6. Spring
7. Upper pressure disc
8. Friction washer
9. Plate
10. Lower pressure disc
11. Nut
12. Upper bracket
13. Bearing seat
14. Dust shield
15. Bearing
16. Bearing race
17. Steering head cup
18. Fork stem
19. Bolt
20. Lock plate
21. Nut
22. Cotter pin
23. Bolt
24. Lower bracket
25. Cover
26. Fork leg

8

1. Remove the front wheel as described in this chapter.

2. Remove the fuel tank as described in Chapter Six.

3. Remove the fork tubes as described in this chapter.

4. Remove the handlebar as described in this chapter.

5A. *Adjustable forks:* Perform the following:

    a. Remove the steering damper adjusting screw by unscrewing it completely. Then remove all parts down to the steering nut.

*NOTE*
*The pressure discs in **Figure 45** and **Figure 46** can be removed by prying them up with a screwdriver.*

    b. Loosen and remove the fork stem nut.

5B. *Non-adjustable forks:* Bend the lockwasher tab down and remove the fork stem nut.

6. Remove the fork bracket and lower the fork stem out of the frame.

7. Remove the dust shields and bearings in order.

8. Inspect the front fork steering unit as described in this chapter.

9. Installation is the reverse of these steps. Note the following.

10. Pack the bearings with bearing grease.

11. Assemble the bearing, bearing cone and dust seal on the steering stem.

12. Install the steering stem and assemble the upper bearing, bearing cone and dust seal.

13. Install the bearing seat onto the fork stem and tighten it until the steering bearings have no noticeable free play.

14. Install the fork stem nut and tighten it securely. Bend the lockwasher tab against the nut.

15. On late 1978 and later models with adjustable forks, note the following:

    a. Install the lower pressure disc so that the pin in the disc engages one of the three holes in the upper bracket.

    b. After installing the lower friction washer on top of the lower pressure disc, install the anchor plate, making sure that the hole in the anchor plate engages with the roll pin on the frame. Install the upper friction washer.

16. Install all parts previously removed.

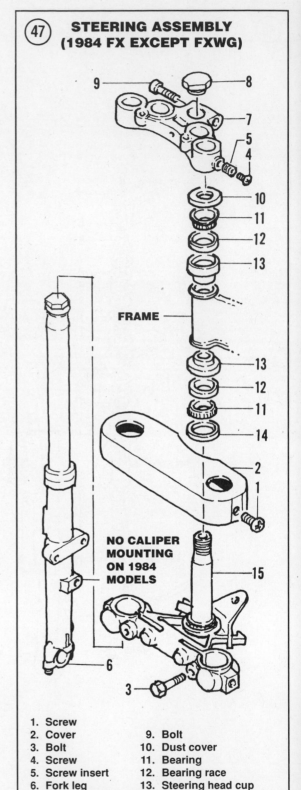

**47 STEERING ASSEMBLY (1984 FX EXCEPT FXWG)**

FRAME

NO CALIPER MOUNTING ON 1984 MODELS

| | |
|---|---|
| 1. Screw | 9. Bolt |
| 2. Cover | 10. Dust cover |
| 3. Bolt | 11. Bearing |
| 4. Screw | 12. Bearing race |
| 5. Screw insert | 13. Steering head cup |
| 6. Fork leg | 14. Dust cover |
| 7. Upper bracket | 15. Fork stem and bracket |
| 8. Nut | |

**FORK BRACKET (1984 FXWG)**

FRAME

NO CALIPER MOUNTING ON 1984 MODELS

1. Bolt
2. Lockwasher
3. Fork cap
4. Washer
5. Oil seal
6. Cap
7. Nut
8. Lockwasher
9. Upper bracket
10. Bearing seat
11. Dust cover
12. Bearing
13. Bearing race
14. Steering head cup
15. Lower dust shield
16. Fork stem and bracket
17. Fork leg

## Disassembly/Reassembly (FX Models)

Refer to **Figure 47** or **Figure 48** for this procedure.

1. Remove the front wheel as described in this chapter.

2. Remove the fuel tank as described in Chapter Six.

3. Remove the fork tubes as described in this chapter.

4. Remove the handlebar as described in this chapter.

5. On models so equipped, bend the lockwasher tab down and remove the fork stem nut. On other models, remove the fork stem nut.

6. Loosen the fork bracket pinch bolt, if so equipped.

7. Remove the fork bracket and lower the fork stem out of the frame.

8. Remove the dust shields and bearings in order.

9. Inspect the front fork steering unit as described in this chapter.

10. Installation is the reverse of these steps. Note the following.

11. Pack the bearings with bearing grease.

12. Assemble the bearing, bearing cone and dust seal on the steering stem.

13. Install the steering stem and assemble the upper bearing, bearing cone and dust seal.

14. Install the bearing seat onto the fork stem and tighten it until the steering bearings have no noticeable free play.

15. Install the fork stem nut and tighten it securely. Bend the lockwasher tab against the nut, if so equipped.

16. Replace all parts previously removed.

## Inspection

1. Clean the bearing races in the steering head and all bearings and bearing cones with solvent.

2. Check for broken welds on the frame around the steering head. If any are found, have them repaired by a competent frame shop or welding service familiar with motorcycle frame repair.

3. Check the bearings and bearing cones for pitting, scratches or discoloration indicating wear or corrosion. Replace them in sets if any are bad.

4. Check the races for pitting, galling and corrosion. If any of these conditions exist, replace the races as described in this chapter.

5. Check the steering stem for cracks and check its race for damage or wear. Replace if necessary.

8

## Steering Head Bearing Outer Race Replacement

The headset and steering stem bearing races are pressed into the frame. Because they are easily bent, do not remove them unless they are worn and require replacement.

To remove a headset race, insert a hardwood stick into the head tube and carefully tap the race out from the inside (**Figure 49**). Tap all around the race so that neither the race nor the head tube is bent. To install a race, fit it into the end of the head tube. Tap it slowly and squarely with a block of wood (**Figure 50**).

## Steering Stem Adjustment

Adjust the steering stem as described under *Front Fork Disassembly/Assembly* for your model in this chapter.

## FRONT FORK

The front suspension consists of a spring-controlled, hydraulically dampened telescopic fork.

Before suspecting major trouble, drain the front fork oil and refill with the proper type and quantity; refer to Chapter Three. If you still have trouble, such as poor damping, a tendency to bottom or top out or leakage around the rubber seals, follow the service procedures in this section.

To simplify fork service and to prevent the mixing of parts, the legs should be removed, serviced and installed individually.

## Removal/Installation

1. Remove the fairing, if so equipped.
2. Prop the bike so that the front wheel clears the ground.
3. Remove the front wheel as described in this chapter.
4. Remove the front fender (**Figure 51**).
5. *Disc brake models:* Remove the brake caliper(s) as described in Chapter Ten. Tie each caliper with wire to the frame to keep tension off the brake hoses.

*NOTE*
*Insert a piece of wood in the caliper(s) in place of the disc. That way, if the*

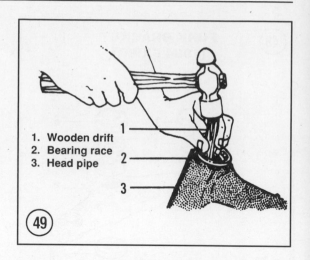

1. Wooden drift
2. Bearing race
3. Head pipe

49

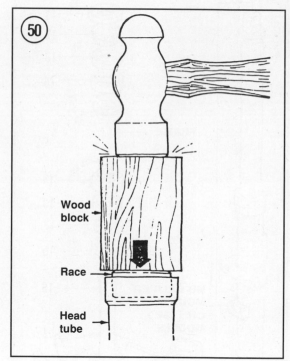

50

Wood block

Race

Head tube

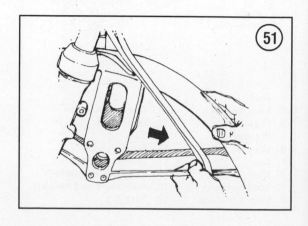

51

*brake lever is inadvertently squeezed, the piston will not be forced out of the caliper. If it does happen, the caliper might have to be disassembled to reseat the piston. By using the wood, bleeding the brake is not necessary when installing the wheel.*

6. Loosen the lower bracket cover screw and lift the cover up, if so equipped.

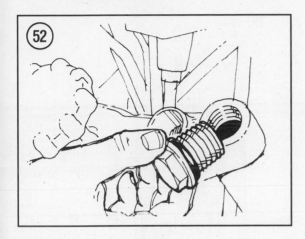

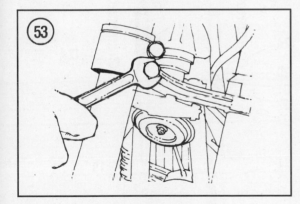

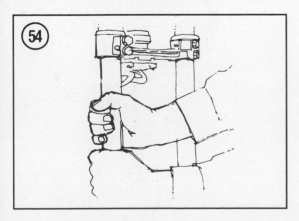

7. Remove the fork cap (**Figure 52**), if necessary.

8. Loosen the upper and lower (**Figure 53**) fork tube pinch bolts. Then rotate the fork tube (**Figure 54**) and remove it.

9. Repeat for the opposite side.

10. Install by reversing these removal steps. Tighten all bolts to specifications (**Table 1** or **Table 2**).

**Front Fork Service Notes**

*WARNING*
*If a fork tube is bent, do not attempt to disassemble it by removing the fork tube cap or plug. The fork spring may be tightly compressed and could cause severe personal injury if the fork cap is removed. Refer service at this point to a qualified Harley-Davidson dealer.*

1. If fork disassembly is required, it is best to loosen, but not remove, the fork cap while the fork tube is still mounted on the bike. Loosen the upper pinch bolt and loosen the fork tube.

2. To measure the correct amount of fork fluid during reassembly, use a plastic baby bottle. These have measurements in fluid ounces (oz.) and cubic centimeters (cc) on the side.

**Disassembly/Reassembly (1966-early 1977 FL Models)**

Refer to **Figure 55** (typical) for this procedure.

1. Remove the front fork as described in this chapter.

2. Remove the fork slider plug (11, **Figure 55**) from the top of the fork tube.

3. Remove the fork spring (12).

4. Hold the fork tube over a drain pan and drain the fork of all fork oil.

5. Remove the damper valve stud locknut (13) at the bottom of the fork slider. Then pull the fork slider tube (14) out of the slider.

6. Remove the snap ring (20) from the top of the slider. Then remove the damper tube lower bushing (17).

7. Discard the damper tube bushing gaskets.

8. Turn the slider over and remove the damper valve assembly (19).

9. Remove snap ring (20). Then remove the following parts in order:
    a. Washer (21).
    b. Felt washer (22).

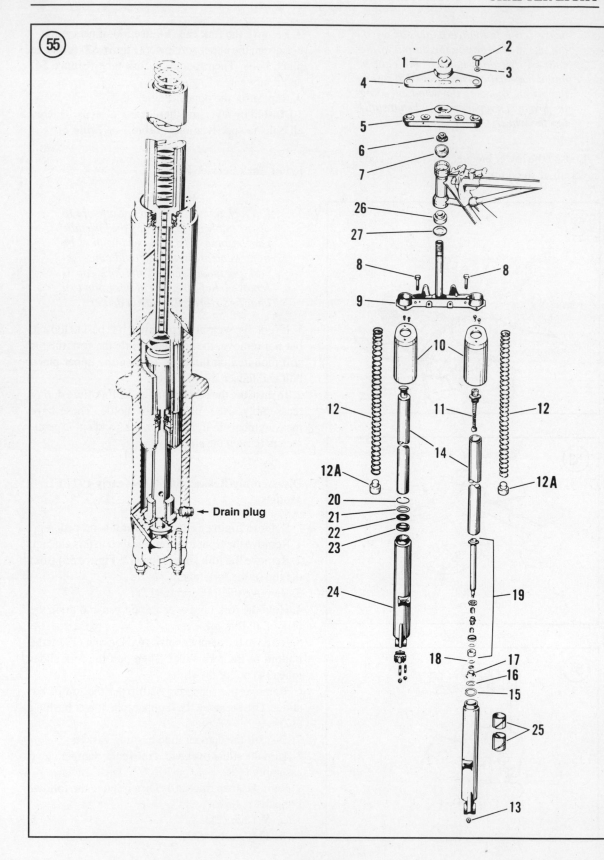

## FRONT FORK
## (1966-EARLY 1977 FL)

1. Fork stem nut
2. Fork upper bracket bolt and valve
3. Tube plug oil seal
4. Fork upper bracket cover
5. Handlebar and fork bracket
6. Head bearing nut
7. Head bearing
8. Fork bracket clamping stud
9. Fork bracket with stem
10. Fork slider cover
11. Fork slider plug
12. Fork spring
12A. Spring spacer (1966)
13. Damper valve stud locknut
14. Fork slider tube
15. Slider tube snap ring
16. Damper tube bushing gasket
17. Damper tube lower bushing
18. Damper valve stud gasket
19. Damper tube valve
20. Snap ring
21. Spring ring washer
22. Upper oil seal felt washer
23. Upper oil seal
24. Slider
25. Slider upper and lower bushing
26. Head bearing
27. Lower head bearing guard

c. Oil seal (23). Remove the oil seal by carefully prying it out of the slider with a screwdriver or small pry bars. Make sure to work around the seal to prevent damaging the fork slider.

10. The slider is equipped with an upper and lower bushing. If the bushings require replacement, refer service to a Harley-Davidson dealer, as special tools are required.

11. Inspect the fork assembly as described in this chapter.

12. Installation is the reverse of these steps. Note the following.

13. Install new oil seals by driving them in with a socket placed on top of the seal. Make sure to drive the oil seals in squarely.

14. When installing snap rings, make sure they seat completely in the machined groove.

15. Refill the fork tube with the correct type and quantity of fork oil as described in Chapter Three.

### Disassembly/Reassembly
### (Late 1977-on FL and 1976-1983 FX)

Refer to **Figure 56**, **Figure 57** or **Figure 58**.

1. Remove the front forks as described in this chapter.

2. Remove the fork tube cap, washer and O-ring.

3. Remove the spring.

4. Pour the oil out of the fork and discard it. Pump the fork several times by hand to expel most of the remaining oil.

5. Remove the Allen bolt and washer (**Figure 59**) at the bottom of the slider.

6. Pull the fork tube out of the slider.

7. Turn the fork tube over and remove the damper rod (**Figure 60**) and small spring.

8. Remove the retaining ring securing the oil seal in the slider.

9. Remove the oil seal by carefully prying it out of the slider with a screwdriver (**Figure 61**) or small pry bars. Make sure to work around the seal to prevent damaging the fork slider.

10. On late 1978-1983 FXWG and FXDG models, remove the washer under the oil seal (11, **Figure 57**).

11. Inspect the fork assembly as described in this chapter.

12. Installation is the reverse of these steps. Note the following.

13. Install new oil seals by driving them in with a socket placed on top of the seal. Make sure to drive the oil seals in squarely.

14. When installing the snap ring, make sure it seats completely in the machined groove.

15. Refill the fork tube with the correct type and quantity of fork oil as described in Chapter Three.

### Disassembly/Reassembly (1971-1975 FX)

Refer to **Figure 62** for this procedure.

1. Remove the front forks as described in this chapter.

2. *1971-1972:* Using Harley-Davidson tool No. 94694-52 (**Figure 63**), loosen and remove the spring retainer (7, **Figure 62**) from the top of the fork tube.

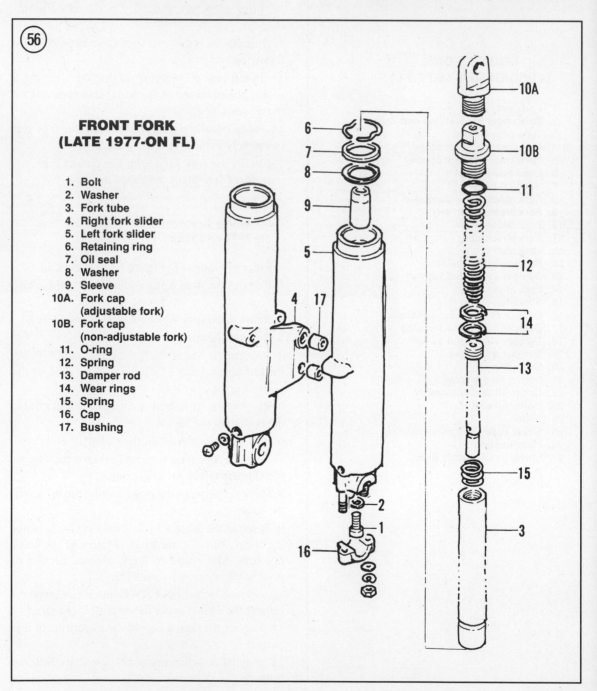

**(56)**

**FRONT FORK
(LATE 1977-ON FL)**

1. Bolt
2. Washer
3. Fork tube
4. Right fork slider
5. Left fork slider
6. Retaining ring
7. Oil seal
8. Washer
9. Sleeve
10A. Fork cap
 (adjustable fork)
10B. Fork cap
 (non-adjustable fork)
11. O-ring
12. Spring
13. Damper rod
14. Wear rings
15. Spring
16. Cap
17. Bushing

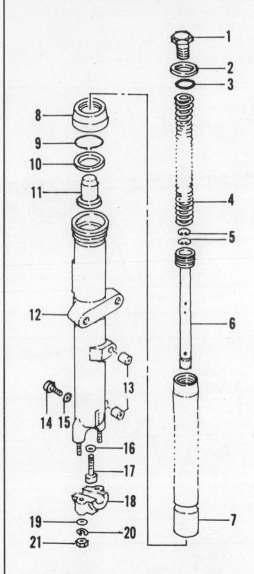

**57**

# FRONT FORK (LATE 1978-1983 FXWG AND FXDG)

| | |
|---|---|
| 1. Fork cap | 11. Washer |
| 2. O-ring | 12. Sleeve |
| 3. Spring | 13. Slider |
| 4. Piston rings | 14. Bushing |
| 5. Damper rod | 15. Washer |
| 6. Spring | 16. Bolt |
| 7. Fork tube | 17. Axle cap |
| 8. Dust boot | 18. Flat washer |
| 9. Retaining ring | 19. Lockwasher |
| 10. Oil seal | 20. Nut |

**58**

# FRONT FORK—1983 AND EARLIER FX MODELS (EXCEPT FXWG/FXDS)

| | |
|---|---|
| 1. Fork cap | 12. Slider |
| 2. Washer | 13. Bushing |
| 3. O-ring | 14. Drain screw |
| 4. Spring | 15. Washer |
| 5. Piston rings | 16. Washer |
| 6. Damper rod | 17. Bolt |
| 7. Fork tube | 18. Axle cap |
| 8. Dust boot | 19. Flat washer |
| 9. Retaining ring | 20. Lockwasher |
| 10. Oil seal | 21. Nut |
| 11. Sleeve | |

**8**

*NOTE*
*A substitute tool can be made by drilling 2 holes in a piece of 1/4 in. flat steel and installing 2 hardened pins the same distance apart as the holes in the spring retainer. Cut the steel long enough so that one end can be used as a handle. See* **Figure 64**.

3. Remove the spring (9, **Figure 62**).

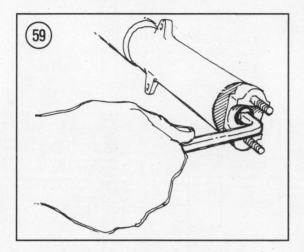

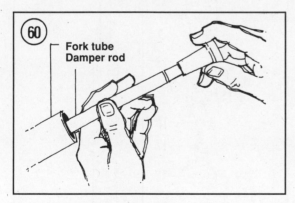

Fork tube
Damper rod

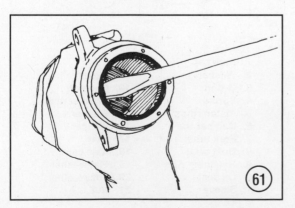

## FRONT FORK
## (1971-1975 FX)

1. Tube cap
1A. Cap washer (1973 and later)
2. Tube breather valve (1972 and earlier)
3. Tube cap seal
4. Pinch bolt
5. Fork boot
5A. Retaining ring (1973 and later)
5B. Retaining washer (1973 and later)
5C. Seal
6. Fork slide
7. Spring retainer (1972 and earlier)
8. Fork tube and shock absorber assembly
9. Fork spring
9A. Spring guide (1973 and later)
10. Fork slider
11. Fork slider bushing (1972 and earlier)
12. Tube end bolt and washer
13. O-ring (1972 and earlier)
14. Fork stem nut
15. Upper bracket pinch bolt
16. Upper bracket
17. Upper bearing shield
18. Upper bearing cone
19. Upper bearing cup
20. Lower bearing cup
21. Lower bearing cone
22. Lower bearing shield
23. Lower bracket and stem
24. Drain screw and washer
25. Cover screw
26. Cover screw insert
27. Cover

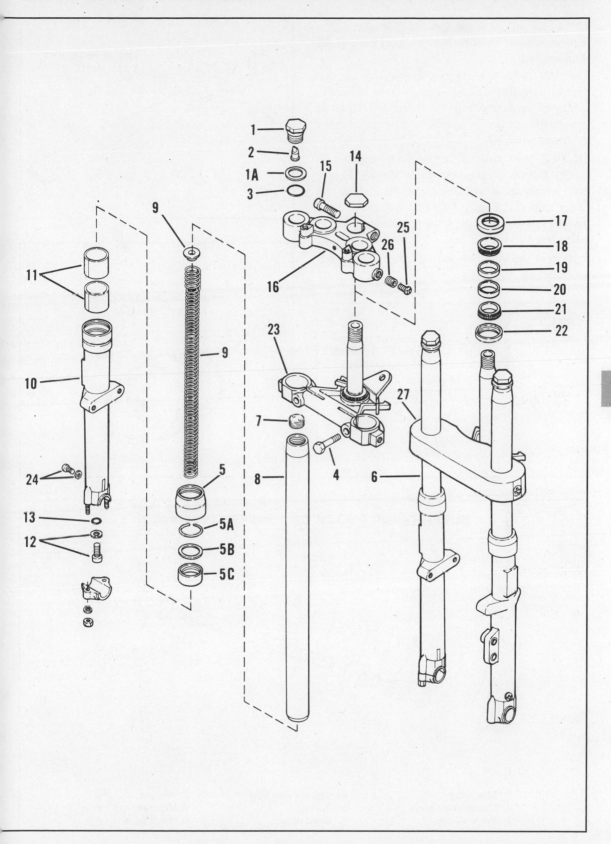

4. Pour the oil out of the fork and discard it. Pump the fork several times by hand to expel most of the remaining oil.

5A. *1971-1972:* Insert a long screwdriver through the top of the fork tube and engage it with the slot in the top of the damper rod (10, **Figure 65**). This will prevent the damper rod from turning when removing the lower Allen bolt.

5B. *1973-1975:* Insert a long ratchet extension with Harley-Davidson extension No. 94556-73 (**Figure 66**) and engage it with the top of the damper rod upper stop (11, **Figure 67**). This will prevent the damper rod from turning when removing the lower Allen bolt.

6. Remove the Allen bolt and washer (12, **Figure 62**) at the bottom of the slider.

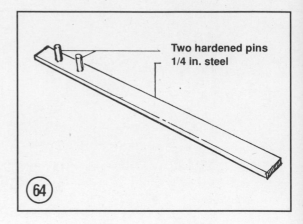

(64) Two hardened pins 1/4 in. steel

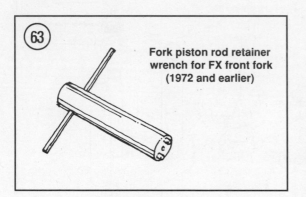

(63) **Fork piston rod retainer wrench for FX front fork (1972 and earlier)**

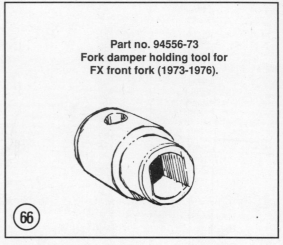

**Part no. 94556-73 Fork damper holding tool for FX front fork (1973-1976).**

(66)

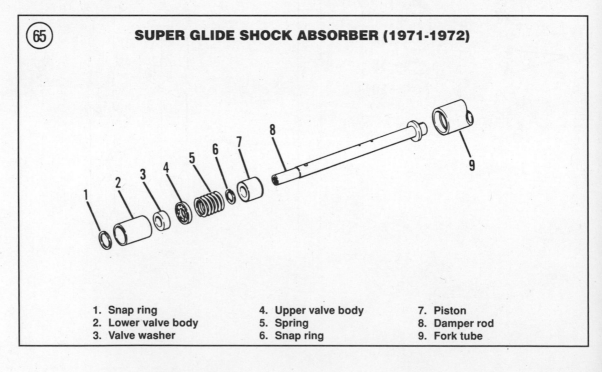

(65) **SUPER GLIDE SHOCK ABSORBER (1971-1972)**

1. Snap ring
2. Lower valve body
3. Valve washer
4. Upper valve body
5. Spring
6. Snap ring
7. Piston
8. Damper rod
9. Fork tube

7. Pull the fork tube out of the slider.

*NOTE*
*Step 8 and Step 9 describe removal of the damper rod assemblies.*

8A. *1971-1972:* Perform the following to disassemble and remove the damper rod assembly. See **Figure 65**.

   a.  Remove the snap ring.
   b.  Remove the lower valve body.
   c.  Remove the valve washer.
   d.  Remove the upper valve body.
   e.  Remove the spring.
   f.  Remove the piston snap ring and remove the piston.
   g.  Remove the damper rod.

8B. *1973-1975:* Perform the following to disassemble and remove the damper rod assembly. See **Figure 67**.

   a.  Remove the snap ring from the fork tube and remove the damper rod assembly.
   b.  Remove the lower piston and lower stop.
   c.  Remove the orifice washer and valve.
   d.  Remove the spring washer and valve body.
   e.  Remove the snap ring and the upper piston.
   f.  Remove the roll pin from the damper rod. Then unscrew the upper stop.

9. Slide the fork boot off of the fork tube.

10. *1973-1975:* Remove the fork seal retaining ring and washer from the slider.

11. Remove the oil seal by carefully prying it out of the slider with a screwdriver or small pry bars (**Figure 61**). Make sure to work around the seal to prevent damaging the fork slider.

12. On 1971 and 1972 models, the slider is equipped with an upper and lower bushing. If the bushings require replacement, refer service to a Harley-Davidson dealer as special tools are required.

13. Inspect the fork assembly as described in this chapter.

14. Installation is the reverse of these steps. Note the following.

15. Install new oil seals by driving them in with a socket placed on top of the seal. Make sure to drive the oil seals in squarely.

16. When installing snap rings, make sure they seat completely in their machined grooves.

17. Refill the fork tube with the correct type and quantity of fork oil as described in Chapter Three.

### Disassembly/Reassembly (1984 FX)

Refer to **Figure 68** or **Figure 69** for this procedure.

1. Remove the fork tube cap, washer and O-ring.
2. Remove the spring.

8

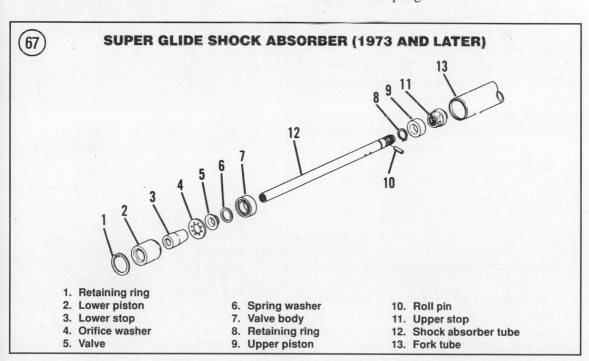

**67  SUPER GLIDE SHOCK ABSORBER (1973 AND LATER)**

1. Retaining ring
2. Lower piston
3. Lower stop
4. Orifice washer
5. Valve
6. Spring washer
7. Valve body
8. Retaining ring
9. Upper piston
10. Roll pin
11. Upper stop
12. Shock absorber tube
13. Fork tube

3. Pour the oil out of the fork and discard it. Pump the fork several times by hand to expel most of the remaining oil.

4. Slip the dust seal off of the lower fork tube and remove it.

5. Remove the retaining ring securing the oil seal in the slider.

6. Remove the Allen bolt and washer (**Figure 59**) at the bottom of the slider.

*NOTE*
*If the Allen bolt is difficult to remove in Step 6, the damper rod is turning as the bolt is turned. Reinstall the fork spring washer and fork tube cap to apply resistance to the damper rod. Then attempt to remove the Allen bolt.*

*NOTE*
*The bushing installed in the slider is an interference fit. When separating the fork tube and slider, the bushing, spacer seal and oil seal will be removed.*

7. While grasping the slider in one hand, work the fork tube in an up-and-down motion to hit the upper bushing installed on the fork tube against the lower bushing in the slider. As the lower bushing works its way free, it will push the oil seal and seal spacer out of the slider. Continue this action until the components are separated.

8. Remove the rebound sleeve from the damper rod.

9. Insert a small rod into the end of the damper rod and push it out of the fork tube. See **Figure 60**.

10. The bushing installed on the fork tube should not be removed unless worn or damaged. To replace it, perform the following:

    a. Wedge a screwdriver into the bushing split to expand the bushing slightly and slide it off of the fork tube.

    b. Install a new bushing by expanding the split as during removal. Expand the new split bushing only enough to fit it over the fork tube.

    c. Seat the new bushing into the groove in the fork tube.

11. Inspect the fork assembly as described in this chapter.

12. Install by reversing these removal steps. Note the following.

13. Install the rebound spring onto the damper rod and install the assembly into the fork tube.

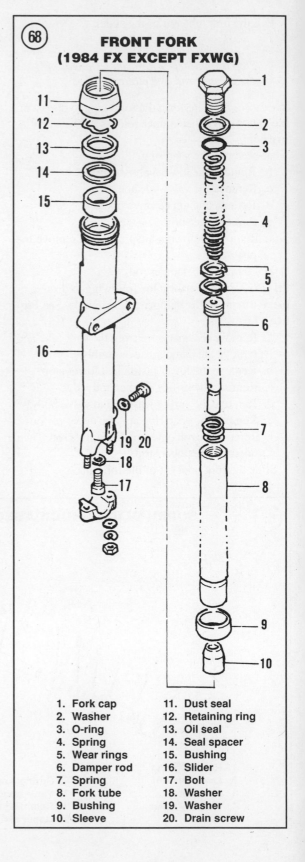

**68**

**FRONT FORK
(1984 FX EXCEPT FXWG)**

| | |
|---|---|
| 1. Fork cap | 11. Dust seal |
| 2. Washer | 12. Retaining ring |
| 3. O-ring | 13. Oil seal |
| 4. Spring | 14. Seal spacer |
| 5. Wear rings | 15. Bushing |
| 6. Damper rod | 16. Slider |
| 7. Spring | 17. Bolt |
| 8. Fork tube | 18. Washer |
| 9. Bushing | 19. Washer |
| 10. Sleeve | 20. Drain screw |

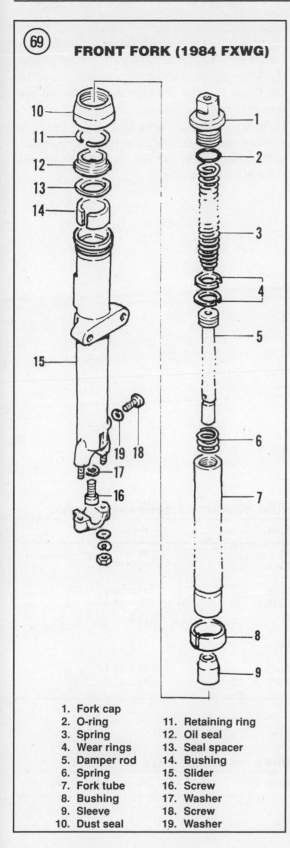

**FRONT FORK (1984 FXWG)**

1. Fork cap
2. O-ring
3. Spring
4. Wear rings
5. Damper rod
6. Spring
7. Fork tube
8. Bushing
9. Sleeve
10. Dust seal
11. Retaining ring
12. Oil seal
13. Seal spacer
14. Bushing
15. Slider
16. Screw
17. Washer
18. Screw
19. Washer

14. Install the spring into the fork tube so that the tapered end faces toward the damper rod.

15. Install the rebound sleeve onto the end of the damper rod.

16. Install the fork tube assembly into the slider. Use the fork spring to keep tension on the damper rod and install the lower Allen bolt and washer. Tighten the bolt securely.

17. Install a new bushing, seal spacer and oil seal over the fork tube.

*NOTE*
*Make sure the flanged surface of the seal spacer faces upward and that the lettering on the oil seal faces upward.*

18. Slide a suitable size hollow tube over the fork tube.

*NOTE*
*The hollow tube must be slightly smaller in diameter than that of the oil seal.*

19. Lightly tap the hollow tube and install the bushing, seal spacer and oil seal. Install the parts until the retaining ring groove in the slider is exposed. Remove the hollow tube.

20. Install the retaining ring. Make sure it seats fully in the groove.

21. Slide the dust seal over the fork tube and seat it in the slider.

22. Refill the fork tube with the correct quantity and weight of fork oil as described in Chapter Three.

**Inspection (All Models)**

1. Thoroughly clean all parts in solvent and dry them.

2. Check fork tube exterior for scratches and straightness. If bent or scratched, it should be replaced.

3. Make sure the small hole in groove at the bottom of the fork tube is clean.

4. Check the slider for dents or exterior damage that may cause the upper fork tube to hang up during riding. Replace if necessary.

5. Replace the oil seal if removed, as removal damages seals.

6. Replace any snap or retaining rings that appear weak or deformed. When installing snap or retaining

rings, compress or expand them only enough to install them and no more.

7. Check the damper rod piston rings for wear or damage. Replace the piston rings by removing them from the damper rod and installing new ones.

8. Check the lower slider for dents or exterior damage that may cause the upper fork tube to hang up during riding conditions. Replace if necessary.

9. Inspect the O-ring in the fork cap. Replace if worn or damaged. When installing a part with an O-ring, make sure the mating surface around the O-ring is free of any foreign material and perfectly smooth to prevent oil leaks.

10. When installing the fork tube into the slider, make sure to guide the tube through the oil seal carefully to prevent damage.

11. Any parts that are worn or damaged should be replaced. Simply cleaning and reinstalling unserviceable components will not improve performance of the front suspension.

### Table 1  FRONT SUSPENSION TIGHTENING TORQUES (1970-EARLY 1978)

|  | ft.-lb. | N·m |
|---|---|---|
| Front axle nut | 50 | 69 |
| Sprocket mounting bolts or nuts | 34-42 | 46.9-57.9 |
| Front fork |  |  |
|   Upper bracket pinch bolts | 22-26 | 30.4-35.9 |
|   Lower bracket pinch bolts | 22-26 | 30.4-35.9 |
|   Slider cap nuts | 11 | 15.2 |
| Handlebar |  |  |
|   Clamp screws | 20 | 27.6 |
|   Riser mount bolts |  |  |
|     FL models | 40-45 | 55.2-62.1 |
|     FX models | 55-70 | 75.9-96.6 |

### Table 2  FRONT SUSPENSION TIGHTENING TORQUES (LATE 1978-1984)

|  | ft.-lb. | N·m |
|---|---|---|
| Front axle nut | 50 | 69 |
| Slider cap nuts | 11 | 15.2 |

# CHAPTER NINE

# REAR SUSPENSION

This chapter includes repair and replacement procedures for the rear wheel, drive chain and belt and rear suspension components.

Tightening torques for 1970-1984 models are listed in **Table 1** (end of chapter). For 1966-1969 models, use the general torque specifications listed in Chapter One.

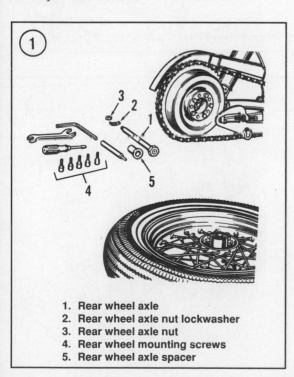

1. Rear wheel axle
2. Rear wheel axle nut lockwasher
3. Rear wheel axle nut
4. Rear wheel mounting screws
5. Rear wheel axle spacer

## REAR WHEEL

### Removal/Installation (1966-1972)

Refer to **Figure 1**.

1. Support the bike so that the rear wheel clears the ground.

2. Remove the rear fender support screws and raise the fender.

3. Remove the wheel-to-brake drum screws (4, **Figure 1**). Rotate wheel to gain access to each screw from behind the axle.

4. Pry the axle nut lockwasher away from the nut. Then loosen and remove the axle nut.

5. Remove the rear axle. If necessary, tap the axle out with an aluminum drift.

6. Remove the axle spacer.

7. Have assistant apply rear brake and remove the rear wheel.

8. If the wheel is going to be off for any length of time, or if it is to be taken to a shop for repair, install the axle spacer on the axle along with the axle nut to prevent losing any parts.

9. Install by reversing these removal steps. Note the following.

10. Make sure the brake drum-to-wheel hub mating surfaces are clean of all dirt and road debris.

11. Tighten the brake drum-to-wheel hub screws to 35 ft.-lb. (48.3 N•m).

12. Adjust the drive chain as described in Chapter Three.

13. Tighten the axle nut to 50 ft.-lb. (69 N•m).

14. Adjust the rear brake as described in Chapter Three.

15. Rotate the wheel several times to make sure it rotates freely and that the brakes work properly.

### Removal/Installation (1973-On)

Refer to **Figure 2** and **Figure 3** (typical) for this procedure.

1. Support the bike so that the rear wheel clears the ground.

2. Loosen the drive chain or drive belt adjusting locknuts and adjuster bolts.

3. Loosen and remove the axle nut and washer.

4. Slide the axle out of the wheel and allow the wheel to drop to the ground.

5. Remove the axle spacer.

6. Lift the drive chain or belt off of the sprocket and remove the rear wheel.

*NOTE*
*On disc brake models, insert a piece of wood or vinyl tubing in the caliper between the brake pads in place of the disc. That way, if the brake pedal is inadvertently depressed, the piston will not be forced out of the cylinder. If this does happen, the caliper might have to be disassembled to reseat the piston and the system will have to be bled. By using the wood or vinyl tubing, bleeding the brake should not be necessary when installing the wheel.*

7. If the wheel is going to be off for any length of time, or if it is to be taken to a shop for repair, install the chain adjusters and axle spacers on the axle along with the axle nut to prevent losing any parts.

*CAUTION*
*On disc brake models, do not set the wheel down on the disc surface, as it may be scratched or warped. Either lean the wheel against a wall or place it on a couple of wood blocks.*

8. Install by reversing these removal steps. Note the following.

9. On disc brake models, *carefully* insert the disc between the pads when installing the wheel.

10. Adjust the drive chain or drive belt as described in Chapter Three.

11. Tighten the axle nut to specifications in **Table 1**.

12. Adjust the rear brake as described in Chapter Three.

13. Rotate the wheel several times to make sure it rotates freely and that the brakes work properly.

### Inspection

Measure the axial and radial runout of the wheel with a dial indicator as shown in **Figure 4**. The maximum allowable axial and radial runout is 0.031 in. If the runout exceeds this dimension, check the

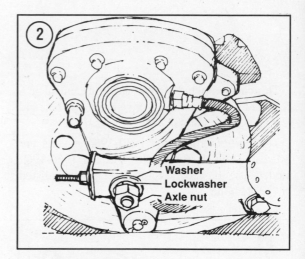

Washer
Lockwasher
Axle nut

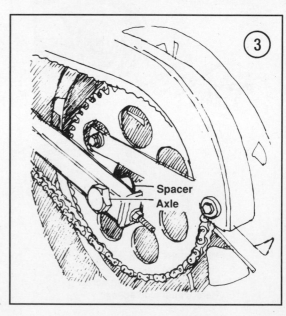

Spacer
Axle

wheel bearings. Some of this condition can be corrected on spoke wheels as described in Chapter Eight. If the wheel bearings are in good condition on cast wheels and no other cause can be found, the wheel will have to be replaced as it cannot be serviced.

Inspect the wheel for signs of cracks, fractures, dents or bends.

> *WARNING*
> *Do not try to repair any damage to a cast wheel as it will result in an unsafe riding condition.*

### Rear Wheel Bearing End Play Check

After tightening the axle nut, check the wheel bearing end play as follows.

1. Mount a dial indicator so that the plunger contacts the wheel hub.

2. Move the rear wheel from side to side and measure the axle end play.

3. The standard end play is 0.004-0.018 in. (0.10-0.46 mm). If the end play does not fall within this range, replace the axle spacer with a longer or shorter one if available. Spacers of various lengths (for some models) are available from Harley-Davidson dealers.

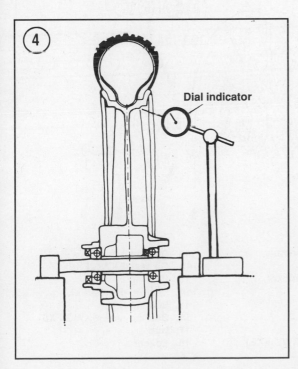

Dial indicator

## REAR HUB

### Disassembly/Inspection/Assembly (1966-early 1978)

The front and rear hubs on these models are interchangeable units and thus identical. To service the rear hub on a 1966-early 1978 model, refer to Chapter Eight under *Front Hub* for your model.

### Disassembly/Inspection/Assembly (Late 1978-on)

Refer to **Figure 5** or **Figure 6**, typical for this procedure.

> *NOTE*
> *The bearing cones and cups on these models are matched pairs. Label all parts so that they may be returned to their original positions.*

1. Remove the rear wheel as described in this chapter.

2. Remove the snap ring. Then remove the washer and spacer.

3. Remove the oil seal by prying it out of the hub.

4. Remove the bearing.

5. Repeat Steps 2-4 for the opposite side.

> *NOTE*
> *The following models do not use a circlip or washer on the left-hand side: 1983 FXDG and 1984 FXSB cast wheels and all FXE and FXWG models with laced wheels.*

6. Check the roller bearing cones and cups for wear, pitting or excessive heat (bluish tint). Replace the bearing cones and cups as a complete set. Replace the bearing cups as described in Step 7. If they do not require replacement, proceed to Step 9.

7. If the bearing cups require replacement, remove them using a blind bearing puller.

8. Remove the center spacer.

9. Clean the inside and outside of the hub with solvent. Dry with compressed air.

10. Installation is the reverse of these steps. Note the following.

11. Blow any dirt or foreign matter out of the hub prior to installing the bearings.

9

12. Pack the bearing cones with grease before installation.

13. Tap the oil seals (**Figure 7**) into the hub to the following specifications:

    a. Laced wheels (except FXWG and FXE): 13/64-7/32 in. (5.16-5.56 mm).

    b. Laced wheels (only FXWG and FXE): 3/16-1/4 in. (4.76-6.35 mm).

    c. Cast wheels (except 1983 FXDG and 1984 FXSB): 13/64-7/32 in. (5.16-5.56 mm).

    d. Cast wheels (only 1983 FXDG and 1984 FXSB): 5/16-21/64 in. (7.94-8.33 mm).

14. If the brake disc was removed, refer to Chapter Ten for correct procedures and tightening torques.

15. After the wheel is installed, the bearing end play should be checked as described in this chapter.

16. If the hub on spoke wheels is damaged, the hub can be replaced by removing the spokes and having a dealer assemble a new hub. If the hub on cast wheels is damaged, the wheel assembly must be replaced; it cannot be repaired.

## DRIVEN SPROCKET ASSEMBLY

### Removal/Installation (1966-1972)

Refer to **Figure 8** for this procedure.

1. Remove the rear wheel as described in this chapter.

2. Remove the rear brake drum as described in Chapter Ten.

3. Place the rear brake drum in a vise.

4. Working from the brake shell side, chisel the heads off of each sprocket-to-brake drum rivet head. Remove the rivets with a drift punch.

5. Separate the rear sprocket from the brake drum.

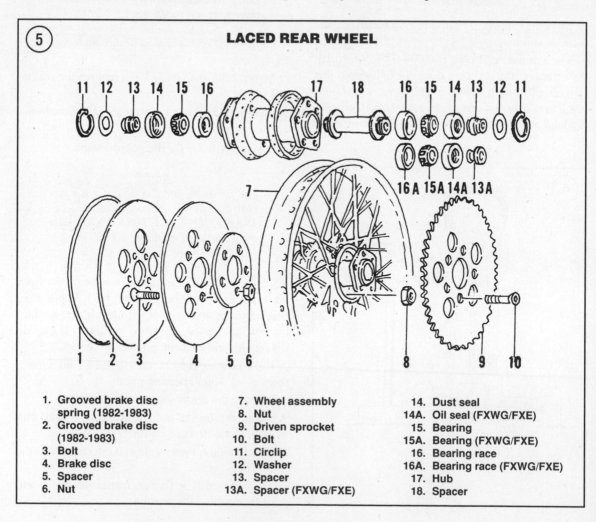

**⑤ LACED REAR WHEEL**

7 — Wheel assembly

16A 15A 14A 13A

| | |
|---|---|
| 1. Grooved brake disc spring (1982-1983) | 7. Wheel assembly |
| 2. Grooved brake disc (1982-1983) | 8. Nut |
| 3. Bolt | 9. Driven sprocket |
| 4. Brake disc | 10. Bolt |
| 5. Spacer | 11. Circlip |
| 6. Nut | 12. Washer |
| | 13. Spacer |
| | 13A. Spacer (FXWG/FXE) |

14. Dust seal
14A. Oil seal (FXWG/FXE)
15. Bearing
15A. Bearing (FXWG/FXE)
16. Bearing race
16A. Bearing race (FXWG/FXE)
17. Hub
18. Spacer

⑥ **CAST REAR WHEEL (ALL MODELS)**

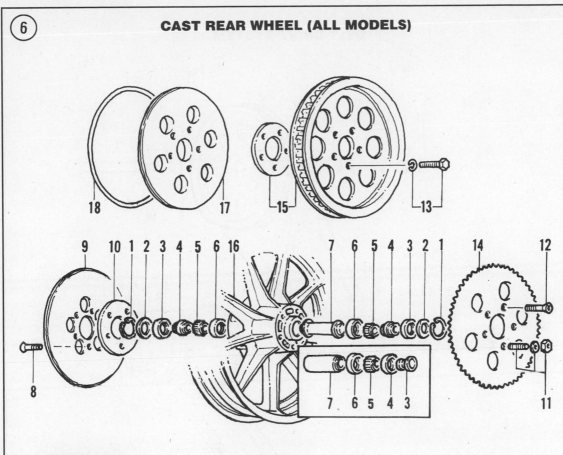

1. Retaining ring
   (1983 and earlier)
2. Washer (1983 and earlier)
3. Spacer
4. Oil seal
5. Bearing
6. Bearing race
7. Spacer
8. Screw
9. Brake disc
10. Spacer
11. Sprocket mounting
    hardware (FL models)
12. Sprocket mounting
    hardware (FX models)
13. Sprocket mounting
    hardware (belt drive)
14. Sprocket (chain drive)
15. Sprocket and spacer
    (belt drive)
16. Rear wheel
17. Brake disc (1982-on FLH)
18. Brake disc spring
    (1982-on FLH)

9

⑦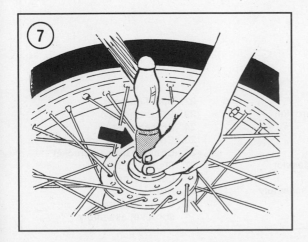

⑧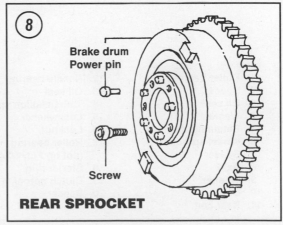

Brake drum
Power pin

Screw

**REAR SPROCKET**

**BELT DRIVE SYSTEM**

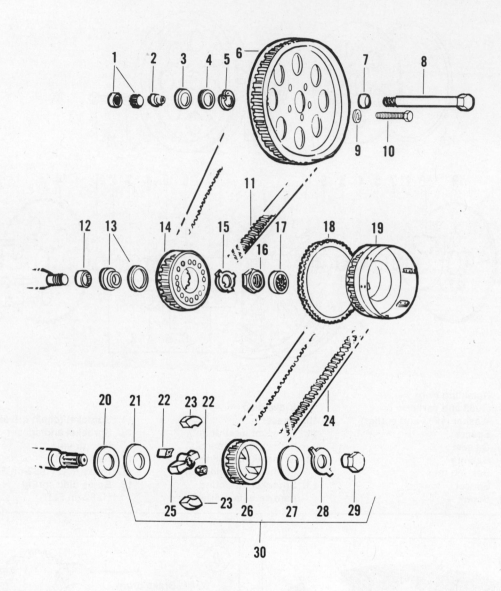

1. Roller bearing
2. Spacer
3. Oil seal
4. Washer
5. Retaining ring
6. Wheel sprocket
7. Spacer
8. Axle
9. Washer
10. Bolt
11. Rear belt
12. Needle bearing
13. Oil seal
14. Transmission sprocket
15. Lockwasher
16. Locknut
17. Roller bearing
    (not on FLH 80 classic)
18. Starter ring
19. Clutch sprocket assembly
20. Spacer
21. Bearing disc
22. Small damper
23. Large damper
24. Primary belt
25. Compensator drive hub
26. Compensating sprocket
27. Flange bearing
28. Lockwasher
29. Nut
30. Compensating
    sprocket assembly

6. Examine the brake drum as described in this chapter. If the drum is okay, proceed to Step 7.

7. Check the brake drum rivet holes. If the holes are elongated or damaged, new rivet holes will have to be drilled in the brake drum. If new holes have to be drilled, proceed to Step 8. If the holes are okay, proceed to Step 9.

8. Drill new rivet holes in the brake drum as follows:

   a. Purchase a new sprocket, if required.

   b. The sprocket holes will be used as a template to drill the new holes.

   c. Place the sprocket over the brake drum.

   d. Align the sprocket holes midway between the original dowel pins and the old rivet holes.

   e. Use a No. 10 drill bit to drill all of the rivet holes.

*NOTE*
*It is important to drill the holes as accurately as possible; clamp the sprocket in place when drilling the holes.*

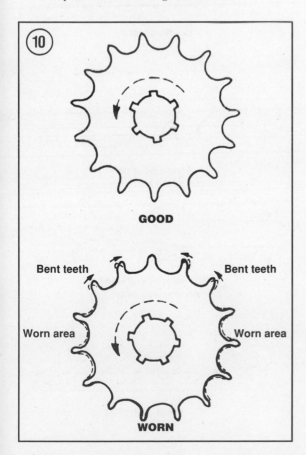

**GOOD**

Bent teeth             Bent teeth

Worn area             Worn area

**WORN**

*NOTE*
*All holes are drilled from the brake shell side.*

   f. Drill the first hole. Then install a new rivet. Do not set rivet heads at this point.

   g. Drill the second hole directly opposite the first hole. Install a new rivet as in Step "f."

   h. Drill the remaining rivet holes.

   i. Remove the 2 rivets and separate the sprocket from the brake drum.

   j. Remove all burrs from the new holes.

9. Drill new dowel pin holes as follows:

   a. Assemble the sprocket and brake drum with 2 or 3 new rivets as previously described. Do not rivet the heads.

   b. Using the sprocket as a template, drill the 4 dowel pin holes with a 3/16 in. (4.76 mm). drill.

   c. Remove the rivets and separate the sprocket from the brake drum.

   d. Remove all burrs from the new holes.

*NOTE*
*The dowel pins and rivets are installed from the brake shell side.*

10. Install sprocket over brake hub. Align the new dowel pin holes and install the new dowel pins.

11. Install the new rivets and set them securely.

### Removal/Installation (1973-on)

Refer to **Figure 9** for this procedure.

1. Remove the rear wheel as described in this chapter.

2. Remove the bolts and nuts securing the sprocket to the hub and remove the sprocket.

3. Remove any sprocket spacer as required.

4. Installation is the reverse of these steps. Tighten the sprocket bolts to specifications in **Table 1**.

### Inspection

Inspect the teeth of the sprocket. If the teeth are visibly worn (**Figure 10**), replace both sprockets and the drive chain or belt. Never replace any one sprocket or chain as a separate item; worn parts will cause rapid wear of the new component. Refer to *Drive Chain Adjustment* and *Drive Belt Adjustment* in Chapter Three.

## DRIVE CHAIN

### Removal/Installation

1. Loosen the rear axle nut, chain adjuster nuts and the anchor bolt (if so equipped). See **Figure 2** and **Figure 3**.

2. Push the rear wheel as far forward in the swing arm as possible.

3. Turn the rear wheel and locate the drive chain master link on the rear sprocket (**Figure 11**).

4. Remove the master link spring clip and separate the chain.

> *NOTE*
> *It may be necessary to use a chain breaking tool to press the connecting link from the side plate. Chain breakers can be purchased at most motorcycle dealerships.*

5. If installing a new drive chain, connect the new chain to the old chain with the old master link. Pull the new chain through the front sprocket. If the original chain is to be reinstalled, tie a piece of wire approximately 30 inches (76.2 cm) long to the drive chain. Pull the chain so that the wire is routed around the front sprocket. Disconnect the wire from the chain so that it can be used to route the chain during installation.

6. Install by reversing these removal steps. Note the following:

    a. Install a new drive chain master link spring clip with the closed end facing in the direction of chain travel (**Figure 11**).

    b. Adjust the drive chain as described in Chapter Three.

    c. Tighten the axle nut to the torque valves in **Table 1**.

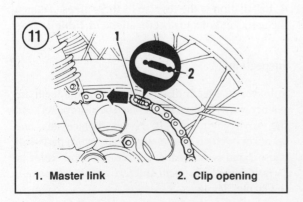

**1. Master link**      **2. Clip opening**

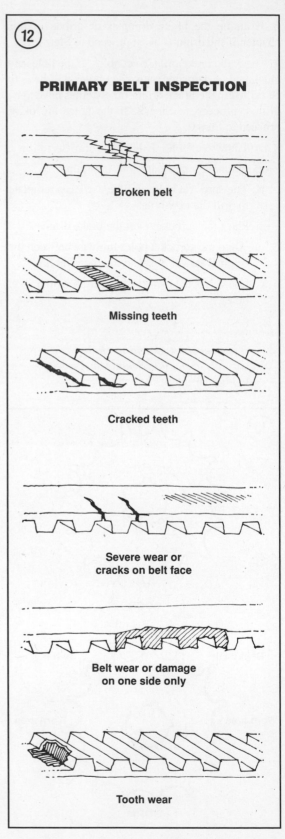

## PRIMARY BELT INSPECTION

**Broken belt**

**Missing teeth**

**Cracked teeth**

**Severe wear or cracks on belt face**

**Belt wear or damage on one side only**

**Tooth wear**

d. Rotate the wheel several times to make sure it rotates smoothly. Apply the brake several times to make sure it operates correctly.

e. Adjust the rear brake as described in Chapter Three.

## Lubrication

For lubrication of the drive chain, refer to Chapter Three.

## DRIVE BELT

### Removal/Installation

Refer to **Figure 9** for this procedure.

1. Remove the rear wheel as described in this chapter.

2. Remove the compensating sprocket and clutch as described in Chapter Five.

3. Remove the primary housing as described in Chapter Five.

4. Remove the swing arm as described in this chapter.

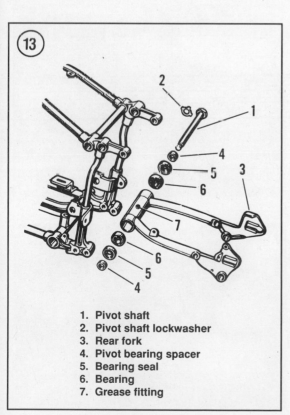

1. Pivot shaft
2. Pivot shaft lockwasher
3. Rear fork
4. Pivot bearing spacer
5. Bearing seal
6. Bearing
7. Grease fitting

5. Remove the drive belt from the sprocket.

> *CAUTION*
> *If the drive belt is going to be reused, never bend the belt sharply as this will weaken the belt and cause premature failure.*

6. Installation is the reverse of these steps. Adjust the drive belt tension as described in Chapter Three.

### Inspection

The drive belt has a built-in polyethylene lubricant coating that burnishes off during break-in. Do not apply lubricants. Inspect the drive belt for wear or damage. Replace any belt that appears questionable. See **Figure 12**.

> *CAUTION*
> *When handling a drive belt, never bend the belt sharply as this will weaken the belt and cause premature failure.*

## WHEEL BALANCING

For complete information refer to *Wheel Balancing* in Chapter Eight.

## TIRE CHANGING

Refer to *Tire Changing* in Chapter Eight.

## REAR SWING ARM

### Removal/Installation

Refer to **Figure 13** for this procedure.

1. Support the bike so that the rear wheel is off the ground.

2. Remove the saddlebags (if so equipped).

3. Remove the mufflers as described in Chapter Six.

4. Remove the rear wheel as described in this chapter.

5. Remove the lower mounting bolts (**Figure 14**) and lockwashers securing the shock absorbers. Do not remove the shock absorber units. Remove the drive chain guard.

6. Remove the rear brake drum or caliper as described in Chapter Ten.

9

*NOTE*
*It is not necessary to disconnect the hydraulic lines. Instead, hang the brake caliper from the frame with wire or bungee cords.*

7. Prior to completing swing arm removal, check its condition by grasping the swing arm on both sides and trying to move it from side to side. If the free play is excessive (approximately 1/32 in. [0.79 mm] of movement) replace the swing arm bushings as described under *Rear Swing Arm Bushing Replacement* in this chapter.

8. Pry back the pivot shaft lockwasher (2, **Figure 13**).

9. Loosen and remove the pivot shaft from the swing arm from the right-hand side.

*NOTE*
*If the pivot shaft is difficult to remove, the primary chain housing (**Figure 15**) must be removed so the pivot shaft (**Figure 16**) can be tapped out from the left-hand side. Primary chain housing removal is described in Chapter Five. Use an aluminum drift or rod and carefully tap the shaft out. Do not use a steel drift or punch as this may damage the pivot shaft threads.*

10. Pull back on the drive chain or drive belt and remove the swing arm.

11. Install by reversing these removal steps. Note the following.

*NOTE*
*Be sure to slide the swing arm through the drive belt or drive chain as it must be on the inside of the swing arm.*

12. Pre-load the swing arm Timken bearings as follows:

   a. After swing arm is installed on bike, install the rear wheel axle through the swing arm at the axle's position. Then attach a spring scale to the axle and lift spring scale to raise the swing arm to a horizontal position.

   b. Note reading on scale. Tighten pivot shaft to pre-load bearings 1-2 lbs. (0.45-0.91 kg). For example, if scale reading is five pounds with swing arm in horizontal position, tighten pivot shaft until scale reads 6-7 lbs. (2.72-3.17 kg) when the swing arm is raised.

   c. Bend the lockwasher tab over the pivot shaft.

13. Grease the swing arm bearings with a hand grease gun through the grease fitting (1966-on models) in the center of the swing arm.

14. Tighten the lower shock absorber nuts securely.

15. Adjust the drive chain or drive belt as described in Chapter Three.

## Rear Swing Arm Bearing Replacement

Bearing replacement requires that the bearings be installed with a press. Refer to **Figure 13**.

1. Secure the swing arm in a vise with soft jaws.

*NOTE*
*Tag all components when removed from the swing arm so they can be reinstalled in their original positions. Bearing components must not be intermixed.*

2. Pull the bearing spacers out of the swing arm.

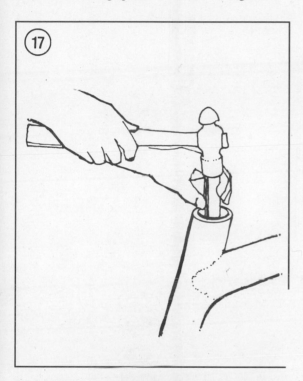

3. Pry the oil seals with a suitable size drift.

4. With your hand, roll the bearings inside the swing arm. If they are rough or damaged, remove them as follows.

5. Drive the bearings out with an aluminum or brass drift from the opposite end (**Figure 17**).

6. Thoroughly clean out the inside of the swing arm with solvent and allow to dry.

*NOTE*
*The new bearings should be installed with a press. If you do not have access to one, take the swing arm and new bearings to a Harley-Davidson dealer or machine shop and have the parts installed.*

7. Pack the new bearings with bearing grease. Press the new bearings into the swing arm.

8. Install 2 new bearing seals.

9. Install the bearing spacers in the bearing seals.

### SHOCK ABSORBERS

The rear shocks are spring controlled and hydraulically damped. Spring preload can be adjusted on all models. Refer to Chapter Three.

### Removal/Installation

Removal and installation of the rear shocks is easier if they are done separately. The remaining unit will support the rear of the bike and maintain the correct relationship between the top and bottom mounts. If both shock absorbers must be removed at the same time, cut a piece of flat metal a few inches longer than the shock absorber and drill two holes in the metal the same distance apart as the bolt holes in a shock absorber. Install the metal support after one shock absorber is removed. This will allow the bike to be easily moved around until the shock absorbers are reinstalled or replaced.

1. Place the bike on the centerstand.

2. Remove the upper and lower bolts (**Figure 18**).

3. Pull the shock off.

4. Install by reversing these removal steps. Torque the upper and lower bolts securely.

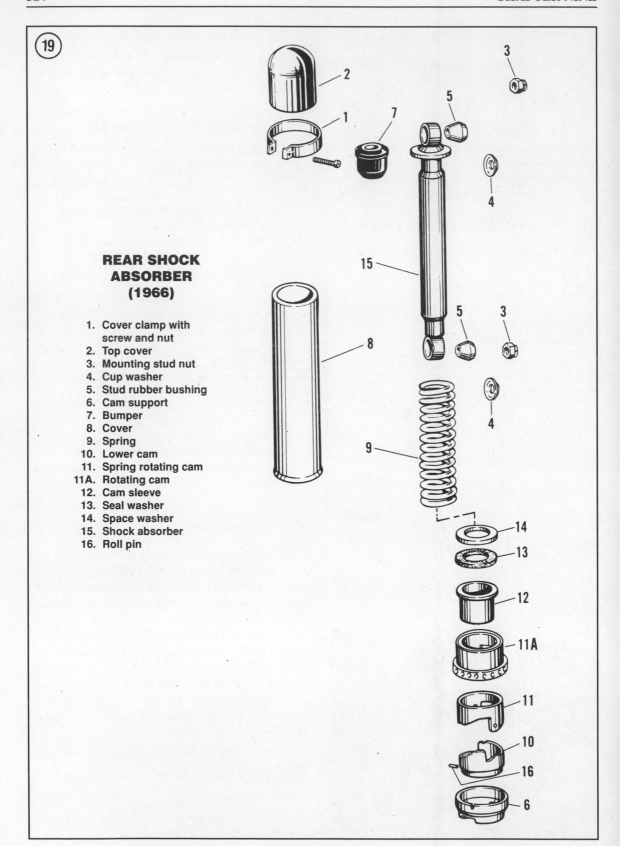

⑲

## REAR SHOCK ABSORBER (1966)

1. Cover clamp with screw and nut
2. Top cover
3. Mounting stud nut
4. Cup washer
5. Stud rubber bushing
6. Cam support
7. Bumper
8. Cover
9. Spring
10. Lower cam
11. Spring rotating cam
11A. Rotating cam
12. Cam sleeve
13. Seal washer
14. Space washer
15. Shock absorber
16. Roll pin

## Disassembly/Assembly (1966)

Refer to **Figure 19** for this procedure.

1. Remove the shock absorber as described in this chapter.

2. Loosen the cover clamp, then slip off the top cover (2, **Figure 19**).

3. Remove the upper and lower rubber bushings.

> *WARNING*
> *Do not attempt to disassemble the shock absorbers without use of the proper spring compression tool.*

4. Using a shock absorber spring compression tool, compress the shock absorber spring enough to turn lower eye of shock absorber 90° into cam support slot (6). **Figure 20** shows the Harley-Davidson shock compressor (97010-52A).

5. Release spring pressure, then remove the shock absorber assembly from tool.

6. Remove the cam support (6).

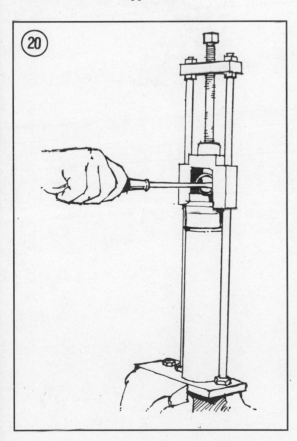

7. Tap the lower end of shock absorber to free the bumper (7) from its retaining flange inside the cover (8).

8. Slide the spring off of the shock absorber.

9. Remove the lower cam (10), spring rotating cam (11 and 11A), cam sleeve (12), seal washer (13), and the spacer washer (14).

10. Remove the bumper (7) by extending the shock shaft and slipping the bumper off the shock.

11. Inspect the shock absorber as described in this chapter.

12. Assembly is the reverse of these steps. Note the following.

13. Lightly grease the cam sleeve and spring rotating cam.

## Disassembly/Assembly (1967-on)

Refer to **Figure 21** (FL) or **Figure 22** (FX) for this procedure.

1. Remove the shock absorber as described in this chapter.

2. Remove the upper and lower rubber bushings.

3. Remove the retaining ring (5, **Figure 21**) if so equipped.

> *WARNING*
> *Do not attempt to disassemble the shock absorbers without use of the proper spring compression tool.*

4. Using a shock absorber spring compression tool, compress the shock absorber spring and remove the upper spring retainer. **Figure 20** shows the Harley-Davidson spring compressor (97010-52A).

5. Release spring pressure, then remove the shock absorber assembly from tool.

6A. *FL models:* Referring to **Figure 21**, remove the following parts in order:

    a. Washer (8).

    b. Spring (9).

    c. Washer (8).

    d. Seal washer (10).

    e. Adjusting cup (11).

    f. Cam assembly (12).

6B. *FX models:* Referring to **Figure 22**, remove the following parts in order:

    a. Spring retainer (4).

    b. Spring seat (5).

    c. Washer (6).

    d. Spring (7).

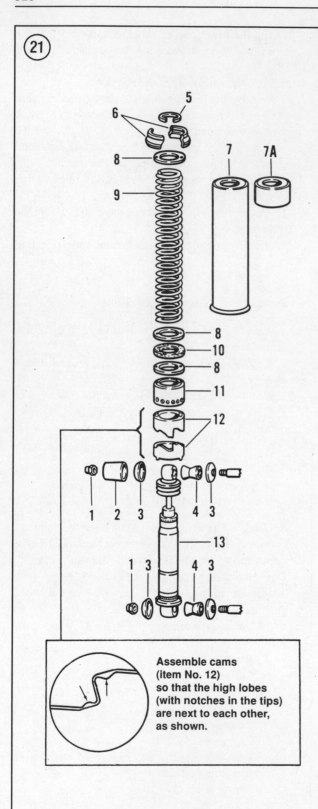

**REAR SHOCK ABSORBER (1967-ON FL)**

1. Mounting stud nut
2. Stud cover
3. Cup washer
4. Stud rubber bushing
5. Retaining ring
6. Split key
7. Cover (long)
7A. Cover (short)
8. Washer
9. Spring
10. Seal washer
11. Adjusting cup
12. Cam
13. Shock absorber unit

Assemble cams (item No. 12) so that the high lobes (with notches in the tips) are next to each other, as shown.

e. 2 washers (6).

f. Spring guide (8).

g. Adjuster spring (9).

h. Adjusting cup (10).

i. Cam (11 and 12).

7. Inspect the shock absorber as described in this chapter.

8. Assembly is the reverse of these steps. Note the following.

9. Lightly grease all cam parts before assembly.

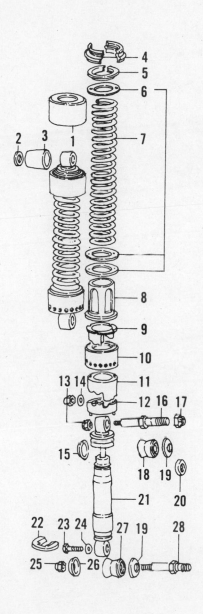

## SHOCK ABSORBER (ALL FX TYPICAL)

1. Cover
2. Plain washer
3. Stud cover (typical)
4. Upper spring retainer
5. Upper spring seat (late 1984)
6. Washers
7. Spring
8. Lower spring guide (late 1984)
9. Spring adjuster (late 1984)
10. Adjuster cup
11. Cam
12. Cam
13. Nut
14. Plain washer
15. Outer washer
16. Upper stud
17. Nut
18. Bushing
19. Outer washer
20. Spacer (not used with safety guards)
21. Shock absorber
22. Cam support
23. Bolt
24. Plain washer
25. Nut
26. Outer washer
27. Bushing
28. Lower stud (1971-early 1972)

9

**Table 1 REAR SUSPENSION TIGHTENING TORQUES (1970-1984)**

|                                      | ft.-lb. | N·m       |
| ------------------------------------ | ------- | --------- |
| Rear axle nut                        | 50      | 69        |
| Sprocket mounting bolts or nuts      |         |           |
| 1970-early 1978                      | 34-42   | 46.9-57.9 |
| Late 1978-on                         |         |           |
| Laced wheels                         | 35      | 48.3      |
| Cast wheels                          | 45-50   | 62.1-69   |

# CHAPTER TEN

# BRAKES

This chapter describes repair and replacement procedures for all brake components.

**Table 1** lists tightening torques for 1970-1984 models. For 1966-1969 models, use the general torque specifications in Chapter One.

> *WARNING*
> *Harley-Davidson specifies DOT 3 brake fluid for models produced prior to September, 1976 and DOT 5 for later models. Mixing the two types of brake fluid can cause brake failure. If you own a 1976 or 1977 model take your frame number to a Harley-Davidson dealer to find out your bike's production date.*

## FRONT DRUM BRAKES

A typical drum brake system consists of a brake drum, camshaft, pivot post, brake shoes and linings. See **Figure 1**.

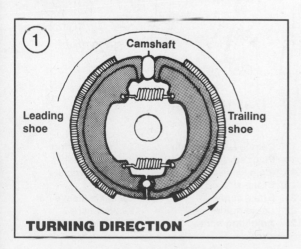

**Removal/Installation (1966-1971 FL)**

Refer to **Figure 2** for this procedure.

1. Remove the front wheel as described in Chapter Eight.

2. Remove the front brake drum bolts (2, **Figure 3**) and remove the brake drum (8, **Figure 3**).

> *CAUTION*
> *Do not touch the brake linings (6, **Figure 2**) during removal. Instead, place a clean shop rag on the linings to protect them from oil and grease.*

> *NOTE*
> *Prior to removing the brake shoes from the backing plate, measure them as described under **Inspection** in this chapter.*

3. Examine the brake shoes for identification marks "L" or "R." These markings represent left- and right-hand brake shoes. If the brake shoes on your bike are not identified, mark the casting with a felt-tipped pen to represent alignment. The brake shoes must be returned to their original positions during installation.

4. Remove the brake shoes from the backing plate by firmly pulling up on the center of each shoe as shown in **Figure 4**.

5. Remove the return springs (7, **Figure 2**) and separate the shoes (6, **Figure 2**).

6. If necessary, remove the cam lever (14, **Figure 2**) assembly as follows:

    a. Remove the cam lever stud lockwasher and washer.

10

b. Remove the cable clevis clamp nut.

c. Operate the front brake lever and slip the cam lever assembly off of its stud.

7. Inspect the front brake assembly as described in this chapter.

8. Installation is the reverse of these steps. Note the following.

9. Assemble the cam lever assembly in the order shown in **Figure 2**. Make sure to use new cotter pins (12, **Figure 2**) for the cam lever stud (16, **Figure 2**) and cam lever clevis pin (17, **Figure 2**).

10. Connect the brake shoes by installing the top brake shoe return spring.

11. Position the brake shoes in the brake side cover. Install the top brake shoe pivot in the pivot stud and the lower brake shoe pivot on the cam lever.

12. Install the lower brake shoe return spring.

*NOTE*
*The brake shoe return spring must be installed in the brake shoe notch nearest the brake side cover.*

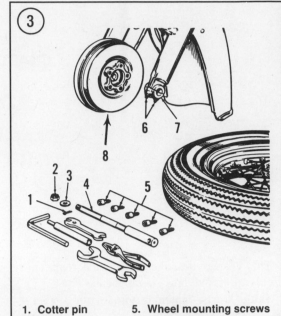

| 1. Cotter pin | 5. Wheel mounting screws |
|---|---|
| 2. Axle nut | 6. Slider cap nuts |
| 3. Flat washer | 7. Slider cap |
| 4. Front axle | 8. Brake drum |

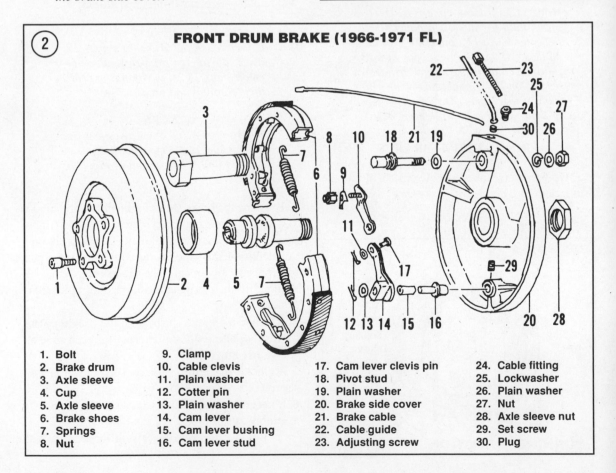

**FRONT DRUM BRAKE (1966-1971 FL)**

| 1. Bolt | 9. Clamp | 17. Cam lever clevis pin | 24. Cable fitting |
|---|---|---|---|
| 2. Brake drum | 10. Cable clevis | 18. Pivot stud | 25. Lockwasher |
| 3. Axle sleeve | 11. Plain washer | 19. Plain washer | 26. Plain washer |
| 4. Cup | 12. Cotter pin | 20. Brake side cover | 27. Nut |
| 5. Axle sleeve | 13. Plain washer | 21. Brake cable | 28. Axle sleeve nut |
| 6. Brake shoes | 14. Cam lever | 22. Cable guide | 29. Set screw |
| 7. Springs | 15. Cam lever bushing | 23. Adjusting screw | 30. Plug |
| 8. Nut | 16. Cam lever stud | | |

13. Slide the brake drum (2, **Figure 2**) over the brake shoes. See **Figure 3**.

14. Install the brake drum bolts (1, **Figure 2**) and tighten securely.

15. Adjust the front brake as described in Chapter Three.

### Removal/Installation (1971-1972 FX)

Refer to **Figure 5** for this procedure.

1. Remove the front wheel as described in Chapter Eight.

2. Disconnect the brake cable at the operating lever and remove the brake assembly.

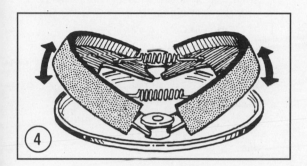

*CAUTION*
*Do not touch the brake linings (8, Figure 5) during removal. Instead place a clean shop rag on the linings to protect them from oil and grease.*

*NOTE*
*Prior to removing the brake shoes from the backing plate, measure them as described under **Inspection** in this chapter.*

3. Examine the brake shoes for identification marks "L" or "R." These markings represent left- and right-hand brakes shoes. If the brake shoes on your bike are not identified, mark the casting with a grease pencil to represent alignment. The brake shoes must be returned to their original positions during installation.

4. Remove the operating shaft nut (2) and lever (3) from the backside of the brake side plate.

5. Tap the operating shaft (4) with a plastic-tipped hammer and remove the brake shoes, operating shaft and shoe pivot stud as one assembly.

6. Slide the shoe pivot stud out from the brake shoes.

7. Remove the brake shoes (8).

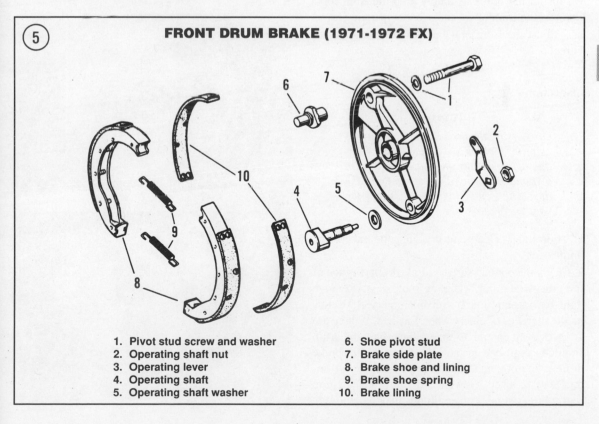

**FRONT DRUM BRAKE (1971-1972 FX)**

| 1. Pivot stud screw and washer | 6. Shoe pivot stud |
|---|---|
| 2. Operating shaft nut | 7. Brake side plate |
| 3. Operating lever | 8. Brake shoe and lining |
| 4. Operating shaft | 9. Brake shoe spring |
| 5. Operating shaft washer | 10. Brake lining |

10

8. Remove the return spring (9) and separate the shoes.

9. Inspect the front brake assembly as described in this chapter.

10. Installation is the reverse of these steps. Note the following.

11. Slide the shoe pivot stud (6) into the brake side plate (7).

12. Install the washer onto the operating shaft (4). Then slide the operating shaft into the brake side plate. Install the operating lever (3) and secure with the nut.

13. Connect the brake shoes by installing the top brake shoe return spring.

14. Position the brake shoes in the brake side cover. Install the top brake shoe pivot in the shoe pivot stud (6) and the lower brake shoe pivot on the operating shaft cam (4).

15. Install the lower brake shoe return spring.

*NOTE*
*The brake shoe return spring must be installed in the brake shoe notch nearest the brake side cover.*

16. Install the brake assembly onto the front fork. Connect the front brake cable to the operating lever.

17. Adjust the front brake as described in Chapter Three.

**Inspection**

*WARNING*
*Do not inhale brake dust. It contains asbestos, which can cause lung damage. Clean brake dust off of parts with a vacuum cleaner. Do not inhale brake dust when emptying or discarding the vacuum cleaner bag.*

1. Thoroughly clean and dry all parts except the linings.

2. Check the contact surface of the drum (**Figure 6**) for cracks, scoring, glaze or roughness. Remove light scoring and glaze with fine emery cloth. Remove all traces of emery when finished. If there are grooves deep enough to snag a fingernail, the drum should be reground and new shoes fitted. If the brake drum is in good condition, standard-size brake linings can be used. However, if the drums require regrinding, oversize brake linings may be required.

Consult your Harley-Davidson dealer for further information.

3. On FL models, check the wheel bearing in the hub on the brake shoe side (**Figure 7**). If the bearing's seal is damaged, it could leak grease and thus contaminate the brake shoes. If necessary, replace the bearing (**Figure 7**) as described in Chapter Eight.

*CAUTION*
*Before installing a brake drum that has been reground, check it for any traces of cuttings or other abrasives left from cutting. If necessary, clean with lacquer thinner. In addition, do not touch the ground surface with your hands as this would contaminate it, thus reducing brake effectiveness.*

*NOTE*
*Oil and grease on the drum surface should be cleaned with a clean rag soaked in lacquer thinner—do not use any solvent that may leave an oil residue.*

*NOTE*
*Riveted brake linings are used on all models.*

4. Inspect the linings for embedded foreign material. Dirt can be removed with a stiff wire brush. Check for traces of oil or grease; if contaminated, the linings must be replaced. Brake lining wear depends upon how close the rivet head is to the brake lining surface. Check each rivet-to-lining area for maximum wear. If necessary, the brake linings can be replaced as follows.

*NOTE*
*Always replace brake linings or shoes in sets of two.*

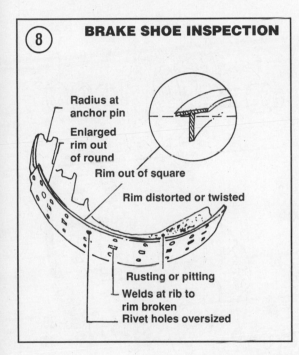

## BRAKE SHOE INSPECTION

Radius at anchor pin

Enlarged rim out of round

Rim out of square

Rim distorted or twisted

Rusting or pitting

Welds at rib to rim broken

Rivet holes oversized

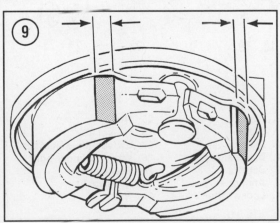

*NOTE*
*Make sure to check the brake drum surface as described in Step 2 before replacing brake linings; oversize linings may be required.*

5. Before replacing brake linings, check the shoe web (**Figure 8**) for cracks or distortion. To check for distortion, lay the web on a flat surface, such as a piece of glass or surface plate. Check how close the web seats along the glass. If the web is twisted at any point, the shoe is bent and must be replaced.
6. Replace riveted brake linings as follows:
  a. Drill out the rivets. Match the drill bit to the rivet head as closely as possible to prevent enlarging the rivet hole in the shoe web.

*CAUTION*
*Do not attempt to punch out the rivets as this may distort the shoe web.*

  b. With a steel brush, clean the brake lining surface thoroughly.
  c. Rivet the new linings to the brake shoes. It is important to start riveting from one end of the lining and work toward the other end to prevent buckling.
  d. Bevel the ends of both brake shoes with a file (**Figure 9**).
7. On early style brakes, check the cam lever stud play on the cam lever. If excessive play is detected, have a Harley-Davidson dealer install a new cam lever bushing.
8. Inspect the brake shoe return springs for wear. If they are stretched, they will not fully retract the brake shoes from the drum, resulting in a power-robbing drag on the drums and premature wear of the linings. Also check springs for rust. Replace as necessary; always replace as a pair.

### REAR DRUM BRAKE
### (1966-1972)

All 1966-1972 models are equipped with a rear hydraulic drum brake assembly (**Figure 10**).
1. Remove the rear wheel as described in Chapter Nine.
2. Remove the brake drum (3, **Figure 10**).

*CAUTION*
*Do not touch the brake linings (12, **Figure 10**) during removal. Instead, place*

10

*a clean shop rag on the linings to pro-
tect them from oil and grease.*

3. The rear brake shoes are manufactured in different widths. The wide shoe is installed in the front (facing front of bike) and the narrow shoe in the rear. The brake shoes should be identified with an "F" or "R" respectively. Mark the brake shoe casting with a felt-tipped pen to represent positioning. The brake

shoes must be returned to their original positions during installation.

4. Remove the upper brake shoe return spring.

5. Remove the anchor stud nut (25) and washer from the back side of the brake side plate (21).

6. Remove the brake shoes from the brake side plate together with the anchor stud (23).

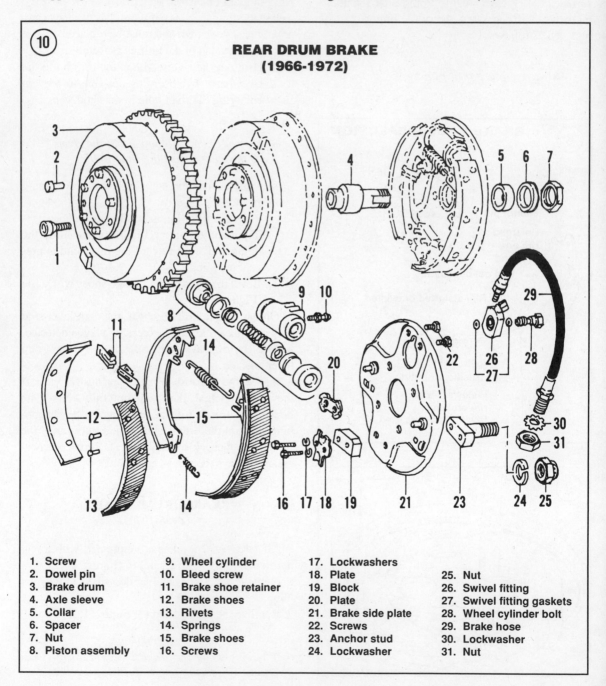

**⑩ REAR DRUM BRAKE
(1966-1972)**

| | | |
|---|---|---|
| 1. Screw | 9. Wheel cylinder | 17. Lockwashers |
| 2. Dowel pin | 10. Bleed screw | 18. Plate |
| 3. Brake drum | 11. Brake shoe retainer | 19. Block |
| 4. Axle sleeve | 12. Brake shoes | 20. Plate |
| 5. Collar | 13. Rivets | 21. Brake side plate |
| 6. Spacer | 14. Springs | 22. Screws |
| 7. Nut | 15. Brake shoes | 23. Anchor stud |
| 8. Piston assembly | 16. Screws | 24. Lockwasher |

| |
|---|
| 25. Nut |
| 26. Swivel fitting |
| 27. Swivel fitting gaskets |
| 28. Wheel cylinder bolt |
| 29. Brake hose |
| 30. Lockwasher |
| 31. Nut |

*NOTE*
*Do not operate the rear brake pedal with the brake shoes removed and the brake cylinder installed. This will force the pistons out of the wheel cylinder and it will have to be rebuilt.*

7. Remove the lower return spring and separate the shoes.

8. Remove the hold-down springs from their posts on the brake side plate.

9. Inspect the brake shoes as described in this chapter.

10. Inspect the wheel cylinder (9) for fluid leakage. If leaking, the wheel cylinder should be removed and rebuilt as described in this chapter.

11. Installation is the reverse of these steps. Note the following.

12. Install the wheel cylinder (9), if removed, as described in this chapter.

13. Install the brake shoes so that the wide shoe faces to the front of the bike and the narrow shoe to the rear. See Step 3.

14. Grease the hold-down springs and pivot posts with a light coat of multipurpose brake grease; avoid getting any grease on the brake plate where the linings come in contact with it.

*WARNING*
*Make sure to use a grease designed to withstand high braking temperatures without thinning. Ordinary multipurpose grease will run onto the brake drum area and contaminate the brake linings.*

15. If the wheel cylinder was removed or the hydraulic hose disconnected, bleed the rear brake as described in this chapter.

**Inspection**

*WARNING*
*Do not inhale brake dust. It contains asbestos, which can cause lung damage. Clean brake dust off of parts with a vacuum cleaner. Do not inhale brake dust when emptying or discarding the vacuum cleaner bag.*

1. Thoroughly clean and dry all parts except the linings.

2. Check the contact surface of the drum for cracks, scoring, glaze or roughness. Remove light scoring and glaze with fine emery cloth. Remove all traces of emery when finished. If there are grooves deep enough to snag a fingernail, the drum should be reground and new shoes fitted. If the brake drum is in good condition, standard-size brake linings can be used. However, if the drum requires regrinding, oversize brake linings may be required. Consult your Harley-Davidson dealer for further information.

*CAUTION*
*Before installing a brake drum that has been reground, check it for any traces of cuttings or other abrasives left from cutting. If necessary, clean with lacquer thinner. In addition, do not touch the ground surface with your hands as this would contaminate it, thus reducing brake effectiveness.*

3. Inspect the linings for embedded foreign material. Dirt can be removed with a stiff wire brush. Check for traces of oil or grease; if contaminated, they must be replaced. Brake lining wear depends upon how close the rivet head is to the brake lining surface. Check each rivet-to-lining area for maximum wear. If necessary, the brake linings can be replaced as follows.

*NOTE*
*Make sure to check the brake drum surface as described in Step 2 before replacing brake linings; oversize linings may be required.*

4. Before replacing brake linings, check the shoe web (**Figure 8**) for cracks or distortion. To check for distortion, lay the web on a flat surface, such as a piece of glass or surface plate. Check how closely the web seats along the glass. If the web is twisted at any point, the shoe is bent and must be replaced.

*NOTE*
*Rear brake shoe linings are either riveted to the brake shoes or bonded. Riveted shoe linings can be removed and replaced. Bonded brake shoes must be replaced as a unit.*

5A. *Riveted brake linings:* Replace riveted brake linings as follows:

a. Drill out the rivets. Match the drill bit to the rivet head as closely as possible to prevent enlarging the rivet hole in the shoe web.

*CAUTION*
*Do not attempt to punch out the rivets, as this may distort the shoe web.*

b. With a steel brush, clean the brake lining surface thoroughly.

c. Rivet the new linings to the brake shoes. It is important to start riveting from one end of the lining and work toward the other end, to prevent the lining from buckling.

d. Bevel the ends of both brake shoes with a file (**Figure 9**).

5B. *Bonded brake linings:* Measure the brake linings with vernier calipers (**Figure 11**). They should be replaced as a set if worn to 0.100 in. (2.54 mm) or less.

6. Inspect the anchor stud pivot area for wear and corrosion. Minor roughness can be removed with fine emery cloth.

7. Check the axle sleeve (4, **Figure 10**) inside diameter for wear or damage. To replace the sleeve, remove the nut (7, **Figure 10**), spacer (6, **Figure 10**) and spacer collar (5, **Figure 10**) from the backside of the brake side plate and remove the sleeve.

8. Replace the hold-down springs if worn or damaged. If installing new brake linings or brake shoes, install new hold-down springs.

9. Inspect the brake shoe return springs (14, **Figure 10**) for wear. If they are stretched, they will not fully retract the brake shoes from the drum, resulting in a power-robbing drag on the drums and premature wear of the linings. Also check springs for rust. Replace as necessary; always replace as a pair.

### Wheel Cylinder Overhaul

Refer to **Figure 10** for this procedure.

1. Disconnect the brake line from the wheel cylinder.
2. Unscrew the bleed screw (10, **Figure 10**).
3. Unbolt the wheel cylinder from the backing plate, then take it off.
4. Refer to 8, **Figure 10**. Remove the piston boots. The pistons, cups and spring can then be removed.
5. Clean the cylinder and pistons in clean brake fluid. Do not clean with gasoline, kerosene or solvent. These leave residue which can cause rubber parts to soften and swell.

6. Check the cylinder bore and pistons for scoring, cracks, corrosion, dirt or excessive wear. Check the pistons for worn brake shoe slots. Replace the cylinder if these conditions are found.
7. Purchase a wheel cylinder overhaul kit.
8. Coat the new piston cups with brake fluid and install them on the pistons.
9. Install the spring and pistons in the cylinder.
10. Install the piston boots on the cylinder.
11. Install the wheel cylinder on the backing plate. Install the bolts and lockwashers. Tighten the bolts securely.
12. Install the bleed valve. Tighten the bleed valve securely.

## DISC BRAKES

Disc brakes are actuated by hydraulic fluid from the master cylinder. The master cylinder is controlled by the hand or foot lever. As the brake pads wear, the brake fluid level drops in the master cylinder reservoir and automatically adjusts for pad wear.

When working on a hydraulic brake system, it is necessary that the work area and all tools be absolutely clean. Any tiny particles of foreign matter or grit on the caliper assembly or the master cylinder can damage the components. Also, sharp tools must not be used inside the caliper or on the caliper piston. If there is any doubt about your ability to correctly and safely carry out major service on the brake components, take the job to a Harley-Davidson dealer or brake specialist.

*WARNING*
*Harley-Davidson specifies DOT 3 brake fluid for models produced prior to September, 1976 and DOT 5 for later*

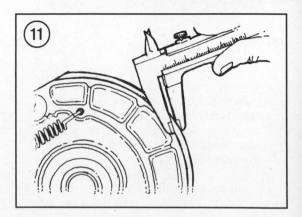

models. Mixing the two types of brake fluids can cause brake failure. If you own a 1976 or 1977 model take your frame number to a Harley-Davidson dealer to find out your bike's production date.

When adding brake fluid use only the specified type from a sealed container. DOT 3 brake fluid will draw moisture which greatly reduces its ability to perform correctly, so it is a good idea to purchase brake fluid in small containers and discard what is not used.

Whenever *any* brake line has been disconnected at either end, the system is considered "opened" and must be bled to remove air bubbles. Also, if the brake feels "spongy," this usually means there are air bubbles in the system and it must be bled. For safe brake operation, refer to *Bleeding the System* in this chapter for complete details.

*CAUTION*
*Disc brake components rarely require disassembly, so do not disassemble unless absolutely necessary. Do not use solvents of any kind on the brake system's internal components. Solvents will cause the seals to swell and distort. When disassembling and cleaning brake components, except brake pads, use new brake fluid.*

## DISC BRAKE PAD REPLACEMENT

### Service Notes

Observe the following service notes before replacing brake pads.
1. Brake pads should be replaced only as a set.
2. Disconnecting the hydraulic brake hose is not required for brake pad replacement. Disconnect the hose only if caliper removal is required.
3. When new pads are installed in the caliper, the master cylinder brake fluid level will rise as the caliper piston is repositioned. Clean the top of the master cylinder of all dirt and foreign matter. Remove the cap and diaphragm from the master cylinder. Then slowly push the caliper pistons into the caliper. Constantly check the reservoir to make sure brake fluid does not overflow. Remove fluid, if necessary, prior to it overflowing. The pistons should move freely. If they don't, and there is evi-

dence of them sticking in the cylinder, the caliper should be removed and serviced as described under *Front Caliper Rebuilding* in this chapter.
4. Push the caliper pistons in all the way to allow room for the new pads.

*WARNING*
*Harley-Davidson specifies DOT 3 brake fluid for models produced prior to September, 1976 and DOT 5 for later models. Mixing the two types of brake fluids can cause brake failure. If you own a 1976 or 1977 model take your frame number to a Harley-Davidson dealer to find out your bike's production date.*

*WARNING*
*After performing any service work to the brake, do not ride the motorcycle until you are sure the brakes are operating correctly with full hydraulic advantage. If necessary, bleed the brake system as described in this chapter.*

### Brake Pad Replacement (Type I)

The Type I disc brake is found on the following models:
   a. 1972-early 1978 FL front and rear (**Figure 12**).
   b. 1973-early 1980 FX front and rear (**Figure 13**).
   c. Late 1980-on FL front and rear (**Figure 14**).
   d. Late 1980 FX rear (**Figure 15**).
1. Remove the brake hose clamp.
2. Remove the caliper housing bolts and washers. Lift the outer caliper housing off. Hang the caliper with a bungee cord.

*CAUTION*
*Do not let the caliper hang by the brake hose.*

3. On front disc brakes, remove the vibration damper clip (if so equipped).
4. Remove the mounting pin and the inner caliper housing.
5. Remove the brake pads and brake pad pins.
6. Inspect pads. Light surface dirt, oil or grease stains may be sanded off. If oil or grease has penetrated the surface, replace the pads. Since brake fluid will ruin the friction material, the pads must be replaced if any

**10**

⑫

**FRONT AND REAR DISC BRAKE
TYPE I (1972-EARLY 1978 FL)**

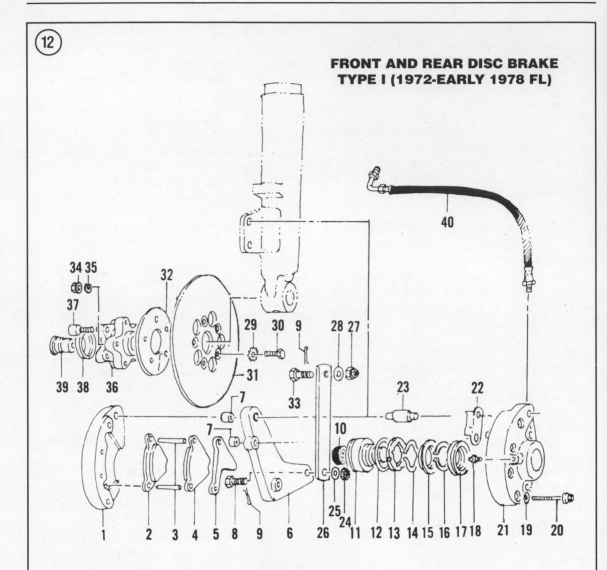

1. Inner caliper half
2. Brake pad
3. Brake pad mounting pins
4. Brake pad
5. Piston shim (late 1978-early 1980 FL, FLH, FLH-80 and Classic)
6. Rear brake mounting bracket (1973-on)
7. Bushing
8. Bolt
9. Cotter pin
10. Insulator (late 1978-early 1980 FL, FLH, FLH-80 and Classic)
11. Piston
12. Seal
13. Friction ring (early models)
14. Spring washer (1972-early 1974 FL, FLH; 1977-early 1980 FL, FLH, FLH-80 and Classic)
15. Backing plate (early models)
16. Circlip (early models)
17. Piston boot
18. Bleed screw
19. Plain washer
20. Bolt
21. Outer caliper half
22. Clip (front brake only) (1973-early 1978 FL, FLH)
23. Mounting pin
24. Nut
25. Plain washer
26. Rear brake torque arm (1973-early 1980 FL, FLH, FLH-80 and Classic)
27. Nut
28. Plain washer
29. Lockwasher
30. Bolt
31. Brake disc
32. Brake disc spacer (late 1976-early 1980 FL, FLH, FLH-80 and Classic)
33. Bolt
34. Nut
35. Lockwasher
36. Brake disc flange (1972 FL, FLH)
37. Bolt
38. Oil seal
39. Front axle spacer (1972 FL, FLH)
40. Brake hose

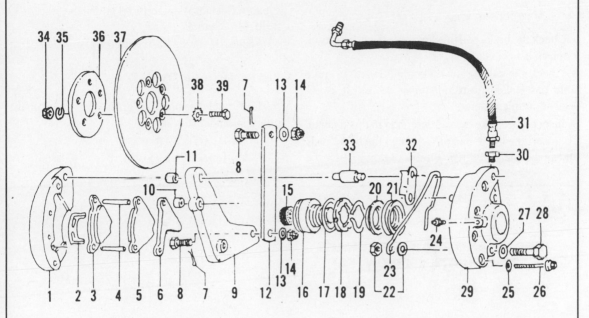

⑬

# FRONT AND REAR DISC BRAKE
## TYPE I (1973-EARLY 1980 FX)

10

1. Inner caliper half
2. Anti-vibration disc
   (rear brake only)
3. Brake pad
4. Brake pad mounting pins
5. Brake pad
6. Shim (late 1978-early 1980)
7. Cotter pin
8. Bolt
9. Caliper mounting bolt
10. Bushing
11. Bushing
12. Anchor arm
13. Plain washer
14. Nut
15. Insulator (late 1978-early 1980)
16. Piston*
17. Seal
18. Friction ring (early models**)
19. Spring washer (1973-early
    1974; 1977-early 1980)
20. Backing plate (early models**)
21. Dust boot

22. Nut (rear brake only)
23. Anti-vibration kit
    (rear brake only)
24. Bleed screw
25. Plain washer
26. Bolt
27. Plain washer
28. Bolt
29. Outer caliper half
30. Brake hose fitting (1973)
31. Brake hose
32. Vibration damper clip
    (1973 front brake only)
33. Mounting pin
34. Nut
35. Lockwasher (early 1973)
36. Brake disc plate
    (late 1976-early 1980)
37. Brake disc
38. Lockwasher
39. Bolt
 * See Harley-Davidson dealer for correct piston
   assembly for your model.
** See Harley-Davidson dealer if used on your model.

brake fluid has touched them. Check the pad for distortion by placing its metal backing on a piece of glass. If the back is bent, replace the pads. If the pads are okay, measure the friction material with a vernier caliper. Replace pads if thickness is 0.062 in. (1.57 mm) or less.

*NOTE*
*Always replace pads as a set.*

7. Check the brake pad pins for damage or distortion. Replace if necessary.

8. Check the caliper for brake fluid leaks. If brake fluid has leaked from the caliper, rebuild it as described in this chapter.

9. Inspect the brake disc as described in this chapter.

10. Insert the brake pad pins through the brake pads and place the pads over the brake disc. Align the inner caliper housing with the brake pad pins and install the housing.

11. Install the mounting pin and damper spring (if used).

12. Position the outer caliper housing next to the brake disc and install the mounting bolts and washers. Lightly coat the bolt threads with Harley-Davidson Pipe Sealant With Teflon. Tighten the bolts to 35 ft.-lb. (48.3 N•m).

13. Secure the brake hose with the clamp.

**Pad Replacement (Type II)**

The Type II disc brake (**Figure 16**) is found on the following models:
  a.  1974-1977 FX (front).
  b.  1974-1977 FXE (front).

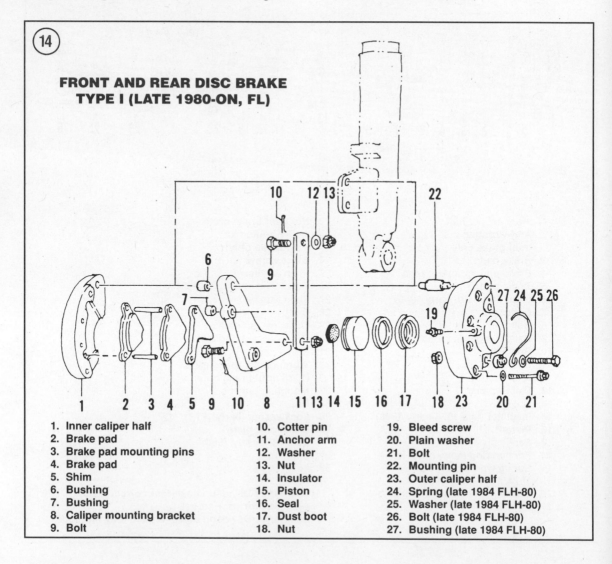

**(14)**

**FRONT AND REAR DISC BRAKE TYPE I (LATE 1980-ON, FL)**

1. Inner caliper half
2. Brake pad
3. Brake pad mounting pins
4. Brake pad
5. Shim
6. Bushing
7. Bushing
8. Caliper mounting bracket
9. Bolt
10. Cotter pin
11. Anchor arm
12. Washer
13. Nut
14. Insulator
15. Piston
16. Seal
17. Dust boot
18. Nut
19. Bleed screw
20. Plain washer
21. Bolt
22. Mounting pin
23. Outer caliper half
24. Spring (late 1984 FLH-80)
25. Washer (late 1984 FLH-80)
26. Bolt (late 1984 FLH-80)
27. Bushing (late 1984 FLH-80)

1. Remove the brake hose clamp.

2. Remove the caliper housing Allen bolts and separate the calipers. Hang the outer caliper housing with a bungee cord.

> *CAUTION*
> *Do not let the caliper hang by the brake hose.*

3. Remove the pressure plate (with outer brake pad attached).

> *NOTE*
> *The inner brake pad is attached to the inner caliper housing.*

4. Inspect pads. Light surface dirt, oil or grease stains may be sanded off. If oil or grease has penetrated the surface, replace the pads. Since brake fluid will ruin the friction material, the pads must be replaced if any brake fluid has touched them. Replace brake pads if worn to indicator groove on bottom of pad.

5. Replace pads as follows:

   a. Brake pads are secured to pressure plate (outer pad) and inner caliper housing (inner pad) by rivets.

   b. Remove pads by drilling out rivets with a 9/64 in. (3.57 mm) drill bit. Drill off rivet head only. Do not drill into the pressure plate or caliper

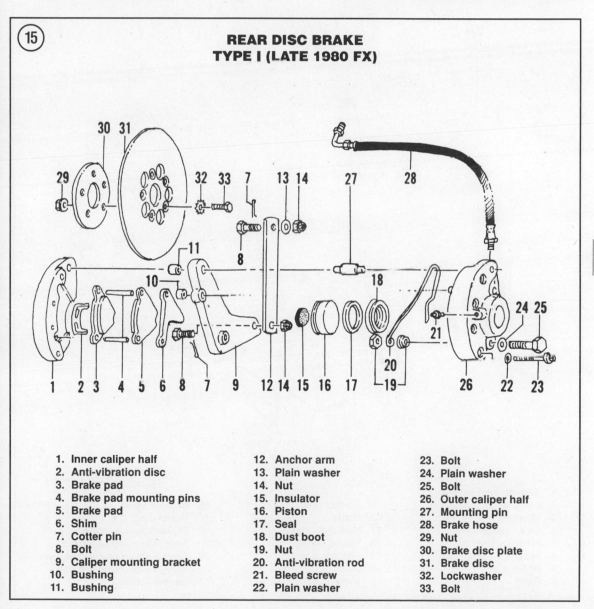

**REAR DISC BRAKE
TYPE I (LATE 1980 FX)**

| | | |
|---|---|---|
| 1. Inner caliper half | 12. Anchor arm | 23. Bolt |
| 2. Anti-vibration disc | 13. Plain washer | 24. Plain washer |
| 3. Brake pad | 14. Nut | 25. Bolt |
| 4. Brake pad mounting pins | 15. Insulator | 26. Outer caliper half |
| 5. Brake pad | 16. Piston | 27. Mounting pin |
| 6. Shim | 17. Seal | 28. Brake hose |
| 7. Cotter pin | 18. Dust boot | 29. Nut |
| 8. Bolt | 19. Nut | 30. Brake disc plate |
| 9. Caliper mounting bracket | 20. Anti-vibration rod | 31. Brake disc |
| 10. Bushing | 21. Bleed screw | 32. Lockwasher |
| 11. Bushing | 22. Plain washer | 33. Bolt |

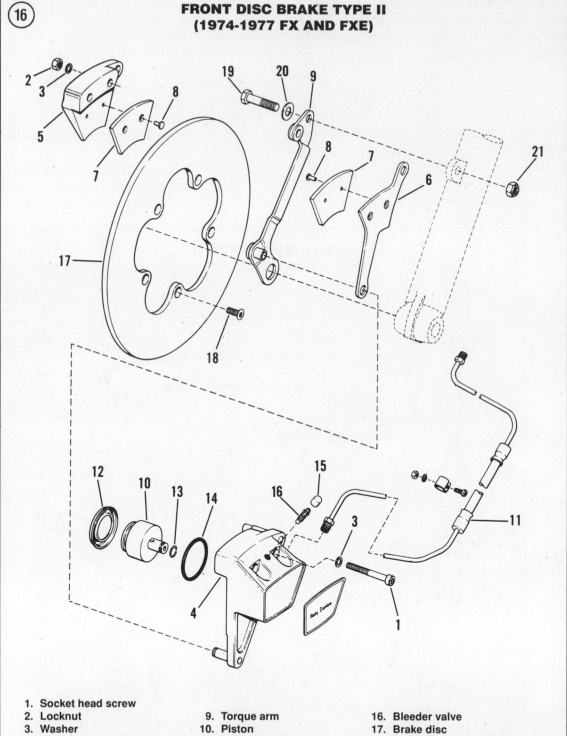

**FRONT DISC BRAKE TYPE II**
**(1974-1977 FX AND FXE)**

1. Socket head screw
2. Locknut
3. Washer
4. Outer caliper half
5. Inner caliper half
6. Pressure plate
7. Brake pad
8. Rivet
9. Torque arm
10. Piston
11. Hydraulic hose
12. Rubber boot
13. Friction ring
14. Seal
15. Bleeder valve cap
16. Bleeder valve
17. Brake disc
18. Brake disc mounting screw
19. Torque arm mounting bolt
20. Torque arm mounting bolt washer
21. Torque arm mounting bolt
    washer locknut

housing as this would not allow seating of new rivets.

c. Check pressure plate for distortion by placing on a piece of glass. Replace pressure plate if necessary.

d. Clean pad mounting surface with a steel brush.

e. Rivet new pads in position with a hollow rivet set.

*NOTE*
*Always replace pads as a set.*

6. Check the caliper for brake fluid leaks. If brake fluid has leaked from the caliper, rebuild it as described in this chapter.

7. Inspect the brake disc as described in this chapter.

8. Install the pressure plate onto the torque arm.

9. Position the inner and brake caliper housings onto the brake disc and install the mounting bolts, washers and nuts. Tighten the bolts to 130 in.-lb. (14.9 N•m). After tightening bolts, make sure the caliper assembly floats on the torque arm.

### Pad Replacement (Type III)

The Type III brake caliper (**Figure 17**) is used on the following models:
a. 1977-1979 FXS (front).
b. 1978-1983 (FX).

1. Remove the 2 nuts and bolts that mount the caliper. Slide the caliper off the disc.

2. Remove the large bolt from the center of the caliper and separate the 2 halves.

3. Hang the outer caliper housing with a bungee cord.

*CAUTION*
*Do not let the caliper hang by the brake hose.*

4. Slide the inner plate, brake pads and outer plate off of the pins.

5. Inspect pads. Light surface dirt, oil or grease stains may be sanded off. If oil or grease has penetrated the surface, replace the pads. Since brake fluid will ruin the friction material, the pads must be replaced if any

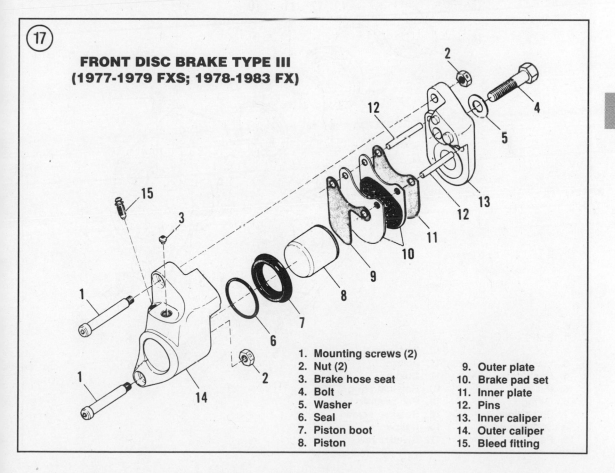

**FRONT DISC BRAKE TYPE III**
**(1977-1979 FXS; 1978-1983 FX)**

1. Mounting screws (2)
2. Nut (2)
3. Brake hose seat
4. Bolt
5. Washer
6. Seal
7. Piston boot
8. Piston
9. Outer plate
10. Brake pad set
11. Inner plate
12. Pins
13. Inner caliper
14. Outer caliper
15. Bleed fitting

10

brake fluid has touched them. Check the pad for distortion by placing its metal backing on a piece of glass. If the back is bent, replace the pads. If the pads are okay, measure the friction material with a vernier caliper. Replace pads if thickness is 0.062 in. (1.57 mm) or less.

*NOTE*
*Always replace pads as a set.*

6. Check the pad pins for damage or distortion. Replace if necessary.

7. Check the caliper for brake fluid leaks. If brake fluid has leaked from the caliper, rebuild it as described in this chapter.

8. Inspect the brake disc as described in this chapter.

9. Install the pad pins in the inner caliper if removed.

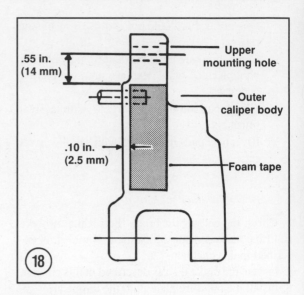

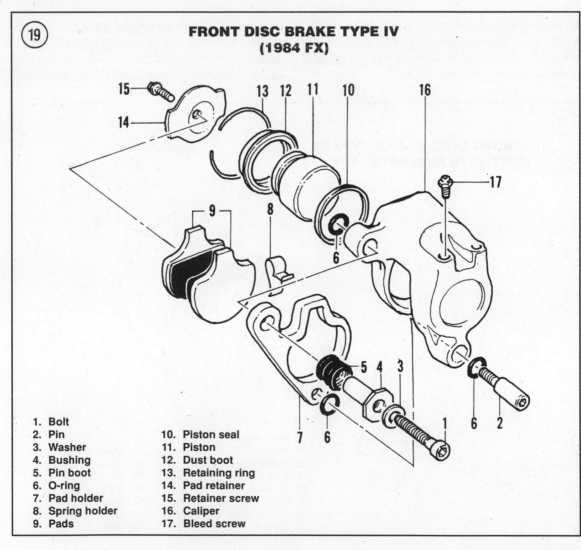

**FRONT DISC BRAKE TYPE IV
(1984 FX)**

1. Bolt
2. Pin
3. Washer
4. Bushing
5. Pin boot
6. O-ring
7. Pad holder
8. Spring holder
9. Pads
10. Piston seal
11. Piston
12. Dust boot
13. Retaining ring
14. Pad retainer
15. Retainer screw
16. Caliper
17. Bleed screw

10. Install the following parts on the pad pins:
   a. Outer plate.
   b. Outer brake pad.
   c. Inner brake pad.
   d. Inner plate.

11. Install the outer caliper housing. Install the large bolt and washer. Tighten the bolts to 45-50 ft.-lb. (62-69 N•m).

> *NOTE*
> *The 1978 FX, FXS and FXE were subject to a safety defect recall for a possible faulty washer under the large bolt that clamps the 2 caliper halves together (5, Figure 17). Inspect the washer closely for cracks. Contact a Harley-Davidson dealer if you suspect the washer is faulty.*

12. Install the caliper over the brake disc. Make sure a brake pad is positioned on both sides of the disc.

13. Install the caliper bolts and secure with *new* locknuts. Tighten the bolts to 115-120 in.-lb. (13.2-13.8 N•m).

> *NOTE*
> *If new locknuts are not installed in Step 13, coat the bolts threads with Loctite Lock N' Seal.*

> *NOTE*
> *Harley-Davidson offers a foam tape to help prevent caliper rattle on 1982-1983 models. Apply the tape as shown in Figure 18.*

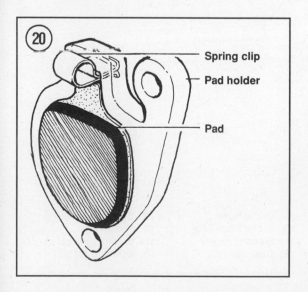

**Spring clip**

**Pad holder**

**Pad**

## Pad Replacement (Type IV)

The Type IV brake disc caliper (**Figure 19**) is found on the following model:
   a. 1984 FX (front).

1. Remove the upper mounting bolt and the lower mounting pin.

2. Pull the caliper assembly away from the brake disc.

3. Remove the screw and remove the pad retainer and the inner brake pad.

4. Remove the outer pad, pad holder and spring clip as an assembly.

> *NOTE*
> *After removing the brake pads, hang the caliper with a bungee cord. Do not allow the caliper to hang by the brake hose.*

5. Push the outer pad free of the spring clip and remove it.

6. Inspect pads. Light surface dirt, oil or grease stains may be sanded off. If oil or grease has penetrated the surface, replace the pads. Since brake fluid will ruin the friction material, the pads must be replaced if any brake fluid has touched them. Check the pad for distortion by placing its metal backing on a piece of glass. If the back is bent, replace the pads. If the pads are okay, measure the friction material with a vernier caliper. Replace pads if thickness is 0.062 in. (1.57 mm) or less.

> *NOTE*
> *Always replace pads as a set.*

7. Place the pad holder on a workbench with the upper mounting bolt hole positioned at the upper right.

8. Install the spring clip at the top of the pad holder as shown in **Figure 20**.

9. Place the brake pad with the insulator backing on top of the spring clip with the lower end of the pad slightly entering the pad holder opening. With the brake pad insulator backing facing toward the pad holder, push the brake pad down until it is held firmly in the pad holder by the spring clip. See **Figure 20**.

10. Insert the outer brake pad/pad holder assembly into the caliper so that the brake pad insulator backing faces against the piston.

**10**

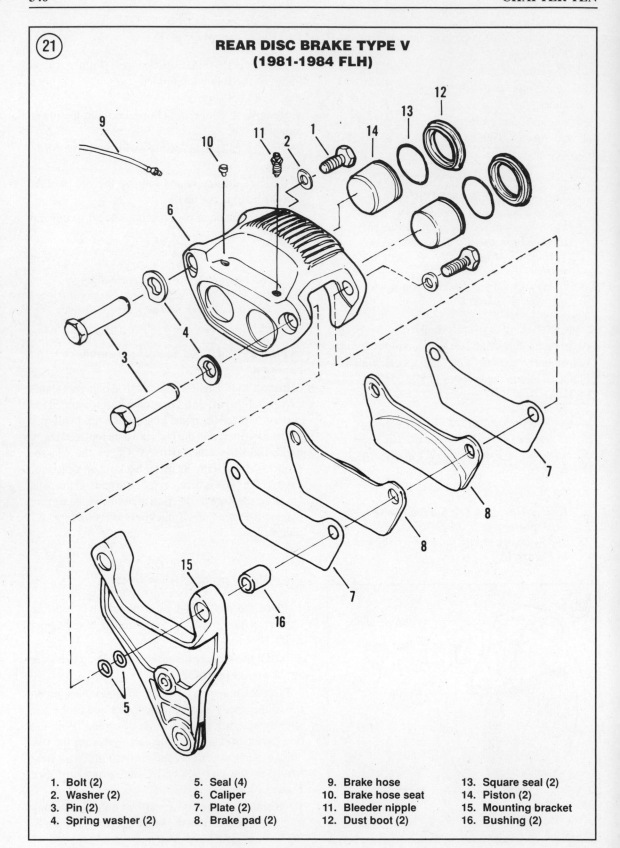

**REAR DISC BRAKE TYPE V
(1981-1984 FLH)**

| | | | |
|---|---|---|---|
| 1. Bolt (2) | 5. Seal (4) | 9. Brake hose | 13. Square seal (2) |
| 2. Washer (2) | 6. Caliper | 10. Brake hose seat | 14. Piston (2) |
| 3. Pin (2) | 7. Plate (2) | 11. Bleeder nipple | 15. Mounting bracket |
| 4. Spring washer (2) | 8. Brake pad (2) | 12. Dust boot (2) | 16. Bushing (2) |

*WARNING*
*The spring clip loop and the brake pad friction material must face away from the piston. Brake failure will occur if the brakes are assembled incorrectly.*

11. Install the inner brake pad (without the insulator backing) in the caliper recessed seat.

12. Insert the pad retainer within the counterbore inside the caliper. Install the self-tapping screw through the pad retainer and thread into the brake pad. Tighten the screw to 55-70 in.-lb. (6.32-8.05 N•m).

13. Coat the lower mounting bolt with Dow Corning Moly 44 grease.

14. Install the caliper over the brake disc and align it with the lower fork tube mounting lugs.

15. Install the upper mounting bolt and washer through the fork lug, pad holder and into the threaded bushing.

16. Install the lower mounting bolt through the caliper, fork lug and into the lower end of the pad holder.

17. Tighten the lower and then the upper mounting bolts to 40-45 ft.-lb. (55.2-62.1 N•m).

18. If it was necessary to reseat the caliper piston, refill the master cylinder reservoir to maintain the correct fluid level. Install the diaphragm and top cap.

## Pad Replacement (Type V)

The Type V brake disc (**Figure 21**) is found on the following models:

    a. 1981-1984 FLH (rear).

1. Remove the brake hose clamp, if so equipped.

2. Remove the 2 bolts and washers that mount the caliper to the bracket. Remove the pins into which the bolts thread and the spring washers and caliper seals.

3. Lift the caliper off of the disc. Then remove the plates and pads.

4. Hang the caliper with a bungee cord.

*CAUTION*
*Do not let the caliper hang by the brake hose.*

5. Inspect pads. Light surface dirt, oil or grease stains may be sanded off. If oil or grease has penetrated the surface, replace the pads. Since brake fluid will ruin the friction material, the pads must be replaced if any brake fluid has touched them. Check the pad for distortion by placing its metal backing on a piece of glass. If the back is bent, replace the pads. If the pads are okay, measure the friction material with a vernier caliper. Replace pads if thickness is 0.062 in. (1.57 mm) or less.

*NOTE*
*Always replace pads as a set.*

6. Check the pad mounting pins for damage or distortion. Replace if necessary.

7. Check the caliper for brake fluid leaks. If brake fluid has leaked from the caliper, rebuild it as described in this chapter.

8. Inspect the brake disc as described in this chapter.

9. Install the plates and brake pads in the caliper. Make sure the plates and pads face in the direction shown in **Figure 21**.

10. Install the caliper housing over the brake disc. Make sure the brake disc separates the brake pads.

11. Install new seals in the mounting bracket as shown in **Figure 21**.

12. Coat the pins with Loctite Anti-seize before installation.

13. Install a spring washer on each pin.

14. Install the pins through the caliper. Make sure to align the pins with the holes in the tops of the plates and brake pads.

15. Install the caliper bolts and washers and tighten to 12-15 ft.-lb. (16.5-20.7 N•m).

## Pad Replacement (Type VI)

The Type VI brake disc (**Figure 22**) is found on the following models:

    a. 1981-1982 FX (rear).

    b. 1983 FXE (rear).

1. Remove the 2 screws and washers that hold the caliper halves together.

2. Remove the caliper halves from the bracket.

3. Remove the damper and sleeve from the mounting bracket.

4. Remove the pins, brake pads, plate, anti-rattle spring and insulator.

5. Hang the caliper with a bungee cord.

*CAUTION*
*Do not let the caliper hang by the brake hose.*

**10**

6. Inspect pads. Light surface dirt, oil or grease stains may be sanded off. If oil or grease has penetrated the surface, replace the pads. Since brake fluid will ruin the friction material, the pads must be replaced if any brake fluid has touched them. Check the pad for distortion by placing its metal backing on a piece of glass. If the back is bent, replace the pads. If the pads are okay, measure the friction material with a vernier caliper. Replace pads if thickness is 0.062 in. (1.57 mm) or less.

*NOTE*
*Always replace pads as a set.*

7. Check the damper and sleeve for damage or distortion. Replace if necessary.
8. Check the caliper for brake fluid leaks. If brake fluid has leaked from the caliper, rebuild it as described in this chapter.
9. Inspect the brake disc as described in this chapter.
10. Install the insulator.
11. Install the brake pad pins, plate and brake pads into the right side caliper housing.
12. Install the anti-rattle spring with its arc to the top and the prongs to the right side of the caliper.
13. Install the damper and sleeve onto the mounting bracket.
14. Install the left side caliper housing onto the mounting bracket.
15. Install the right side caliper housing onto the mounting bracket.

16. Coat the mounting screws with Loctite Anti-seize and install them with their lockwashers. Tighten the screws to 30-35 ft.-lb. (41.4-48.3 N•m).

**Pad Replacement (Type VII)**

The Type VII brake disc (**Figure 23**) is found on the following models:
   a. 1983-on FX (rear).
   b. 1983-on FLHS (rear).
1. Remove the 2 caliper housing Allen bolts and washers.
2. Pull the caliper assembly away from the caliper frame.
3. Remove the brake pads.
4. Remove the pad spring from inside the caliper.
5. Hang the caliper with a bungee cord.

*CAUTION*
*Do not let the caliper hang by the brake hose.*

6. Inspect pads. Light surface dirt, oil or grease stains may be sanded off. If oil or grease has penetrated the surface, replace the pads. Since brake fluid will ruin the friction material, the pads must be replaced if any brake fluid has touched them. Check the pad for distortion by placing its metal backing on a piece of glass. If the back is bent, replace the pads. If the pads

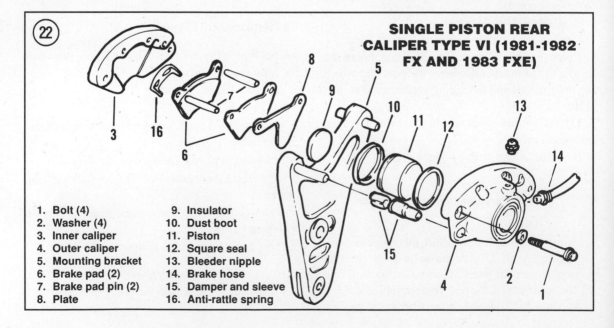

㉒ **SINGLE PISTON REAR CALIPER TYPE VI (1981-1982 FX AND 1983 FXE)**

1. Bolt (4)
2. Washer (4)
3. Inner caliper
4. Outer caliper
5. Mounting bracket
6. Brake pad (2)
7. Brake pad pin (2)
8. Plate
9. Insulator
10. Dust boot
11. Piston
12. Square seal
13. Bleeder nipple
14. Brake hose
15. Damper and sleeve
16. Anti-rattle spring

are okay, measure the friction material with a vernier caliper. Replace pads if thickness is 0.062 in. (1.57 mm) or less.

*NOTE*
*Always replace pads as a set.*

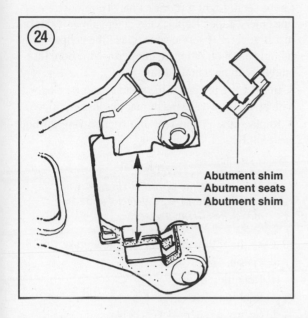

**Abutment shim**
**Abutment seats**
**Abutment shim**

7. Check the abutment shim in the caliper frame (**Figure 24**). If it is worn or damaged, replace it as follows:

   a. Pry the abutment shim away from the caliper.

   b. Remove all adhesive from the caliper surface where the abutment shim is located.

   c. Clean the abutment shim surface with denatured alcohol.

   d. Apply silicone sealant to the abutment shim surface on the caliper and install a new abutment shim. Hold the shim in position by installing the brake pads in the bracket.

   e. Allow the silicone sealant to dry thoroughly before completing brake pad installation.

   f. Check that the brake pads slide freely in the bracket.

8. Check the caliper for brake fluid leaks. If brake fluid has leaked from the caliper, rebuild it as described in this chapter.

9. Check the brake disc as described in this chapter.

10. Refer to **Figure 25**. Install the pad spring into the top of the caliper so that the spring's long tab extends above the piston. Hook the spring's short tab above the ridge on the caliper casting opposite the piston.

10

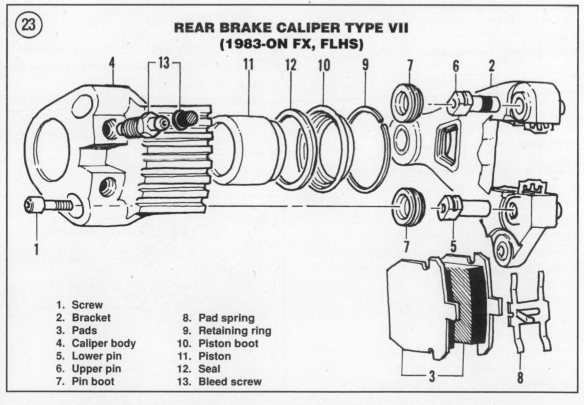

**REAR BRAKE CALIPER TYPE VII**
**(1983-ON FX, FLHS)**

1. Screw
2. Bracket
3. Pads
4. Caliper body
5. Lower pin
6. Upper pin
7. Pin boot
8. Pad spring
9. Retaining ring
10. Piston boot
11. Piston
12. Seal
13. Bleed screw

11. Install the brake pads on the bracket. Then install the caliper body over the brake pads and onto the bracket. Make sure the upper and lower pins do not move when installing the caliper body.

12. Install the caliper screws and tighten to 12-15 ft.-lb. (16.5-20.7 N•m).

## BRAKE CALIPER

### Removal/Installation

1. Remove the brake pads as described in this chapter.

2. Disconnect the caliper brake line at the caliper. Plug the end of the line to prevent brake fluid from dripping onto other motorcycle parts.

3. Remove the caliper(s).

4. Installation is the reverse of these steps.

5. Bleed the brakes as described in this chapter.

### Caliper Overhaul (Type I)

Refer to **Figures 12-15** for this procedure.

1. Remove the brake pads as described in this chapter.

2. Disconnect the caliper brake line. Plug the end of the line to prevent brake fluid from dripping onto the motorcycle.

3. Remove the piston boot.

> *WARNING*
> *During the next step, the piston may shoot out like a bullet. Keep your fingers out of the way. Wear shop gloves and apply compressed air gradually.*

4. Place a rag or piece of wood in the path of the piston (**Figure 26**). Blow the piston out with compressed air directed through the hydraulic hole fitting. Use a service station air hose if you don't have a compressor.

5. Remove the retaining ring (if used) from the piston bore. Then remove the following parts in order:

   a. Backing plate.
   b. Wave spring (1972-1974).
   c. Friction ring.
   d. O-ring.

6. Check the mounting bracket bushings for wear and damage. Replace them if necessary.

> *NOTE*
> *If the bushings are tight, have a Harley-Davidson dealer or machine shop replace the bushings for you.*

7. Inspect the cylinder bore in the outer caliper housing. Replace the outer caliper if wear or damage can be seen. Light dirt or rust may be removed with fine emery paper. Replace the outer caliper if dirt or rust is severe. If serviceable, clean the caliper with rubbing alcohol. After cleaning, rinse the bore thoroughly with new brake fluid.

8. Inspect the piston. If the piston can't be cleaned with brake fluid and a rag, replace it (if so equipped).

9. Replace the piston seal and friction ring during reassembly.

> *NOTE*
> *Never reuse the piston O-ring. Very minor damage or age deterioration can make the O-ring useless.*

10. Coat the following parts with new brake fluid:

   a. Piston.
   b. Retaining ring.
   c. Backing plate.
   d. Spring washers (1972-1974).
   e. Friction ring.
   f. Seal.

11. Install the piston boot into the caliper bore so that the small hole in the boot faces downward when the caliper is installed on the motorcycle.

12. Install the remaining ring, if so equipped. Make sure the ring seats in the caliper groove completely.

13. Install the backing plate, spring washer (if equipped) and friction ring (if so equipped).

14. Install seal on the piston.

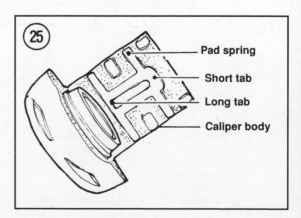

㉕  Pad spring — Short tab — Long tab — Caliper body

15. Align the piston with the caliper bore and install it. Press the piston in all the way.

16. Install the brake pads as described in this chapter.

## Caliper Overhaul (Type II)

Refer to **Figure 16** for this procedure.

1. Remove the brake pads as described in this chapter.

2. Disconnect the caliper brake line. Plug the end of the line to prevent brake fluid from dripping onto the motorcycle.

3. Remove the rubber boot (12, **Figure 16**).

> *WARNING*
> *The piston may shoot out like a bullet. Keep your fingers out of the way. Wear shop gloves and apply compressed air gradually.*

4. Place a rag or piece of wood in the path of the piston (**Figure 26**). Blow the piston out with compressed air directed through the hydraulic hole fitting. Use a service station air hose if you don't have a compressor.

5. Remove the friction ring (13, **Figure 16**) from the end of the piston.

6. Remove the seal (14, **Figure 16**) from the piston bore.

7. Check the caliper housing for damage. Replace if necessary.

8. Inspect the cylinder bore in the outer caliper housing. Replace the outer caliper if wear or damage can be seen. Light dirt or rust may be removed with fine emery paper. Replace the outer caliper if dirt or rust is severe. If serviceable, clean the caliper with

rubbing alcohol. After cleaning, rinse the bore thoroughly with new brake fluid.

9. Inspect the piston (10, **Figure 16**). If the piston can't be cleaned with brake fluid and a rag, replace it.

10. Replace the piston boot, piston seal and friction ring during reassembly.

> *NOTE*
> *Never reuse the piston seal. Very minor damage or age deterioration can make the seal useless.*

11. Coat the following parts with new brake fluid:
   a. Piston (10, **Figure 16**).
   b. Piston boot (12, **Figure 16**).
   c. Seal (14, **Figure 16**).
   d. Friction ring (13, **Figure 16**).

12. Install the seal in the piston bore groove.

13. Install the friction ring onto the piston.

14. Align the piston with the piston bore and install. Rotate the piston during the installation to prevent damaging the friction ring.

15. Install the piston boot. Make sure both boot lips mesh with their individual caliper grooves.

16. Align the piston with the caliper bore and install it. Press the piston in all the way.

17. Install the brake pads as described in this chapter.

## Caliper Overhaul (All Calipers Except Type I and Type II)

Refer to the appropriate drawings for your model when performing this procedure. See **Figure 17**, **Figure 19**, **Figure 21**, **Figure 22** or **Figure 23**.

1. Remove the brake pads as described in this chapter.

2. Disconnect the caliper brake line. Plug the end of the line to prevent brake fluid from dripping onto the motorcycle.

3. On Type IV (**Figure 19**) and Type VII (**Figure 23**) calipers, remove the retaining ring with a small screwdriver.

4. Remove the front piston boot, if so equipped.

> *WARNING*
> *During the next step, the piston may shoot out like a bullet. Keep your fingers out of the way. Wear shop gloves and apply compressed air gradually.*

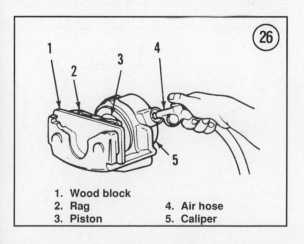

26

1. Wood block
2. Rag
3. Piston
4. Air hose
5. Caliper

5. Place a rag or piece of wood in the path of the piston (**Figure 26**). Blow the piston out with compressed air directed through the hydraulic hole fitting. Use a service station air hose if you don't have a compressor.

6. Remove the piston or seal.

7. Check the caliper housing for damage. Replace if necessary.

8. Inspect the cylinder bore in the outer caliper housing. Replace the outer caliper if wear or damage can be seen. Light dirt or rust may be removed with fine emery paper. Replace the outer caliper if dirt or rust is severe. If serviceable, clean the caliper with rubbing alcohol. After cleaning, rinse the bore thoroughly with new brake fluid.

9. Inspect the piston. If the piston can't be cleaned with brake fluid and a rag, replace it.

10. Replace the piston boot and O-ring during reassembly.

11. Coat the following parts with new brake fluid:

   a. Piston.

   b. Piston boot.

   c. O-ring.

12. Install the piston O-ring.

13. Align the piston with the piston bore and install. Press the piston in all the way.

14. Install the retaining ring, if used.

15. Install the front piston boot, if so equipped.

16. Install the brake pads as described in this chapter.

> *NOTE*
> *Never reuse the piston O-ring. Very minor damage or age deterioration can make the O-ring useless.*

## BRAKE DISC

### Removal/Installation

1. Remove the front or rear wheel as described in Chapter Eight or Chapter Nine.

2. Remove the bolts securing the disc (**Figure 27**) to the hub and remove it.

3. Installation is the reverse of these steps. Tighten the disc nuts or bolts to the specifications in **Table 1**.

### Inspection

1. Check the disc for cracks, rust or scratches. Replace the disc if cracked. Light rust may be removed with medium emery paper.

2. Measure disc thickness with a micrometer at 12 equal points (**Figure 28**). Measure at the brake pad contact point. Minimum thickness for stock disc units is stamped on the brake disc. Replace the disc if thinner than the minimum.

## FRONT MASTER CYLINDER

### Removal/Installation (1972-1981)

Refer to **Figure 29** for this procedure.

1. Attach a hose to the front caliper bleed valve (**Figure 30**). Then open the bleed valve with a wrench and drain the brake fluid by operating the hand lever. Discard the brake fluid.

> *CAUTION*
> *Brake fluid will damage paint and chrome part. Wipe up any spilled fluid*

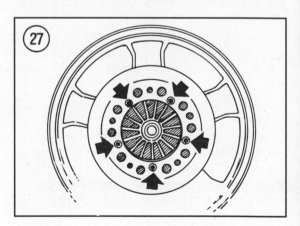

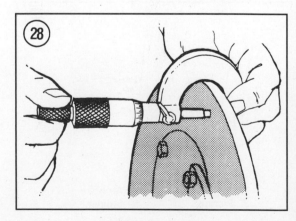

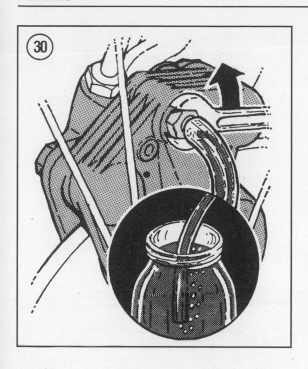

immediately, then wash the area with soap and water.

2. Loosen the front brake hose (5, **Figure 29**) at the front of the caliper.

3. Disconnect the brake light switch wires at the master cylinder.

4. Remove the master cylinder mounting bolts and remove the master cylinder (1, **Figure 29**).

5. Installation is the reverse of these steps. Note the following.

6. Install the master cylinder mounting bolts. Tighten the bolts securely.

7. Connect the stop light switch wire connector at the master cylinder (if so equipped).

8. Connect the brake and reservoir supply hoses at the master cylinder.

*WARNING*
*Harley-Davidson specifies DOT 3 brake fluid for models produced prior*

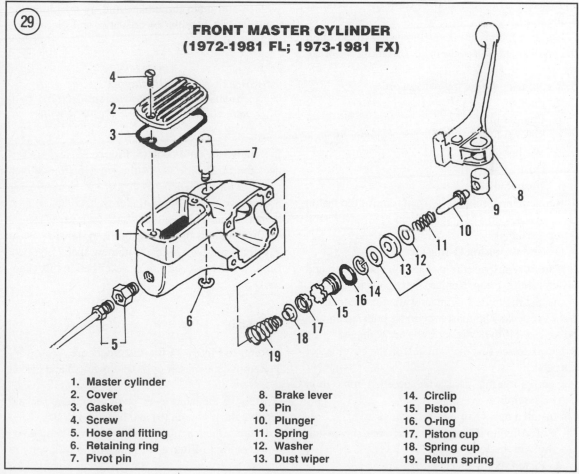

**FRONT MASTER CYLINDER**
**(1972-1981 FL; 1973-1981 FX)**

1. Master cylinder
2. Cover
3. Gasket
4. Screw
5. Hose and fitting
6. Retaining ring
7. Pivot pin
8. Brake lever
9. Pin
10. Plunger
11. Spring
12. Washer
13. Dust wiper
14. Circlip
15. Piston
16. O-ring
17. Piston cup
18. Spring cup
19. Return spring

10

*to September, 1976 and DOT 5 for later models. Mixing the two types of brake fluids can cause brake failure. If you own a 1976 or 1977 model take your frame number to a Harley-Davidson dealer to find out your bike's production date.*

9. Fill the master cylinder with new brake fluid. Bleed brake system as described under *Bleeding the System* in this chapter.

10. Install the master cylinder gasket and cover after bleeding the brakes.

*WARNING*
*Do not ride the motorcycle until you are sure the brakes are working properly.*

### Overhaul (1972-1981)

Refer to **Figure 29** for this procedure.

1. Remove the master cylinder cover (2, **Figure 29**) and gasket (3). Pour and discard any remaining brake fluid.

2. Remove the pivot pin retaining ring (6) and pull the pivot pin (7) out of the caliper. Remove the brake lever.

3. Remove the following parts in order:
   a. Pin (9).
   b. Plunger (10).
   c. Spring (11).
   d. Washer (12).
   e. Dust wiper (13).
   f. Washer (12).

4. Remove the circlip (14). Then remove the piston (15), piston cup (17), spring cup (18) and piston return spring (19).

5. Discard the piston O-ring (16).

6. Carefully inspect the piston cup, spring cup and piston return spring. Replace these parts if required.

7. Inspect the master cylinder walls for scratches or wear grooves. The master cylinder housing should be replaced if the cylinder walls are damaged.

8. Check to see that the vent hole in the cover is not plugged.

9. Coat all internal parts with specified brake fluid before assembly.

10. Install a new O-ring (16) on the piston.

11. Referring to **Figure 29**, assemble the piston assembly in the following order:
   a. Piston return spring (19).

   b. Spring cup (18).
   c. Piston cup (17).
   d. Piston (15).
   e. Circlip (14).

12. Referring to **Figure 29**, install the plunger assembly as follows:
   a. Washer (12).
   b. Dust wiper (13).
   c. Washer (12).
   d. Spring (11).
   e. Plunger (10).
   f. Pin (9).

13. Lightly grease the pivot pin (7). Then install the hand lever and pivot pin (7). Secure the pivot pin with a new circlip (6).

### Removal/Installation (1982-on)

Refer to **Figure 31** for this procedure.

1. Attach a hose to the front caliper bleed valve (**Figure 30**). Then open the bleed valve with a wrench and drain the brake fluid by operating the hand lever. Close the bleed valve and discard the brake fluid.

*CAUTION*
*Brake fluid will damage paint and chrome parts. Wipe up any spilled fluid immediately, then wash the area with soap and water.*

2. Remove the union bolt (5, **Figure 31**) and disconnect the brake line (7) at the front of the caliper. Remove the 2 washers (6).

3. Remove the master cylinder-to handlebar clamp bolts (13) and remove the master cylinder (1).

4. Installation is the reverse of these steps.

5. Fill the master cylinder with new specified brake fluid. Bleed brake system as described in this chapter.

### Overhaul

Refer to **Figure 31** for this procedure.

1. Remove the master cylinder cover (2) and gasket (3).

2. Remove the hand lever pivot pin circlip (8) and remove the pivot pin (9) and hand lever (10).

3. Remove the reaction pin (23) from the hand lever.

4. Referring to **Figure 31**, remove the following in order:

a. Pushrod and switch actuator (22).

b. Dust boot (21).

c. Piston and O-ring (19 and 20).

d. Backup disc (18).

*WARNING*
*All 1984 FL and FX models do not use the backup disc (18, **Figure 31**) because of a design change with the master cylinder piston. Use of the backup disc on these models may cause brake drag or brake lockup. See your Harley-Davidson dealer.*

e. Cup (17).

f. Stop (16).

g. Spring (15).

5. If damaged, remove the grommet and sight glass from the rear side of the master cylinder housing.

6. Carefully inspect the O-ring, cup, dust boot and grommet for softening or wear. Replace these parts if required.

7. Inspect the master cylinder walls for scratches or wear grooves. The master cylinder housing should be replaced if the cylinder walls are damaged.

8. Check to see that the vent hole in the cover is not plugged.

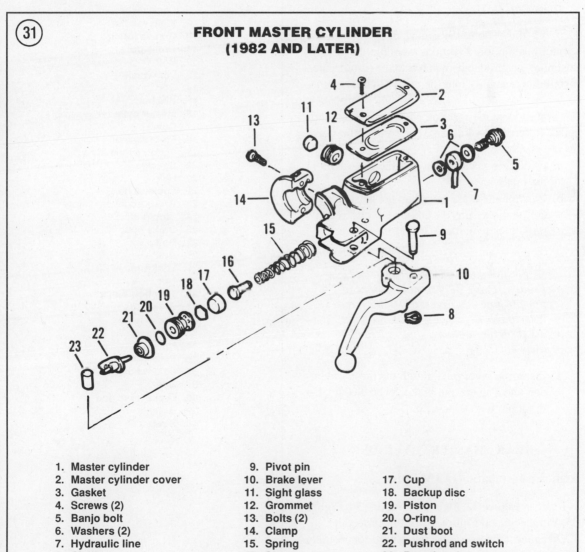

**(31)**

**FRONT MASTER CYLINDER
(1982 AND LATER)**

10

| | | |
|---|---|---|
| 1. Master cylinder | 9. Pivot pin | |
| 2. Master cylinder cover | 10. Brake lever | 17. Cup |
| 3. Gasket | 11. Sight glass | 18. Backup disc |
| 4. Screws (2) | 12. Grommet | 19. Piston |
| 5. Banjo bolt | 13. Bolts (2) | 20. O-ring |
| 6. Washers (2) | 14. Clamp | 21. Dust boot |
| 7. Hydraulic line | 15. Spring | 22. Pushrod and switch |
| 8. Circlip | 16. Stop | 23. Reaction pin |

9. Coat all internal parts in specified brake fluid before assembly.

10. Install the grommet and sight glass if removed.

11. Referring to **Figure 31**, install the following parts in order:

    a. Spring (15).

    b. Stop (16).

    c. Cup (17).

    d. Backup disc (18).

> *WARNING*
> *The backup disc (18, **Figure 31**) is not used on 1984 FL and FX models. Installation of the backup disc on these models may cause brake drag or brake lockup.*

12. With the piston (19, **Figure 31**) facing in the correct installation direction, assemble the O-ring, dust boot, pushrod and switch onto the piston. Then install this assembly into the master cylinder housing.

13. Lightly coat the reaction pin with Loctite Antiseize. Referring to **Figure 31**, assemble the brake lever as follows:

    a. Install the reaction pin into the large hole in the brake lever.

    b. Position the brake lever into the master cylinder, making sure the end of the pushrod fits into the hole in the reaction pin.

> *NOTE*
> *Make sure the pushrod and switch (22, **Figure 31**) are fully seated in the reaction pin hole (23, **Figure 31**). If the hand lever binds or is not smooth in action, disassemble the parts and reassemble them correctly.*

    c. Insert the pivot pin through the master cylinder and engage the brake lever. Secure the pivot pin with the circlip.

## REAR MASTER CYLINDER

### Removal/Installation (All Models)

Refer to **Figure 32, Figure 33, Figure 34, Figure 35, Figure 36** and **Figure 37** for this procedure.

1. Disconnect the stop light switch wire connector at the master cylinder (if so equipped).

2. Disconnect the brake pedal to master cylinder rod.

**REAR BRAKE MASTER CYLINDE**
**(1966-EARLY 1979 FL)**

1. Clevis pin
2. Plunger (1966-1969)
3. Washer
4. Cotter pin
5. Nut
6. Plunger (1970-early 1979)
7. Nut
8. Lockwasher
9. Plain washer
10. Grease fitting
11. Brake pedal
12. Bolt
13. Lockwasher
14. Cap
15. Washer
16. Master cylinder assembly
17. Cover
18. Plain washer
19. Piston assembly
20. Spacers
21. Bolt
22. Lockwasher
23. Brake pedal assembly
24. Stop plate
25. Brake hose
26. Bolt
27. Bolt
28. Lockwasher
29. Nut
30. Junction block
31. Brake hose
32. Brake hose
33. Spring
34. Nut
35. Plain washer
36. Lockwasher
37. Bolt
38. Brake hose pad
39. Bracket
40. Strap

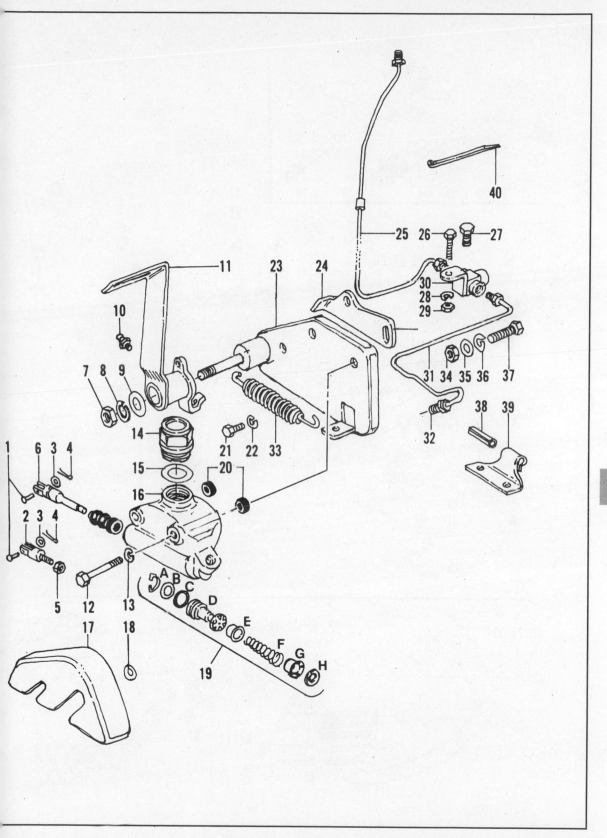

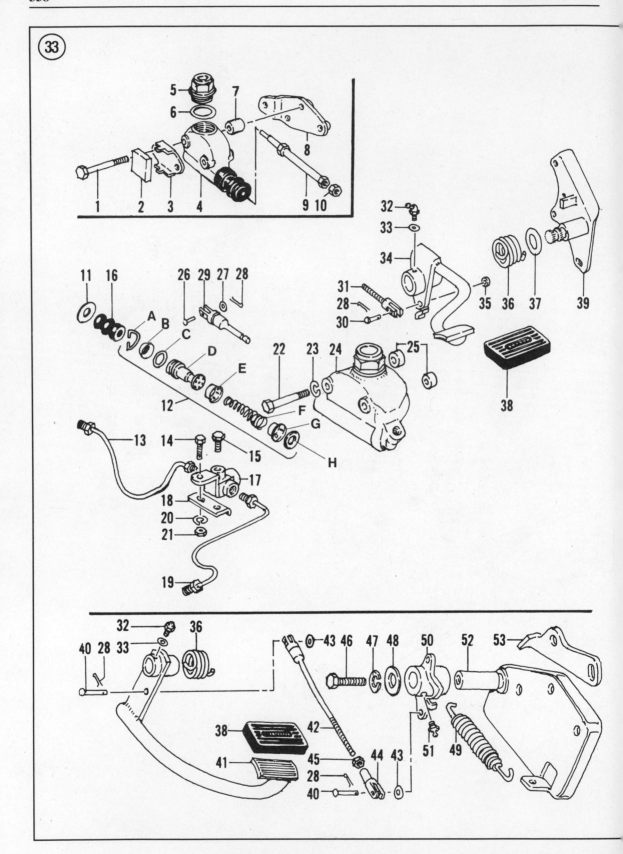

## REAR BRAKE MASTER CYLINDER
## (1971-EARLY 1979 FX)

1. Bolt
2. Insulator
3. Clamp
4. Master cylinder·
   (1973-early 1979)
5. Cap
6. Washer
7. Spacer
8. Plate
9. Brake rod (late
   1972-early 1979)
10. Nut
11. Rubber washer
    (late 1972-early 1979)
12. Piston assembly
13. Brake hose
14. Bolt
15. Bolt
16. Boot
17. Junction block
    (1971-early 1972 FX)
18. Plate
19. Brake hose
20. Lockwasher
21. Nut
22. Bolt
23. Lockwasher
24. Master cylinder
    (1971-1972)
25. Spacers (1971-early 1972)
26. Clevis pin
27. Plain washer
28. Cotter pin

29. Plunger (1971-early 1972)
30. Clevis pin
31. Brake rod (late
    1972-early 1979)
32. Grease fitting
33. Plain washer
34. Brake pedal (late
    1972-early 1979)
35. Plain washer
36. Spring
37. Plain washer
38. Foot lever rubber
39. Foot lever bracket
    (late 1972-early 1979)
40. Clevis pin
41. Brake pedal
    (1971-early 1979)
42. Plunger (early style)
43. Plain washer
44. Brake rod clevis
    (1971-early 1972)
45. Nut
46. Bolt
47. Lockwasher
48. Plain washer
49. Spring (1971-early 1972)
50. Brake pedal support
51. Grease fitting
52. Brake pedal plate
    (1971-early 1972)
53. Stop plate
    (1971-early 1972)

10

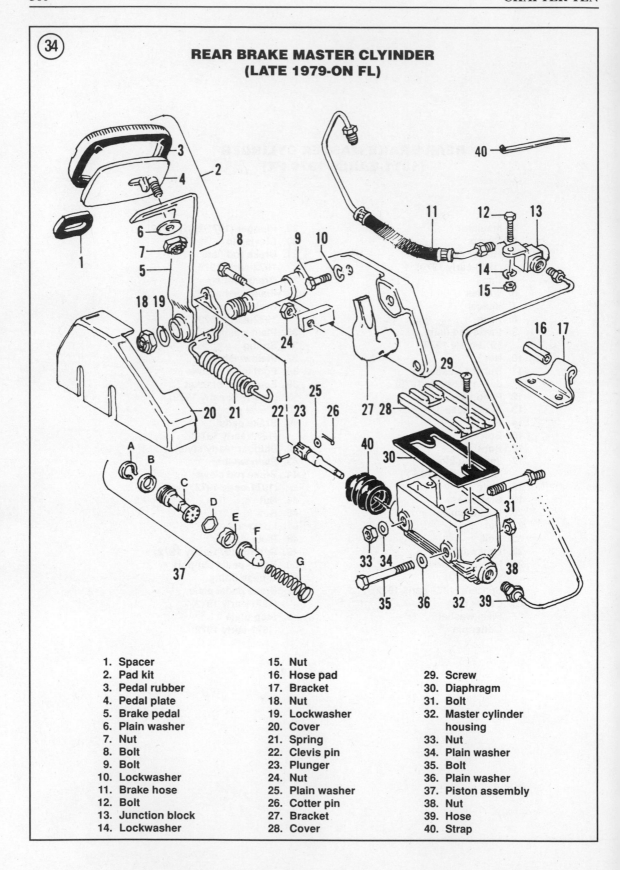

**REAR BRAKE MASTER CLYINDER
(LATE 1979-ON FL)**

1. Spacer
2. Pad kit
3. Pedal rubber
4. Pedal plate
5. Brake pedal
6. Plain washer
7. Nut
8. Bolt
9. Bolt
10. Lockwasher
11. Brake hose
12. Bolt
13. Junction block
14. Lockwasher

15. Nut
16. Hose pad
17. Bracket
18. Nut
19. Lockwasher
20. Cover
21. Spring
22. Clevis pin
23. Plunger
24. Nut
25. Plain washer
26. Cotter pin
27. Bracket
28. Cover

29. Screw
30. Diaphragm
31. Bolt
32. Master cylinder housing
33. Nut
34. Plain washer
35. Bolt
36. Plain washer
37. Piston assembly
38. Nut
39. Hose
40. Strap

3. Disconnect the brake and reservoir supply hoses at the master cylinder and drain the reservoir.

*WARNING*
*Discard all brake fluid.*

4. Remove the master cylinder mounting bolts and remove the master cylinder assembly.
5. Installation is the reverse of these steps.
6. Bleed the rear brake as described in this chapter.
7. Adjust the rear brake pedal as described in Chapter Three.

## Overhaul (1966-early 1979)

It is not necessary to remove the master cylinder to remove the piston assembly. Refer to **Figure 32** (FL) or **Figure 33** (FX).

1. Disconnect the brake line at the master cylinder and drain the reservoir of all brake fluid.
2A. *1966-1969:* Remove the plunger clevis pin. Then loosen the master cylinder plunger locknut and unscrew the lever clevis.
2B. *1970-early 1979:* Remove the plunger clevis pin.
3. Pull the plunger out of the master cylinder.
4. Pull the rubber boot off the end of the master cylinder and remove the piston assembly in the order shown in **Figure 32** or **Figure 33**.
5. On 1972 and earlier models with drum brake, remove the valve and valve seat.
6. Carefully inspect the O-ring, cup, dust boot and grommet for softening or wear. Replace these parts each time the master cylinder is disassembled.
7. Inspect the master cylinder walls for scratches or wear grooves. The master cylinder housing should be replaced if the cylinder walls are damaged. Do not hone the cylinder walls.
8. Check to see that the vent hole in the cover is not plugged.
9. Check the plunger assembly for wear, bending or damage. Replace it if necessary.
10. Coat all internal parts in specified brake fluid before assembly.
11. Referring to **Figure 32** or **Figure 33**, install the piston assembly into the master cylinder bore.

*NOTE*
*On 1972 and earlier models with drum brake, make sure to order parts 41773-58 (valve and seat) and install them in*

the order shown in 19(G, H), **Figure 32** or 12(G, H), **Figure 33**.

12. Install the stop wire in the end of the master cylinder housing. Make sure the wire seats in the groove.
13. Install the rubber boot and plunger.
14. Reconnect the clevis pin.
15. Bleed the rear brake as described in this chapter.
16. Adjust the rear brake as described in Chapter Three.

## Overhaul (Late 1979-on FL and Late 1979-on FX Except 1983-1984 FXWG)

Refer to **Figure 34** (FL), **Figure 35** (FX except FXWG) or **Figure 36** (1980-1982 FXWG).

1. Remove the master cylinder assembly as described in this chapter.
2. Remove the cover and gasket. Drain and discard the brake fluid from the master cylinder.
3. Pull the rubber boot off the end of the master cylinder and remove the retaining ring. Then remove the piston assembly in the order shown in **Figure 34**, **Figure 35** or **Figure 36**.

*WARNING*
*The 1984 FL models do not use the backup disc (37D, **Figure 34**) because of a design change to the master cylinder piston. Use of the backup disc on these models may cause brake drag or brake lockup.*

4. Remove the O-ring from the piston.
5. Carefully inspect the O-ring, cup and dust boot for softening or wear. Replace these parts whenever the master cylinder is disassembled.
6. Inspect the master cylinder walls for scratches or wear grooves. The master cylinder housing should be replaced if the cylinder walls are damaged. Do not hone the cylinder walls.
7. Check to see that the vent hole in the cover is not plugged.
8. Check the plunger assembly for wear, bending or damage. Replace it if necessary.
9. Coat all internal parts in specified brake fluid before assembly.
10A. Referring to **Figure 34**, install the following parts in order:
   a. Spring (37G).

**10**

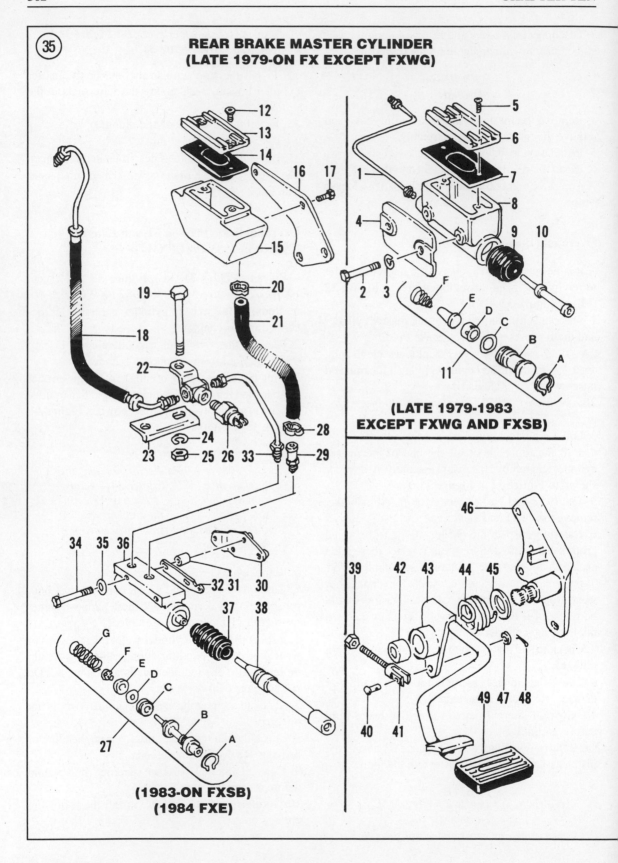

35

# REAR BRAKE MASTER CYLINDER
## (LATE 1979-ON FX EXCEPT FXWG)

(LATE 1979-1983
EXCEPT FXWG AND FXSB)

(1983-ON FXSB)
(1984 FXE)

1. Brake hose
2. Bolt
3. Lockwasher
4. Plate
5. Screw
6. Cover
7. Diaphragm
8. Master cylinder
   housing
9. Rubber boot
10. Brake rod
11. Piston assembly
12. Screw
13. Cover
14. Diaphragm
15. Master cylinder
    housing
16. Plate
17. Bolt
18. Brake hose
19. Bolt
20. Clip
21. Brake hose
22. Junction block
    (all models)
23. Plate
24. Lockwasher
25. Nut
26. Stoplight switch
    (all models)
27. Piston assembly
    (1983-on FXSB; 1984 FXE)
28. Clip
29. Bleed scew
30. Bracket
31. Spacer
32. Mounting plate
33. Brake hose
34. Bolt
35. Washer
36. Master cylinder
    housing
37. Rubber boot
38. Brake rod
39. Nut
40. Clevis pin
41. Brake rod clevis
42. Bushing
43. Brake pedal
44. Spring
45. Plain washer
46. Bracket
47. Washer
48. Cotter pin
49. Brake pedal rubber

b. Spring seat (37F).
c. Piston cup (37E).
d. Backup disc (37D).

*WARNING*
*The backup disc (37D, **Figure 34**) is not used on 1984 FL models. Installation of the disc on this model may cause brake drag or brake lockup.*

e. Install the O-ring (B) on the piston (C).
f. Insert the piston assembly into the master cylinder. Install the circlip (A) in the end of the master cylinder housing.

10B. *Internal reservoir models:* Referring to **Figure 35**, install the following parts in order:
   a. Spring (11F).
   b. Spring seat (11E).
   c. Piston cup (11D).
   d. Install the O-ring (11C) on the piston (11B).
   e. Insert the piston assembly into the master cylinder. Install the circlip (11A) in the end of the master cylinder housing.

10C. *Separate reservoir models:* Referring to **Figure 35**, install the following parts in order:
   a. Spring (27G).
   b. Stop (27F).
   c. Piston cup (27E).
   d. Piston cup (27D).
   e. Install the seal (27C) on the piston (27B). Insert the piston assembly into the master cylinder. Install the circlip (27A) in the end of the master cylinder housing.

10D. Referring to **Figure 36**, install the following parts in order:
   a. Spring (35F).
   b. Spring seat (35E).
   c. Piston cup (37D).
   d. Install the O-ring (B) on the piston (C).
   e. Insert the piston assembly into the master cylinder. Install the circlip (A) in the end of the master cylinder housing.

11. Install the boot onto the master cylinder.
12. Install the master cylinder as described in this chapter.

**Overhaul (1983-on FXWG)**

Refer to **Figure 37** for this procedure.
1. Remove the master cylinder assembly as described in this chapter.

**10**

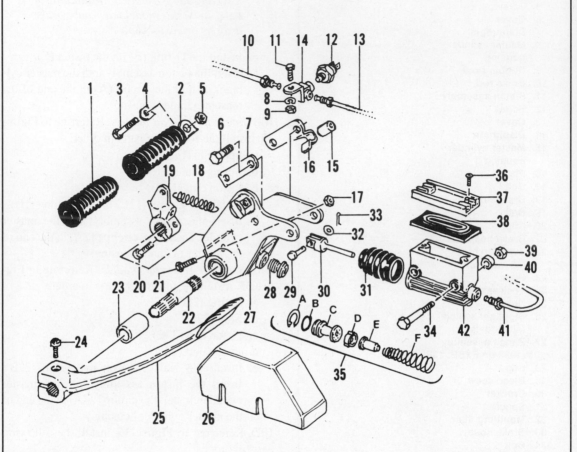

## REAR BRAKE MASTER CYLINDER
## (1980-1982 FXWG)

1. Brake pedal rubber
2. Brake pedal
3. Bolt
4. Spring washer
5. Nut
6. Bolt
7. Plate
8. Lockwasher
9. Nut
10. Brake hose
11. Bolt
12. Stoplight switch
13. Brake hose
14. Junction block
15. Hose pad
16. Bracket
17. Nut
18. Spring
19. Brake lever clevis
20. Bolt
21. Bolt
22. Brake pedal shaft
23. Bearing
24. Screw
25. Brake pedal
26. Cover
27. Bracket assembly
28. Plug
29. Clevis pin
30. Plunger
31. Rubber boot
32. Washer
33. Cotter pin
34. Bolt
35. Piston assembly
36. Screw
37. Cover
38. Diaphragm
39. Nut
40. Spacer
41. Brake hose
42. Master cylinder

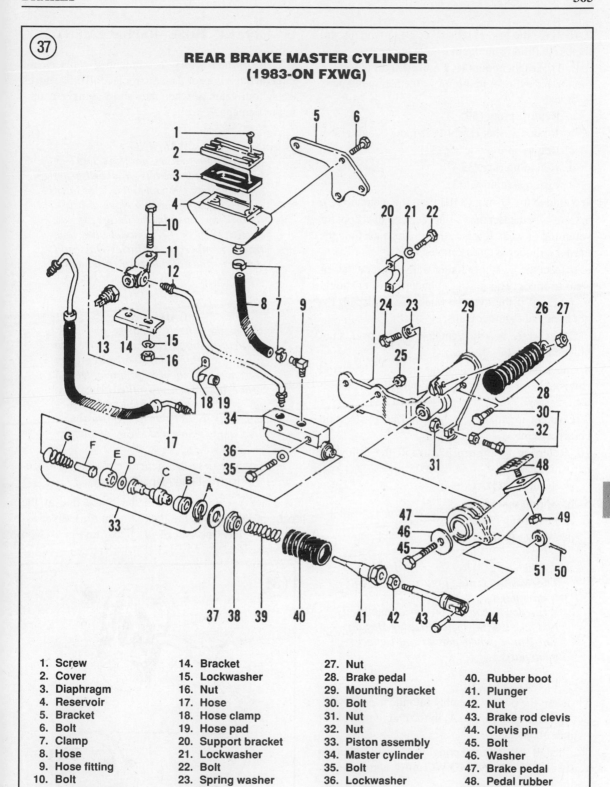

## REAR BRAKE MASTER CYLINDER
## (1983-ON FXWG)

| | | |
|---|---|---|
| 1. Screw | 14. Bracket | 27. Nut |
| 2. Cover | 15. Lockwasher | 28. Brake pedal |
| 3. Diaphragm | 16. Nut | 29. Mounting bracket |
| 4. Reservoir | 17. Hose | 30. Bolt |
| 5. Bracket | 18. Hose clamp | 31. Nut |
| 6. Bolt | 19. Hose pad | 32. Nut |
| 7. Clamp | 20. Support bracket | 33. Piston assembly |
| 8. Hose | 21. Lockwasher | 34. Master cylinder |
| 9. Hose fitting | 22. Bolt | 35. Bolt |
| 10. Bolt | 23. Spring washer | 36. Lockwasher |
| 11. Junction block | 24. Bolt | 37. Washer |
| 12. Hose | 25. Nut | 38. Washer |
| 13. Stoplight switch | 26. Brake pedal | 39. Spring |

| | |
|---|---|
| 40. Rubber boot | |
| 41. Plunger | |
| 42. Nut | |
| 43. Brake rod clevis | |
| 44. Clevis pin | |
| 45. Bolt | |
| 46. Washer | |
| 47. Brake pedal | |
| 48. Pedal rubber | |
| 49. Nut | |
| 50. Cotter pin | |
| 51. Plain washer | |

**10**

2. Remove the cover and gasket. Drain and discard the brake fluid from the master cylinder.

3. Pull the rubber boot (40, **Figure 37**) off the end of the master cylinder and remove the following parts in order:

    a. Return spring (39).

    b. Return washer (early 1983 models only) (38).

    c. Return spring washer (37).

    d. Retaining ring (33A).

    e. Piston assembly (33).

4. Remove the O-ring (33B) from the piston (33C).

5. Carefully inspect the O-ring, cup and dust boot for softening or wear. Replace these parts each time the master cylinder is disassembled.

6. Inspect the master cylinder walls for scratches or wear grooves. The master cylinder housing should be replaced if the cylinder walls are damaged. Do not hone the cylinder walls.

7. Check to see that the vent hole in the cover is not plugged.

8. Check the plunger (41) on the end of the brake rod clevis (43) for wear, bending or damage. Replace it if necessary.

9. Coat all internal parts in specified brake fluid before assembly.

10. Referring to **Figure 37**, install the following parts in order:

    a. Spring (33G).

    b. Spring seat (33F).

    c. Piston cup (33E).

    d. Washer (33D).

*NOTE*
*The washer (33D, **Figure 37**) is not used on early 1983 FXWG models. Write down your engine serial number and take it with you to a Harley-Davidson dealer when purchasing replacement parts.*

    e. Piston cup (33B).

11. Insert the piston assembly into the master cylinder. Install the circlip (33A) in the end of the master cylinder housing.

12. Install the return spring washer (37), return washer (38) (early 1983 FXWG only), and the return spring (39).

13. Install the boot onto the master cylinder.

14. Install the master cylinder as described in this chapter.

## BRAKE HOSE REPLACEMENT

There is no factory-recommended replacement interval but it is a good idea to replace all brake hoses every four years or when they show signs of cracking or damage.

*WARNING*
*Harley-Davidson specifies DOT 3 brake fluid for models produced prior to September, 1976 and DOT 5 for later models. Mixing the two types of brake fluids can cause brake failure. If you own a 1976 or 1977 model take your frame number to a Harley-Davidson dealer to find out your bike's production date.*

*CAUTION*
*Cover the wheels, fenders, fuel tank, swing arm and drive chain or belt with a heavy cloth or plastic tarp to protect it from the accidental spilling of DOT 3 brake fluid. Wash any spilled brake fluid off of any painted or plated surface immediately, as it will destroy the finish. Use soapy water and rinse completely.*

This section describes general replacement procedures for the front and rear brake hoses.

1. Place a container under the brake line at the caliper. Disconnect the brake line at the caliper assembly. Place the end of the brake line in a clean

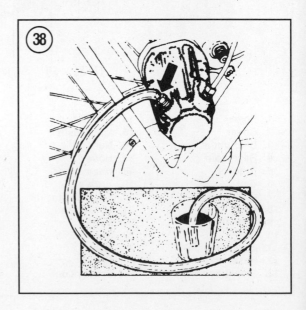

container. Operate the brake lever to drain all brake fluid from the master cylinder.

2. Disconnect and remove all brake hoses, bolts and washers.

3. Install new hoses, sealing washers and union bolts (where used) in the reverse order of removal. Be sure to install new sealing washers in the correct positions.

4. Tighten all union bolts to torque specifications in **Table 1**.

5. Refill the master cylinder with specified type of fresh brake fluid only. Bleed the brake as described in this chapter.

### BLEEDING THE SYSTEM

This procedure is necessary only when the brakes feel spongy, there is a leak in the hydraulic system, a component has been replaced or the brake fluid is being replaced.

1. Flip off the dust cap from the brake bleeder valve.

2. Connect a length of clear tubing to the bleeder valve on the caliper. Place the other end of the tube into a clean container. Fill the container with enough fresh brake fluid to keep the end submerged. The tube should be long enough so that a loop can be made higher than the bleeder valve to prevent air from being drawn into the caliper during bleeding. **Figure 38** shows a typical setup for bleeding disc brakes. **Figure 39** shows the bleed valve for rear drum brakes.

3. Clean the top of the master cylinder of all dirt and foreign matter. Remove the cap and diaphragm. Fill the reservoir to about 3/8 in. (9.52 mm) from the top. Insert the diaphragm to prevent the entry of dirt and moisture.

4. Slowly apply the brake lever several times. Hold the lever in the applied position and open the bleeder valve about 1/2 turn. See **Figure 38** or **Figure 39**. Allow the lever to travel to its limit. When this limit is reached, tighten the bleeder screw. As the brake fluid enters the system, the level will drop in the master cylinder reservoir. Maintain the level at about 3/8 in. (9.52 mm) from the top of the reservoir to prevent air from being drawn into the system.

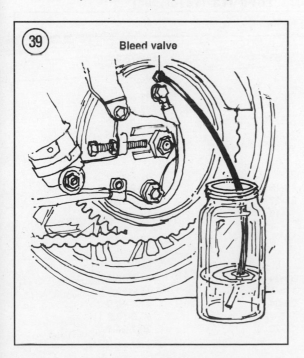

(39) Bleed valve

5. Continue to pump the lever and fill the reservoir until the fluid emerging from the hose is completely free of air bubbles. If you are replacing the fluid, continue until the fluid emerging from the hose is clean.

> NOTE
> If bleeding is difficult, it may be necessary to allow the fluid to stabilize for a few hours. Repeat the bleeding procedure when the tiny bubbles in the system settle out.

6. Hold the lever in the applied position and tighten the bleeder valve. Remove the bleeder tube and install the bleeder valve dust cap.
7. If necessary, add fluid to correct the level in the master cylinder reservoir. It must be above the level line.

8. Install the cap and tighten the screws.

9. Test the feel of the brake lever. It should feel firm and should offer the same resistance each time it's operated. If it feels spongy, it is likely that air is still in the system and it must be bled again. When all air has been bled from the system, and the brake fluid level is correct in the reservoir, double-check for leaks and tighten all fittings and connections.

> WARNING
> Before riding the motorcycle, make certain that the brake is operating correctly by operating the lever several times. Then make the test ride a slow one at first to make sure the brake is operating correctly.

### Table 1 BRAKE TIGHTENING TORQUES (1970-1984)

| | ft.-lb. | N·m |
|---|---|---|
| Wheel mount bolts (drum brake) | | |
| 1970-early 1978 | 50-55 | 69-75.9 |
| Brake disc screws | | |
| 1970-early 1978 | | |
| 1972-1973 | 35 | 48.3 |
| 1974-1977 | 10 | 13.8 |
| Early 1978 | | |
| 16 in. wheel | 21-27 | 29-37.3 |
| 19 in. wheel | 16-19 | 22-26.2 |
| Late 1978-on | | |
| Laced 16 and 21 in. wheels | 23-27 | 31.7-37.3 |
| Laced 19 in. wheels | 16-19 | 22-26.2 |
| Cast 16 in. wheels | 23-27 | 31.7-37.3 |
| Cast 19 in. wheels | 14-16 | 19.3-22 |
| Rear brake anchor nuts | | |
| 1970-early 1978 | 50 | 69 |
| Brake caliper | | |
| 1970-early 1978 | | |
| FL models | 30-34 | 41.4-46.9 |
| FX models 1978 (front) | 115-120 in.-lb. | 13.2-13.8 |
| FX 1970-1977 | 30-34 | 41.4-46.9 |
| Late 1978-on | 115-120 in.-lb. | 13.2-13.8 |
| Union bolts | 15-20 | 20.7-27.6 |

# INDEX

11

11

# WIRING
# DIAGRAMS

## 1966-1967 ELECTRA-GLIDE (FL)

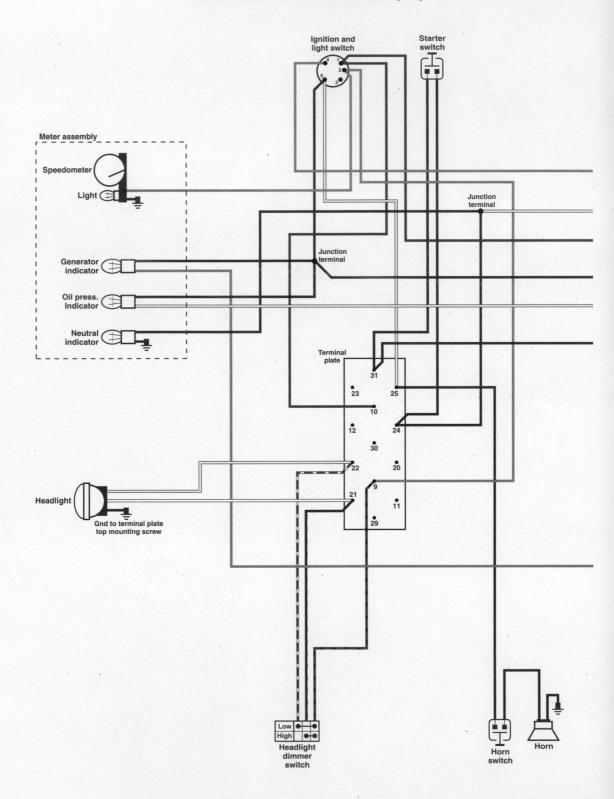

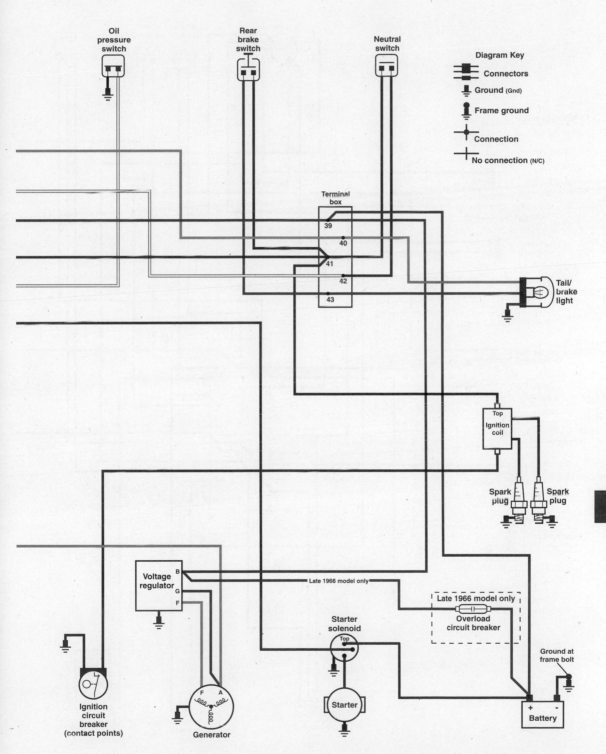

Oil
pressure
switch

Rear
brake
switch

Neutral
switch

**Diagram Key**

Connectors

Ground (Gnd)

Frame ground

Connection

No connection (N/C)

Terminal
box

39

40

41

42

43

Tail/
brake
light

Top
Ignition
coil

Spark
plug

Spark
plug

Voltage
regulator

B

G

F

Late 1966 model only

Late 1966 model only

Overload
circuit breaker

Ground at
frame bolt

Starter
solenoid

Top

Ignition
circuit
breaker
(contact points)

F   A

Generator

Starter

+   –

Battery

12

## 1968-1969 ELECTRA-GLIDE (FL)

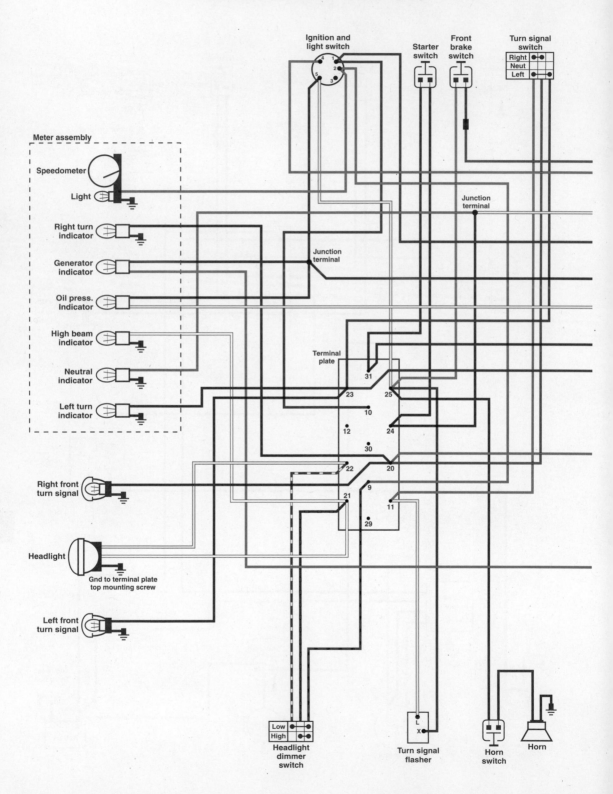

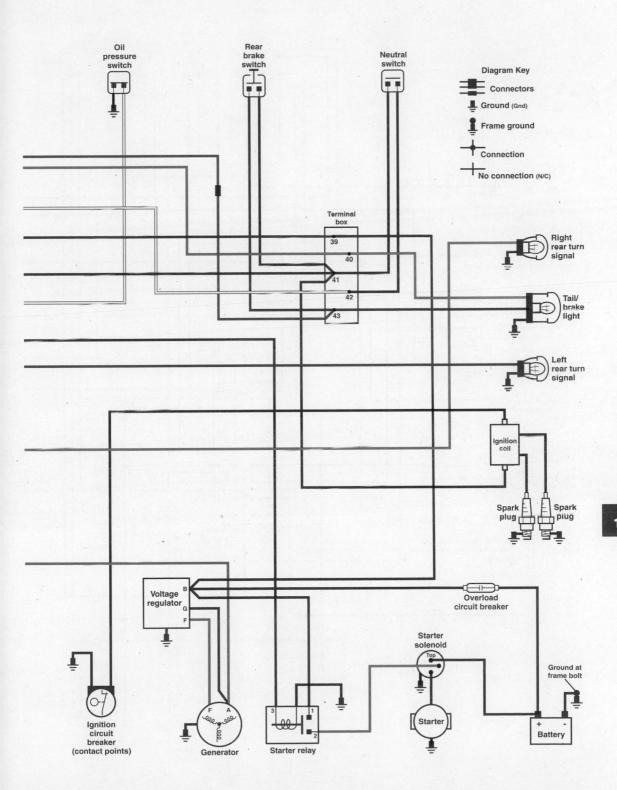

12

## 1970-1971 ELECTRA-GLIDE (FL)

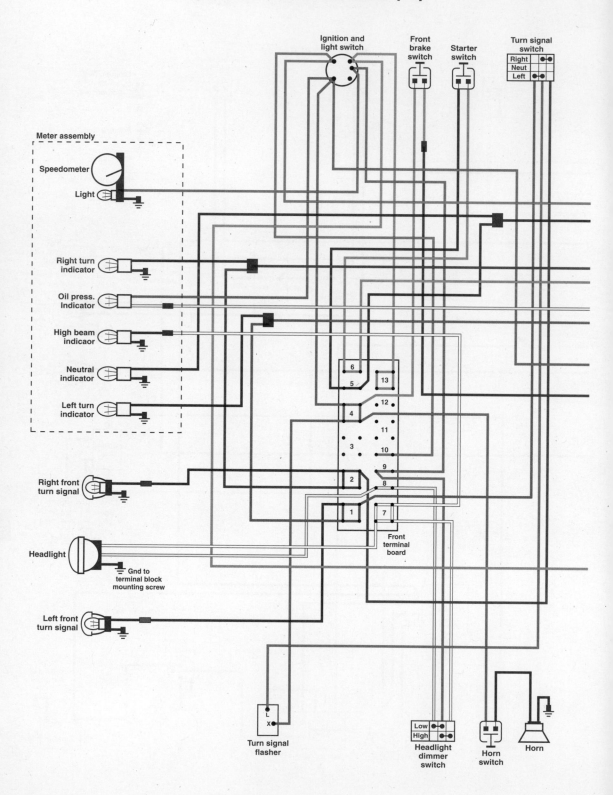

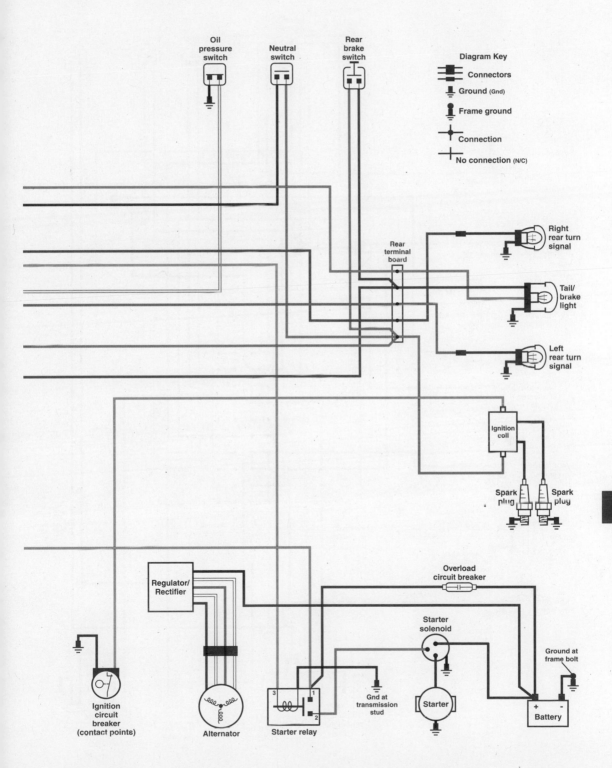

Oil pressure switch

Neutral switch

Rear brake switch

**Diagram Key**

Connectors

Ground (Gnd)

Frame ground

Connection

No connection (N/C)

Right rear turn signal

Tail/ brake light

Left rear turn signal

Rear terminal board

Ignition coil

Spark plug

Spark plug

Overload circuit breaker

Starter solenoid

Ground at frame bolt

Starter

Regulator/ Rectifier

Ignition circuit breaker (contact points)

Alternator

Starter relay

Gnd at transmission stud

Battery

12

# 1972 ELECTRA-GLIDE (FL)

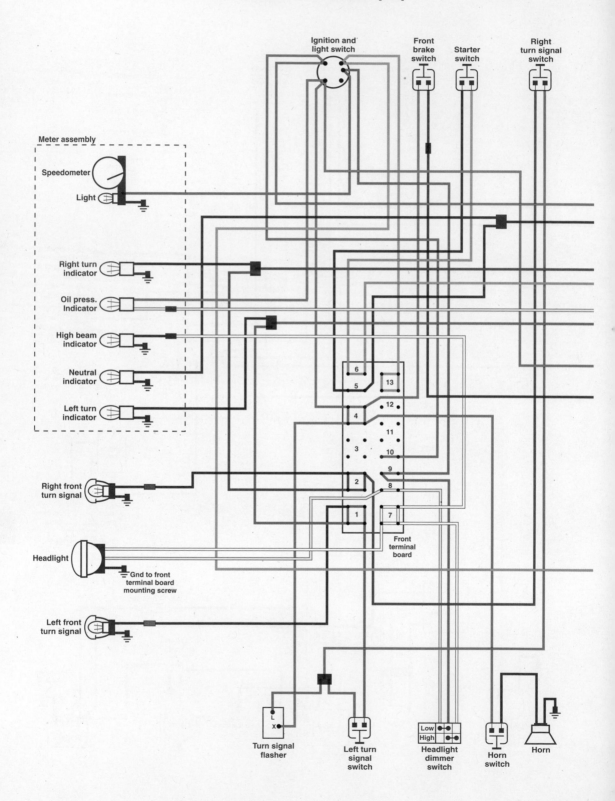

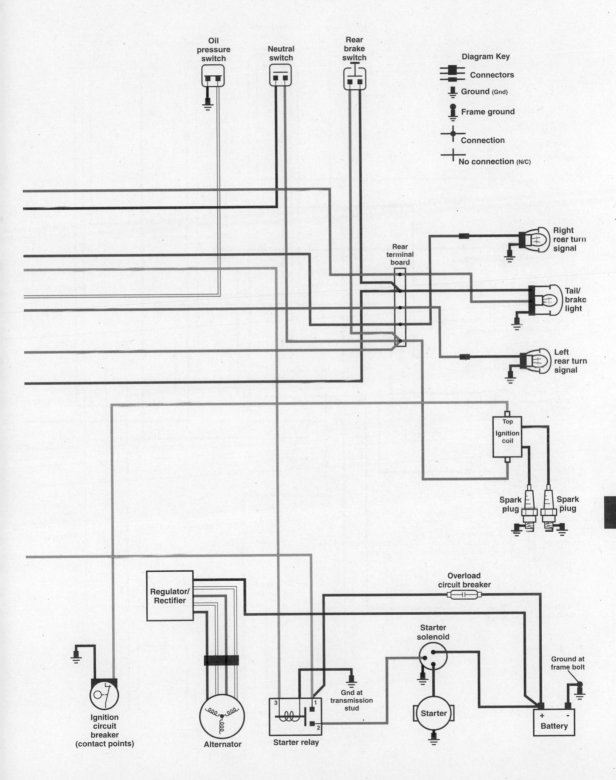

**Diagram Key**

Connectors

Ground (Gnd)

Frame ground

Connection

No connection (N/C)

Oil pressure switch

Neutral switch

Rear brake switch

Rear terminal board

Right rear turn signal

Tail/brake light

Left rear turn signal

Top Ignition coil

Spark plug

Spark plug

Regulator/ Rectifier

Overload circuit breaker

Starter solenoid

Ground at frame bolt

Ignition circuit breaker (contact points)

Alternator

Starter relay

Gnd at transmission stud

Starter

Battery

**12**

## 1971 SUPER GLIDE (FL)

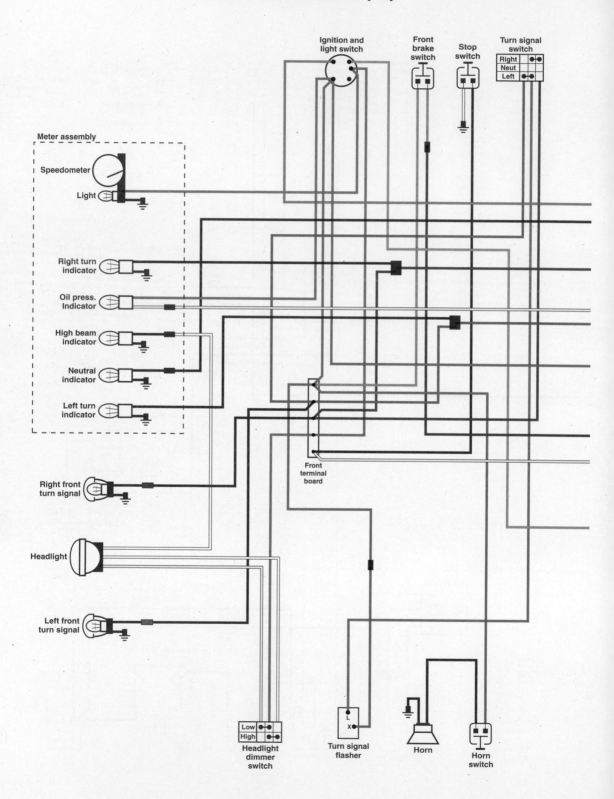

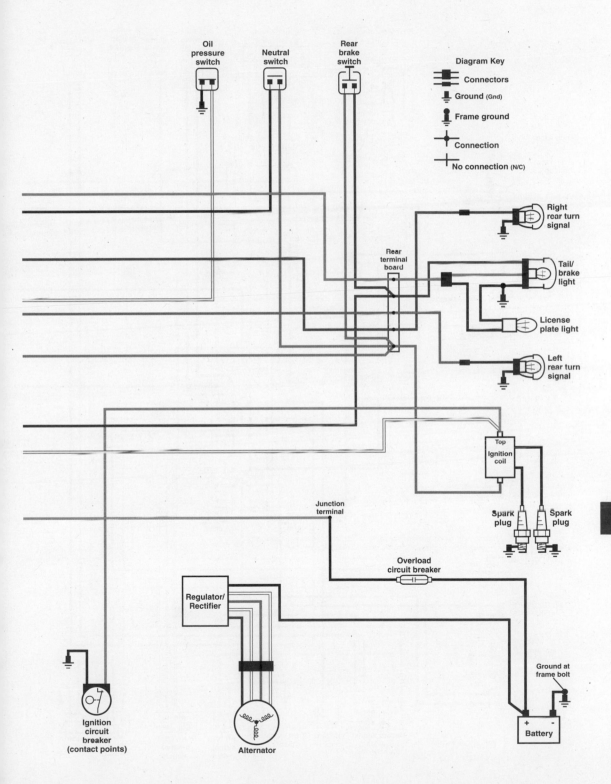

Oil pressure switch

Neutral switch

Rear brake switch

Diagram Key

Connectors

Ground (Gnd)

Frame ground

Connection

No connection (N/C)

Right rear turn signal

Rear terminal board

Tail/ brake light

License plate light

Left rear turn signal

Top Ignition coil

Junction terminal

Spark plug

Spark plug

Overload circuit breaker

Regulator/ Rectifier

Ground at frame bolt

Ignition circuit breaker (contact points)

Alternator

+ Battery -

12

## 1973-1974 FL/FLH 1200

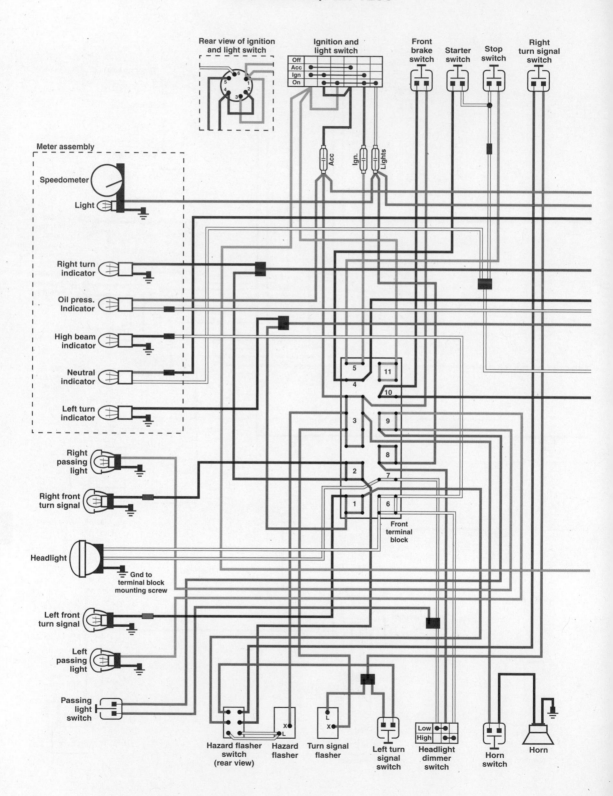

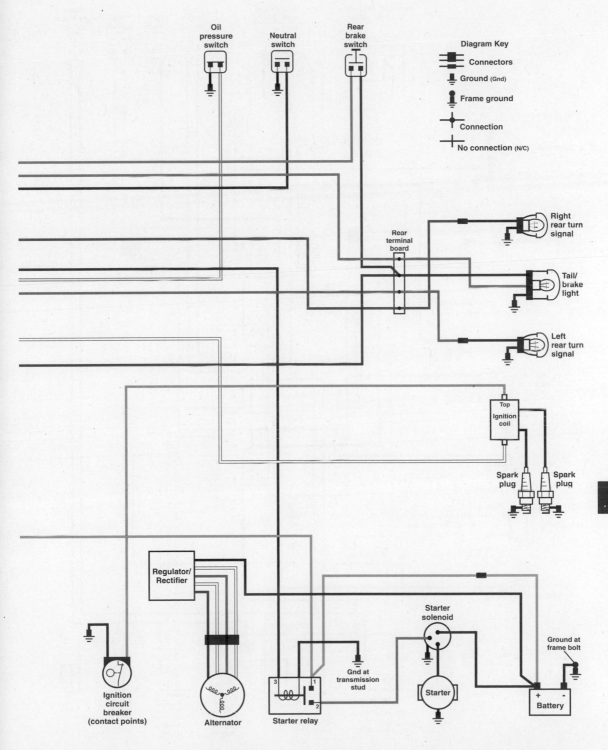

## 1975 FL/FLH 1200

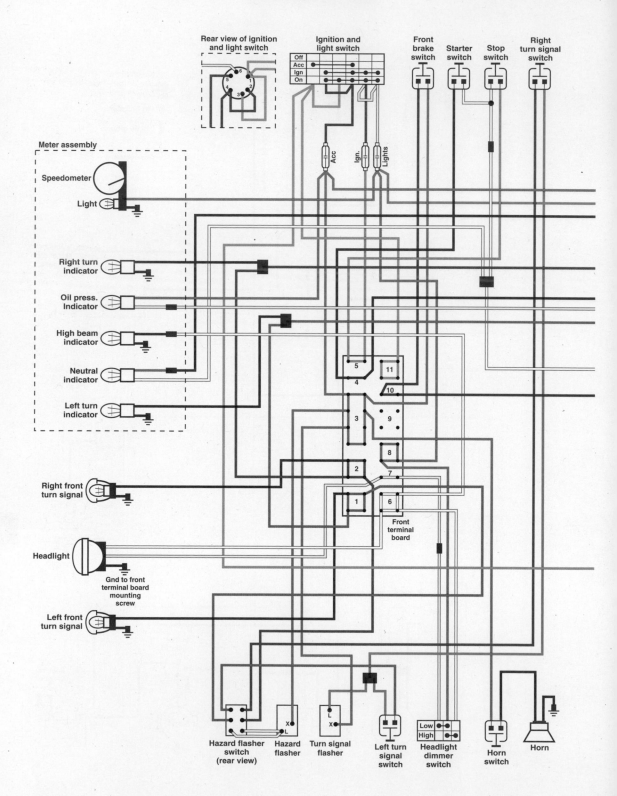

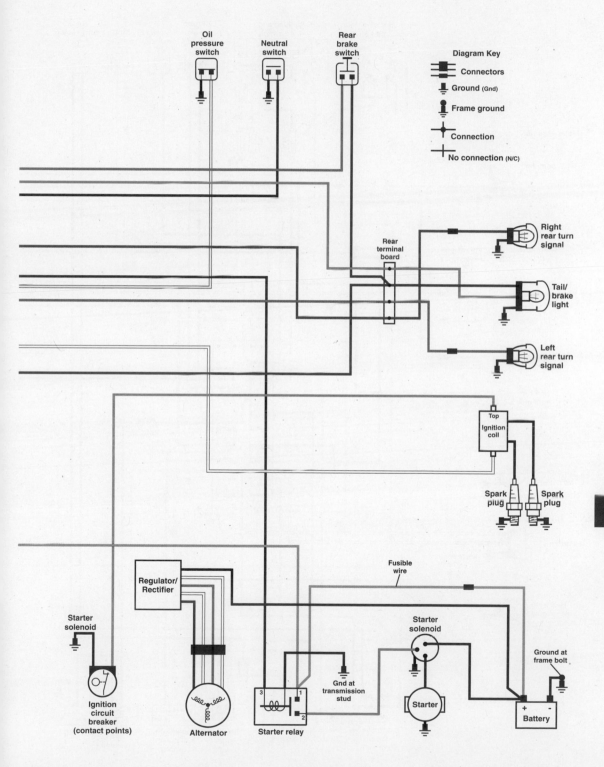

Oil pressure switch

Neutral switch

Rear brake switch

**Diagram Key**

Connectors

Ground (Gnd)

Frame ground

Connection

No connection (N/C)

Rear terminal board

Right rear turn signal

Tail/brake light

Left rear turn signal

Top Ignition coil

Spark plug

Spark plug

Fusible wire

Regulator/Rectifier

Starter solenoid

Starter solenoid

Ground at frame bolt

Gnd at transmission stud

Starter

Ignition circuit breaker (contact points)

Alternator

Starter relay

Battery

**12**

## 1976-1977 FL/FLH 1200

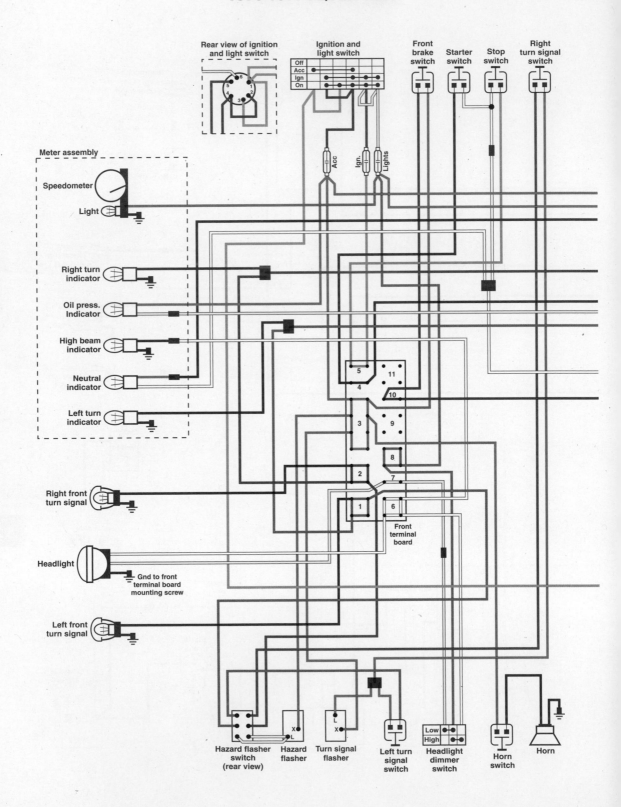

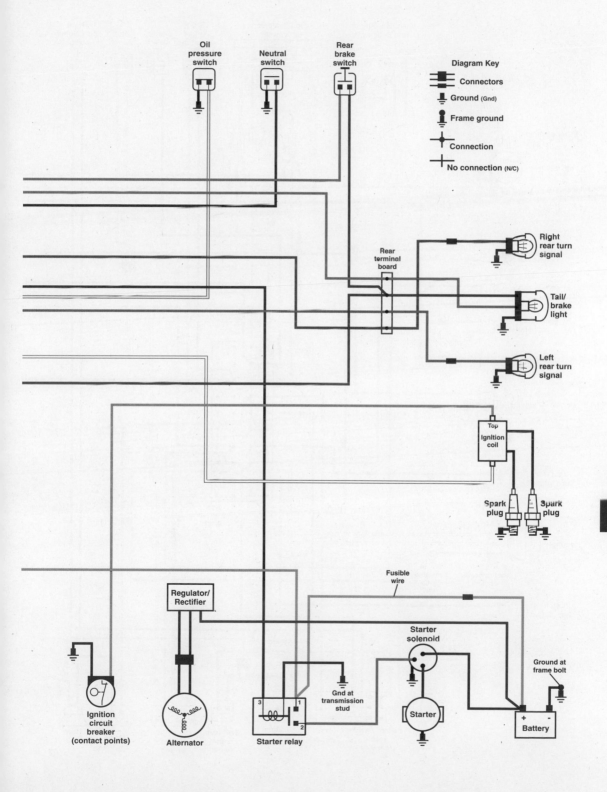

**Oil pressure switch**

**Neutral switch**

**Rear brake switch**

**Diagram Key**

Connectors

Ground (Gnd)

Frame ground

Connection

No connection (N/C)

**Rear terminal board**

**Right rear turn signal**

**Tail/ brake light**

**Left rear turn signal**

**Top Ignition coil**

**Spark plug**

**Spark plug**

**12**

**Fusible wire**

**Regulator/ Rectifier**

**Starter solenoid**

**Ground at frame bolt**

**Starter**

**Gnd at transmission stud**

**Ignition circuit breaker (contact points)**

**Alternator**

**Starter relay**

**+ - Battery**

## 1978-1979 FL/FLH

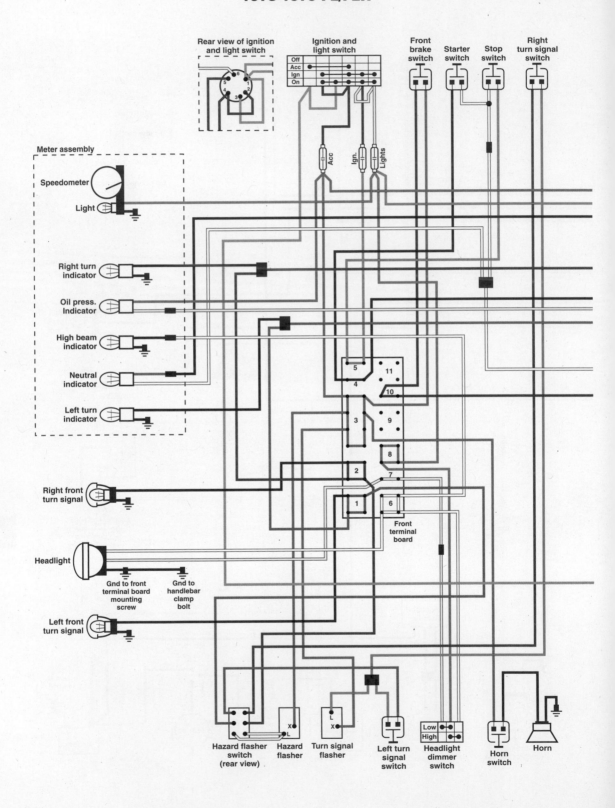

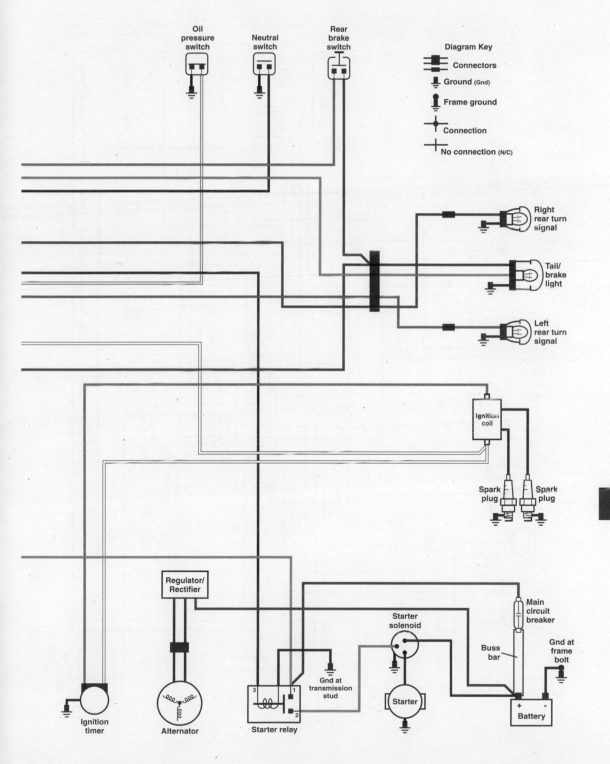

## 1980-1984 FL/FLH

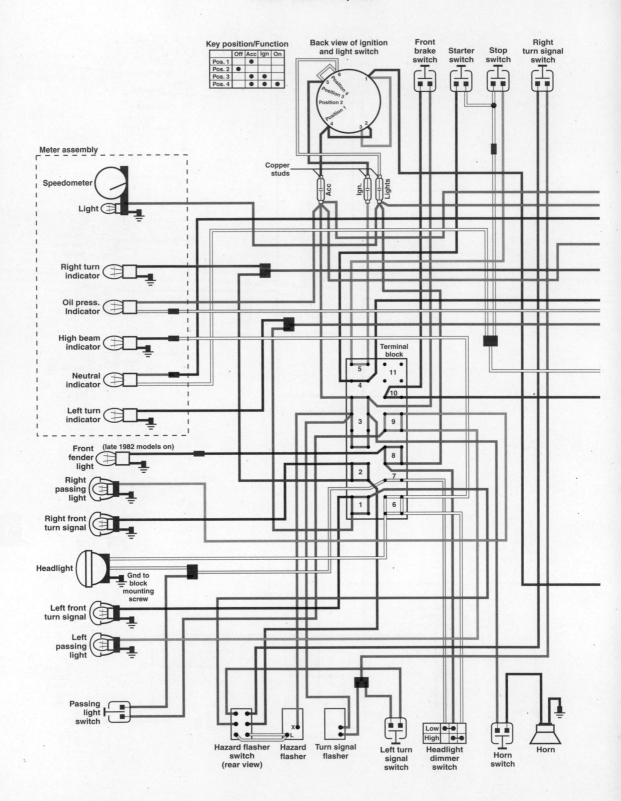

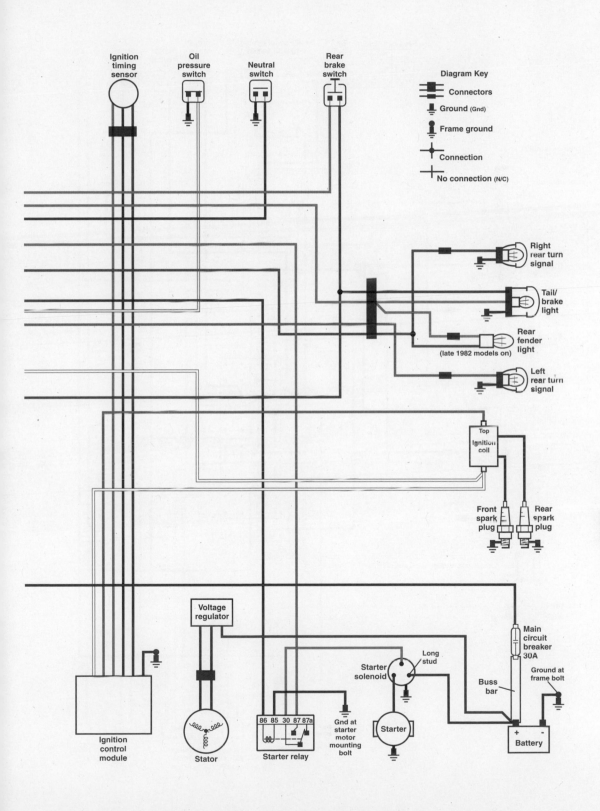

Ignition
timing
sensor

Oil
pressure
switch

Neutral
switch

Rear
brake
switch

Diagram Key

Connectors

Ground (Gnd)

Frame ground

Connection

No connection (N/C)

Right
rear turn
signal

Tail/
brake
light

Rear
fender
light

(late 1982 models on)

Left
rear turn
signal

Top
Ignition
coil

Front
spark
plug

Rear
spark
plug

Voltage
regulator

Main
circuit
breaker
30A

Long
stud

Starter
solenoid

Buss
bar

Ground at
frame bolt

86 85 30 87 87a

Gnd at
starter
motor
mounting
bolt

Starter

+ −
Battery

Ignition
control
module

Stator

Starter relay

12

## 1981-1982 FLHS

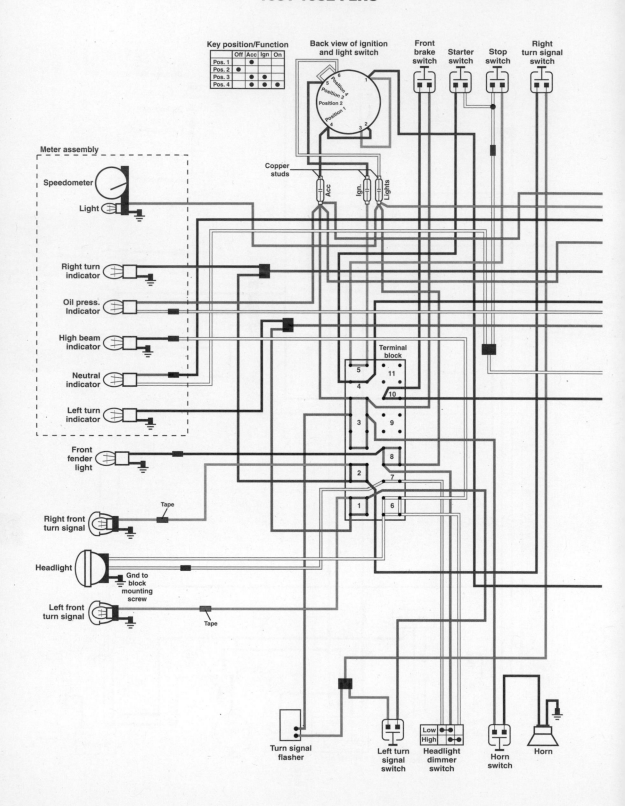

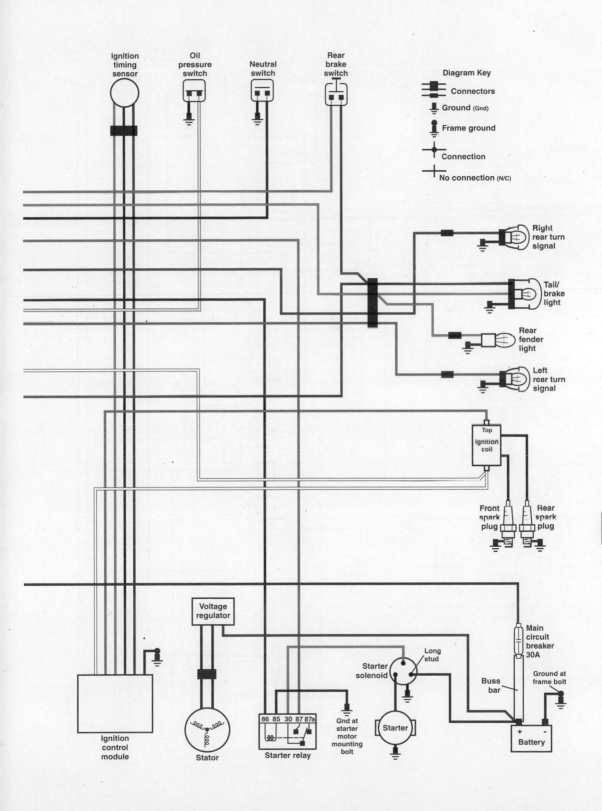

12

## 1982-1983 FLHS

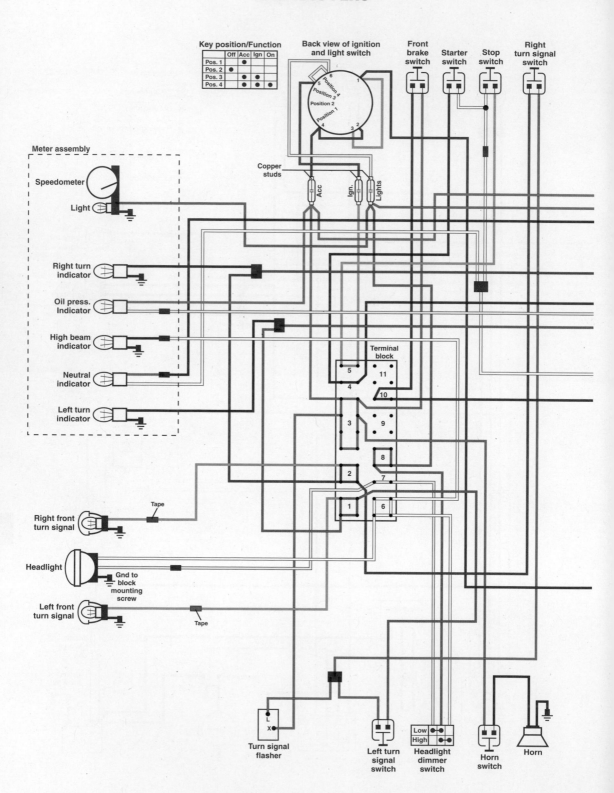

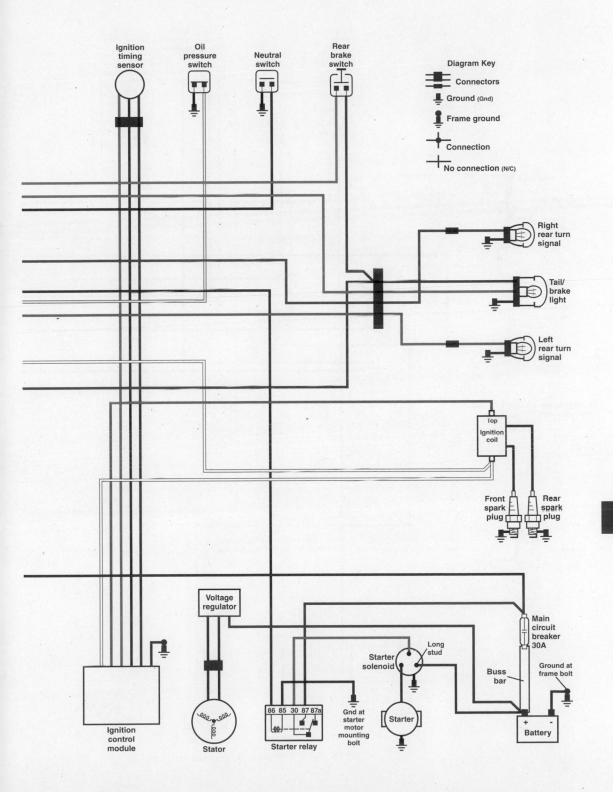

## 1973 FX 1200

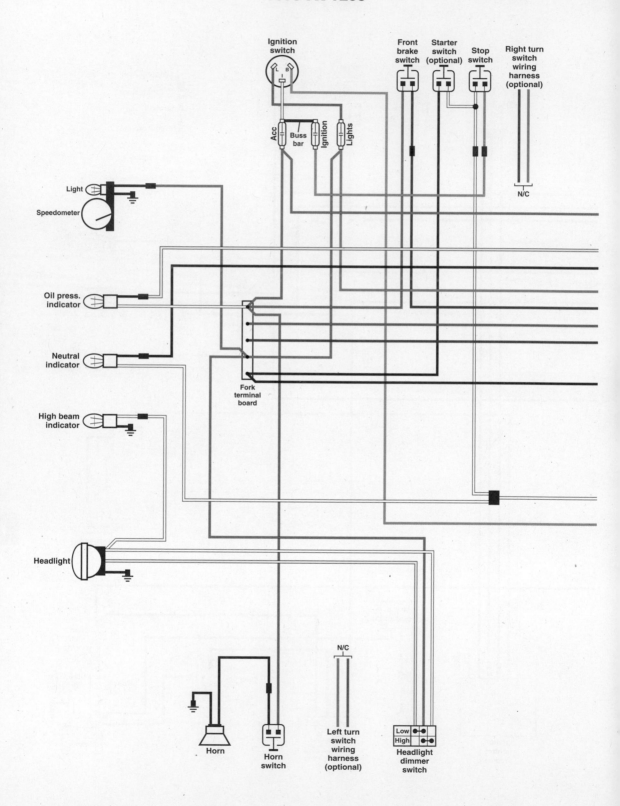

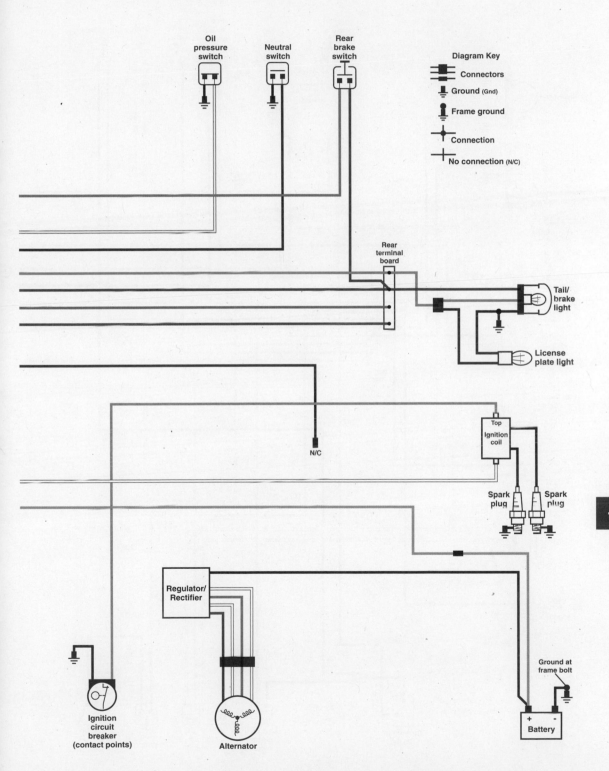

Oil pressure switch

Neutral switch

Rear brake switch

**Diagram Key**

Connectors

Ground (Gnd)

Frame ground

Connection

No connection (N/C)

Rear terminal board

Tail/ brake light

License plate light

N/C

Top Ignition coil

Spark plug

Spark plug

Regulator/ Rectifier

Ground at frame bolt

Ignition circuit breaker (contact points)

Alternator

+ −
Battery

**12**

## 1974 FX/FXE 1200

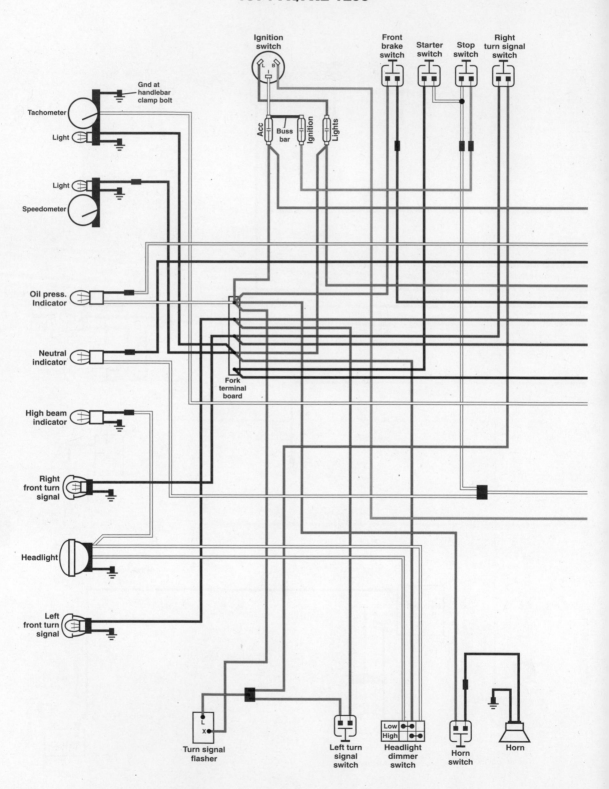

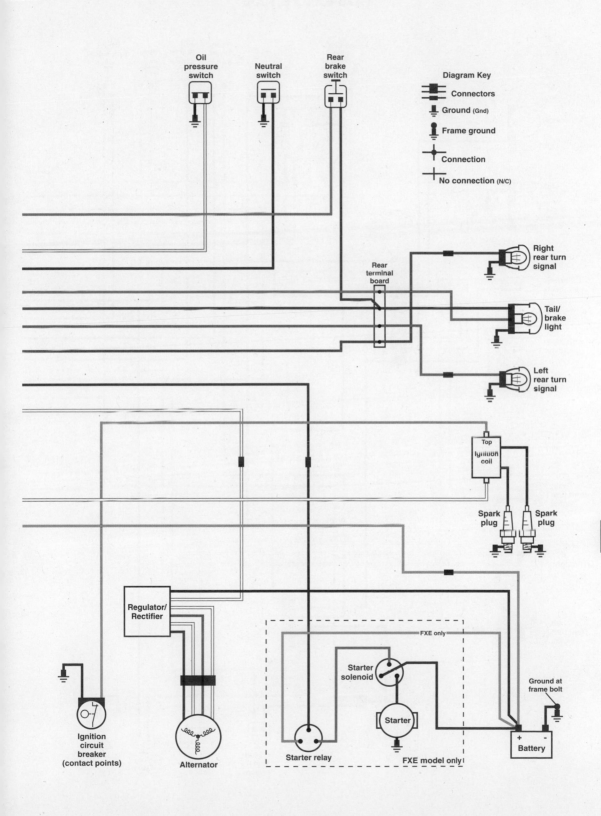

Oil pressure switch

Neutral switch

Rear brake switch

**Diagram Key**

Connectors

Ground (Gnd)

Frame ground

Connection

No connection (N/C)

Right rear turn signal

Rear terminal board

Tail/ brake light

Left rear turn signal

Top
Ignition coil

Spark plug

Spark plug

Regulator/ Rectifier

FXE only

Starter solenoid

Ground at frame bolt

Ignition circuit breaker (contact points)

Alternator

Starter relay

Starter

FXE model only

Battery

12

## 1975 FX/FXE 1200

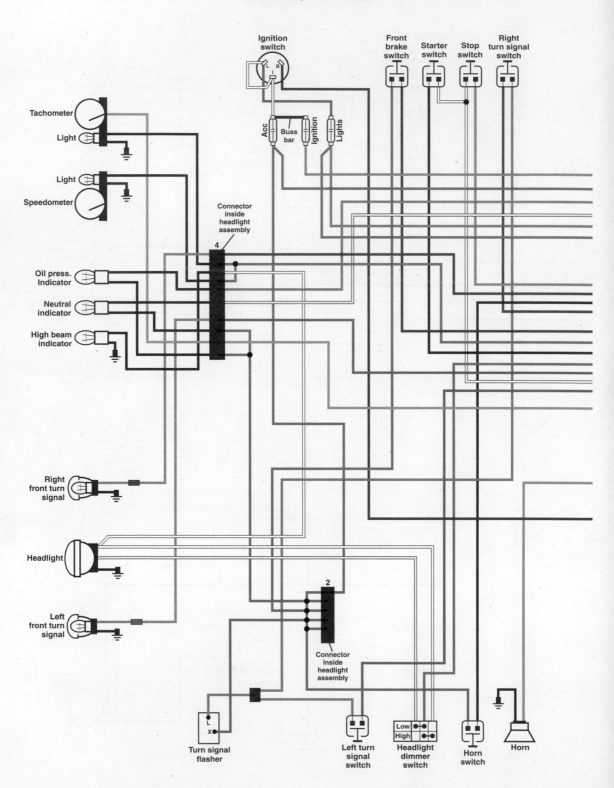

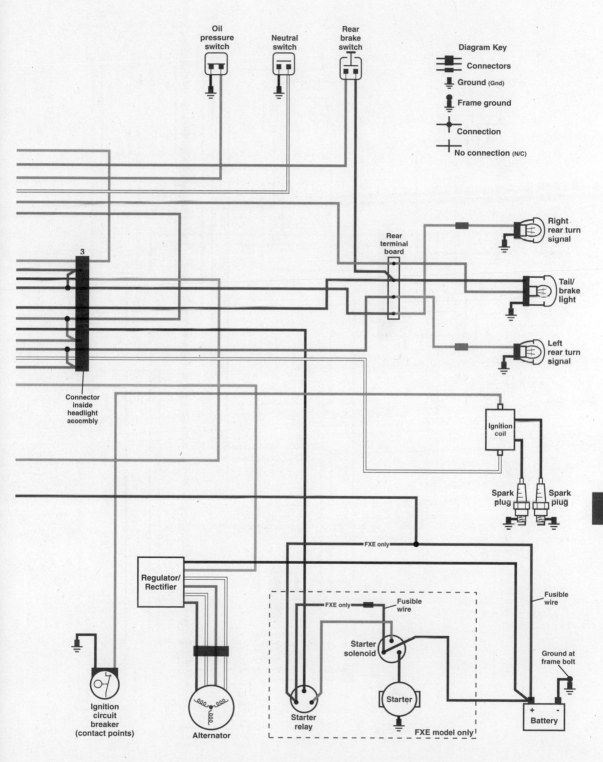

Oil pressure switch

Neutral switch

Rear brake switch

Diagram Key

Connectors

Ground (Gnd)

Frame ground

Connection

No connection (N/C)

Rear terminal board

Right rear turn signal

Tail/ brake light

Left rear turn signal

3

Connector inside headlight assembly

Ignition coil

Spark plug

Spark plug

**12**

FXE only

Regulator/ Rectifier

Fusible wire

FXE only

Fusible wire

Starter solenoid

Ground at frame bolt

Ignition circuit breaker (contact points)

Alternator

Starter relay

Starter

FXE model only

+ − Battery

## 1976-1977 FX/FXE 1200

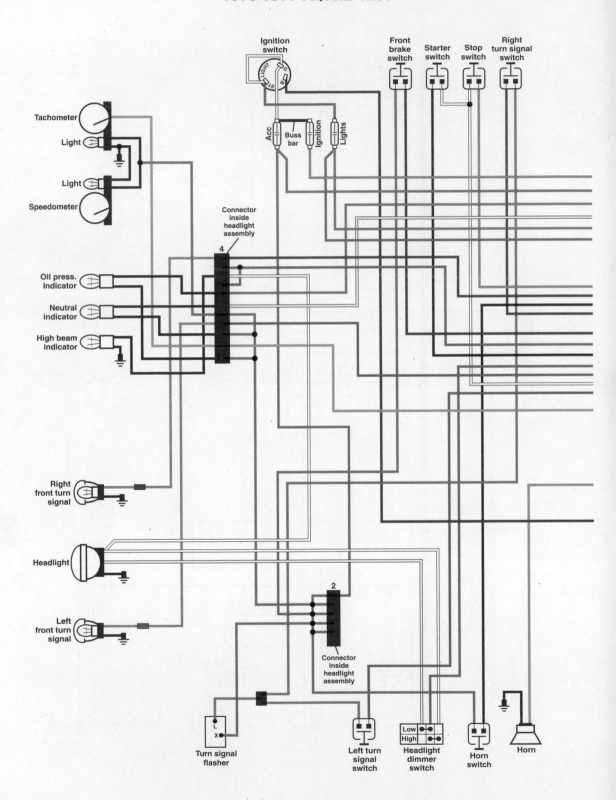

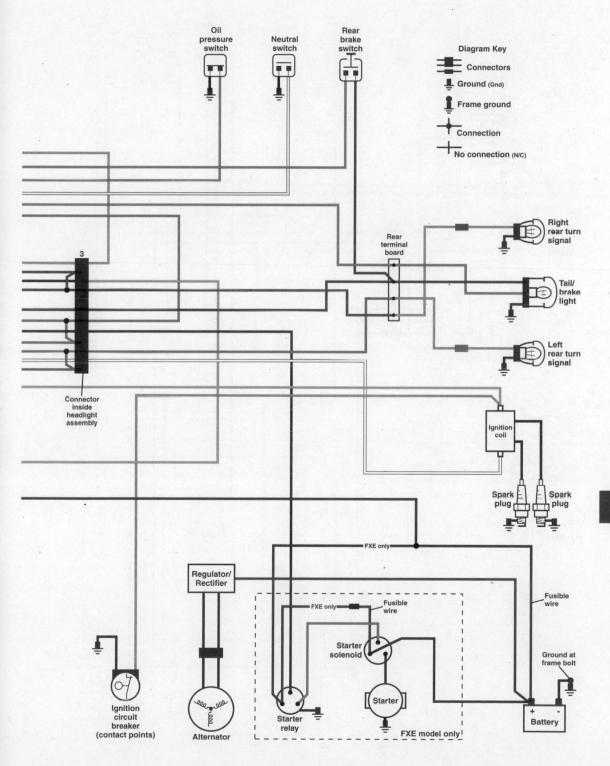

## 1978-1979 FX/FXE

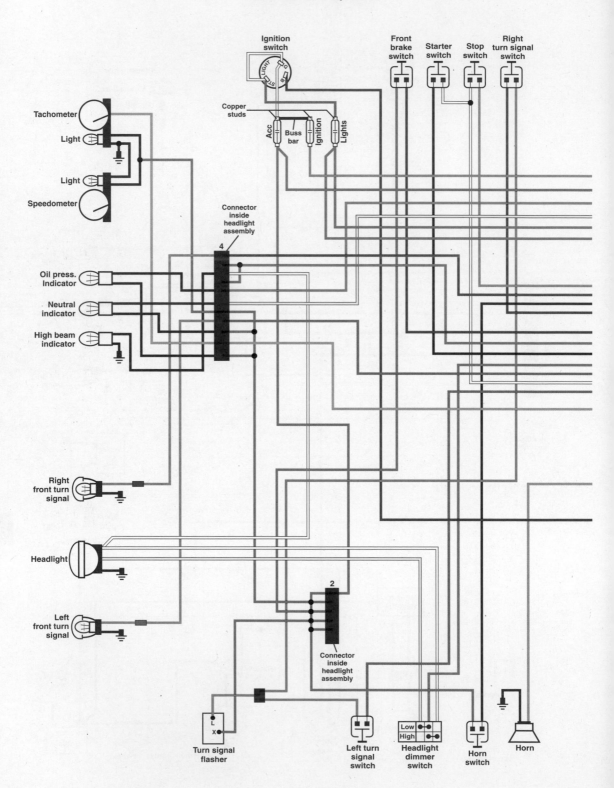

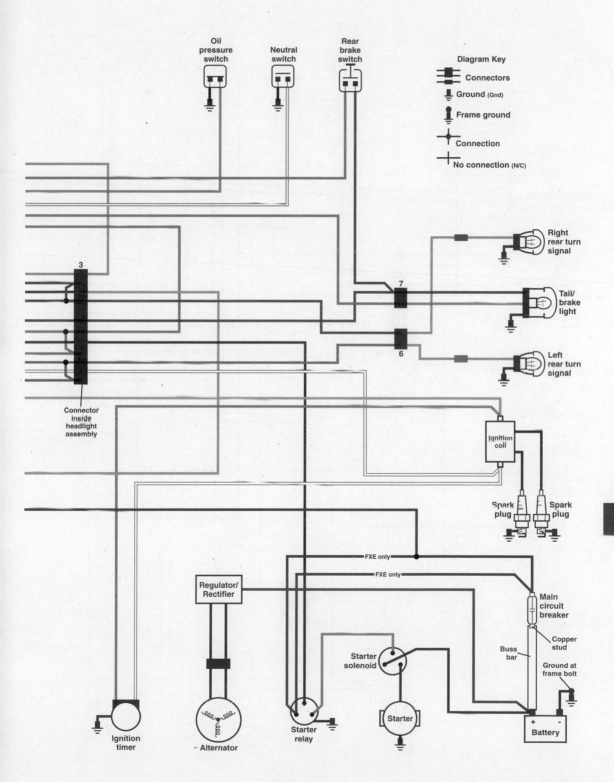

Oil pressure switch

Neutral switch

Rear brake switch

**Diagram Key**
- Connectors
- Ground (Gnd)
- Frame ground
- Connection
- No connection (N/C)

Right rear turn signal

Tail/brake light

Left rear turn signal

3

7

6

Connector inside headlight assembly

Ignition coil

Spark plug

Spark plug

FXE only

FXE only

Regulator/Rectifier

Main circuit breaker

Copper stud

Starter solenoid

Buss bar

Ground at frame bolt

Ignition timer

Alternator

Starter relay

Starter

Battery

+ −

**12**

**1980-1982 FXE**

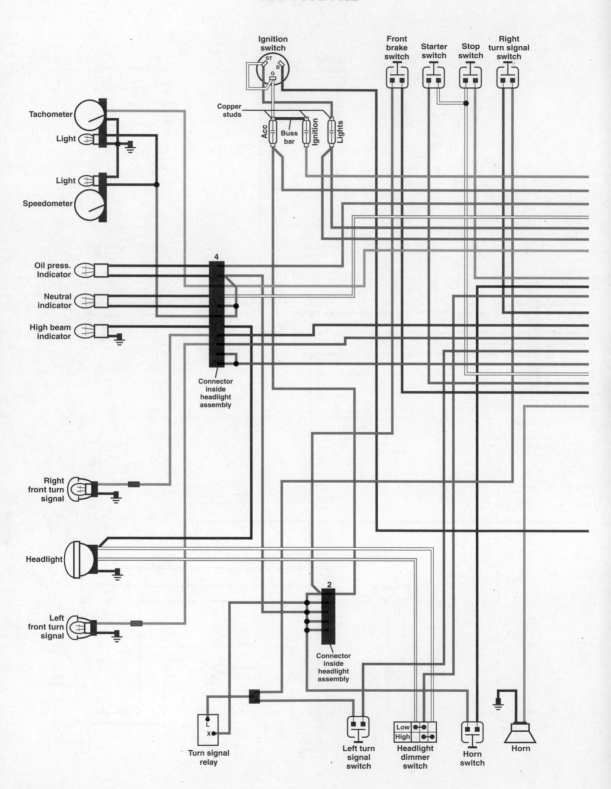

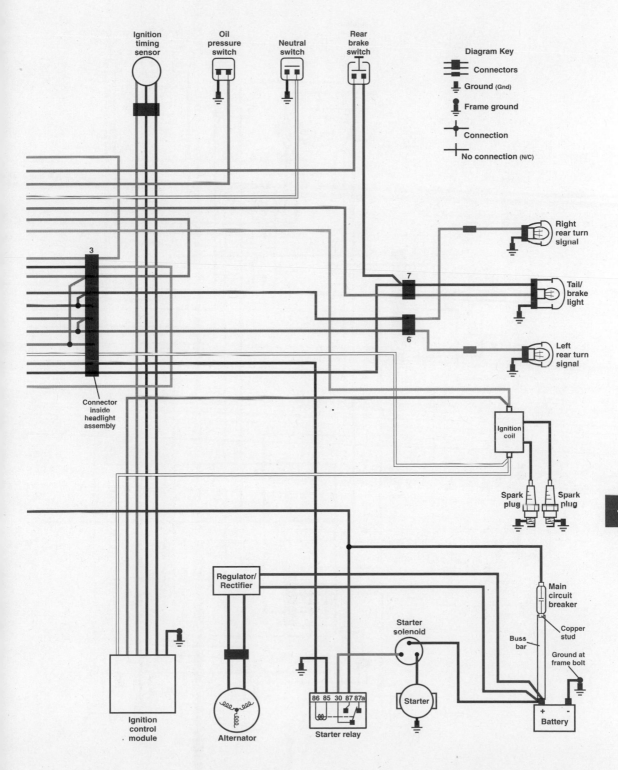

Ignition timing sensor

Oil pressure switch

Neutral switch

Rear brake switch

Diagram Key

Connectors

Ground (Gnd)

Frame ground

Connection

No connection (N/C)

3

Connector inside headlight assembly

7

6

Right rear turn signal

Tail/ brake light

Left rear turn signal

Ignition coil

Spark plug

Spark plug

12

Regulator/ Rectifier

Starter solenoid

Main circuit breaker

Copper stud

Buss bar

Ground at frame bolt

Ignition control module

Alternator

| 86 | 85 | 30 | 87 | 87a |

Starter relay

Starter

Battery

## 1983-1984 FXE

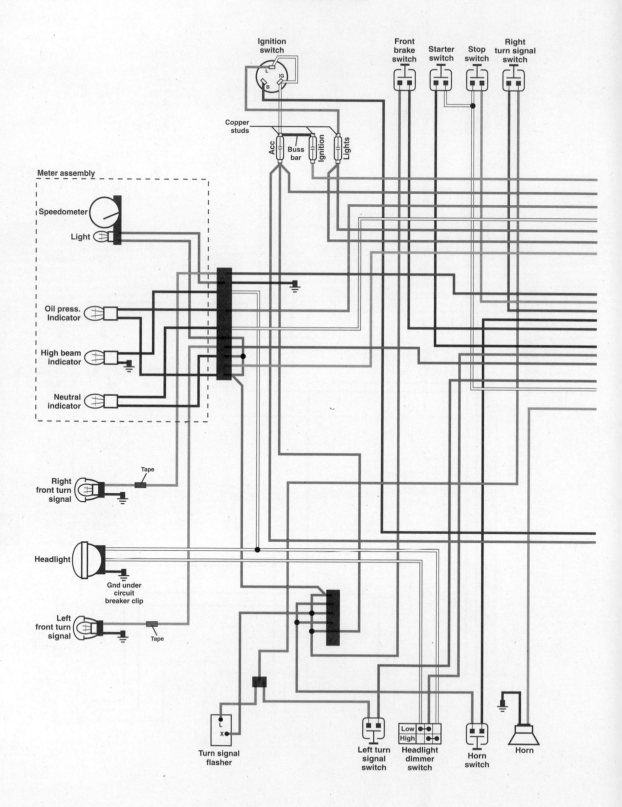

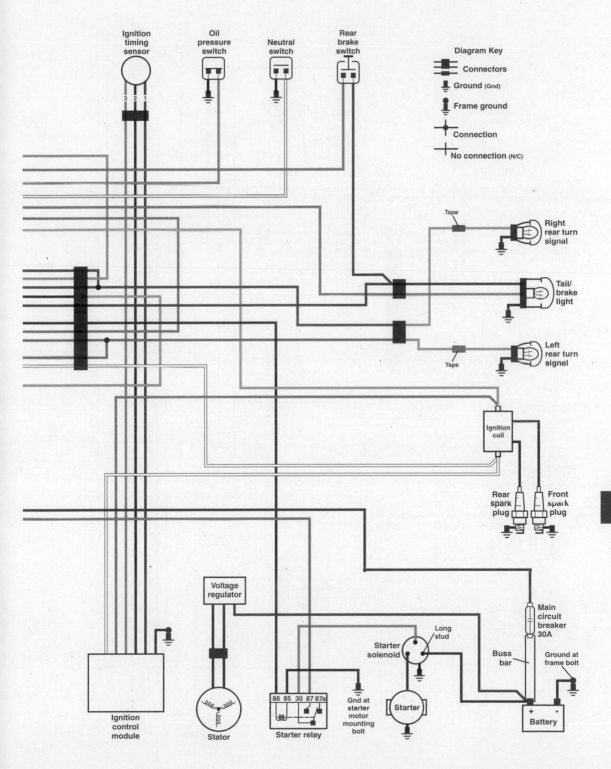

**Ignition timing sensor**

3 2 1

**Oil pressure switch**

**Neutral switch**

**Rear brake switch**

**Diagram Key**

Connectors

Ground (Gnd)

Frame ground

Connection

No connection (N/C)

Tape

**Right rear turn signal**

**Tail/ brake light**

**Left rear turn signal**

Tape

**Ignition coil**

**Rear spark plug**

**Front spark plug**

**12**

**Voltage regulator**

**Long stud**

**Main circuit breaker 30A**

**Starter solenoid**

**Buss bar**

**Ground at frame bolt**

**Ignition control module**

**Stator**

86 85 30 87 87a

**Starter relay**

**Gnd at starter motor mounting bolt**

**Starter**

**+ −**
**Battery**

## 1978-1979 FXS

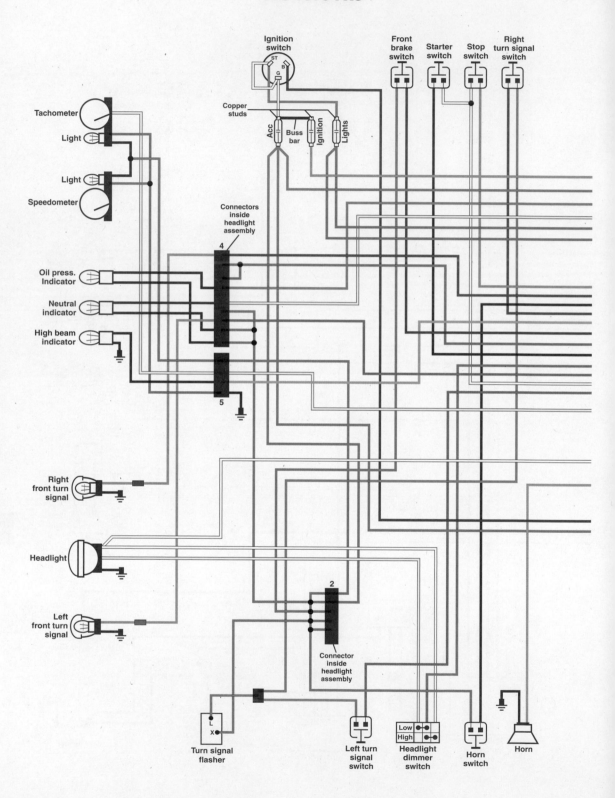

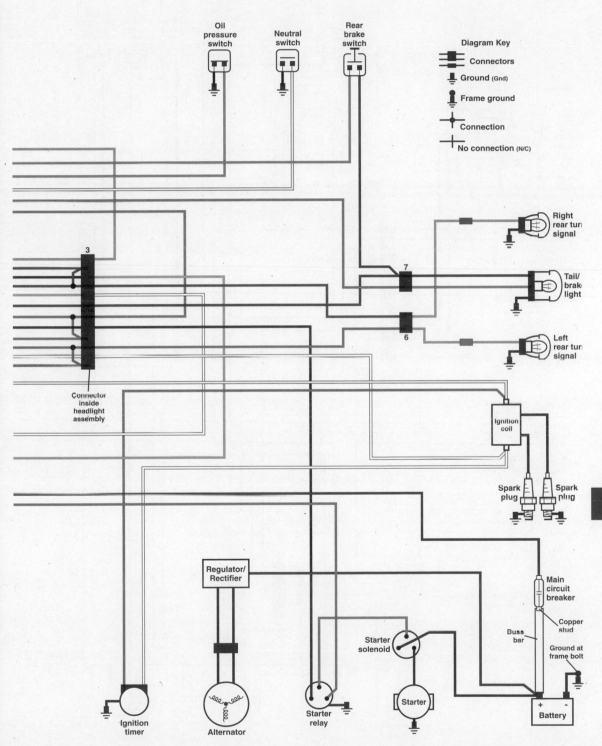

Oil pressure switch

Neutral switch

Rear brake switch

**Diagram Key**

Connectors

Ground (Gnd)

Frame ground

Connection

No connection (N/C)

Right rear turn signal

Tail/brake light

Left rear turn signal

3

7

6

Connector inside headlight assembly

Ignition coil

Spark plug

Spark plug

Regulator/ Rectifier

Main circuit breaker

Copper stud

Buss bar

Ground at frame bolt

Starter solenoid

Starter

Ignition timer

Alternator

Starter relay

Battery

**12**

## 1980-1984 FXS/FXEF/FXB

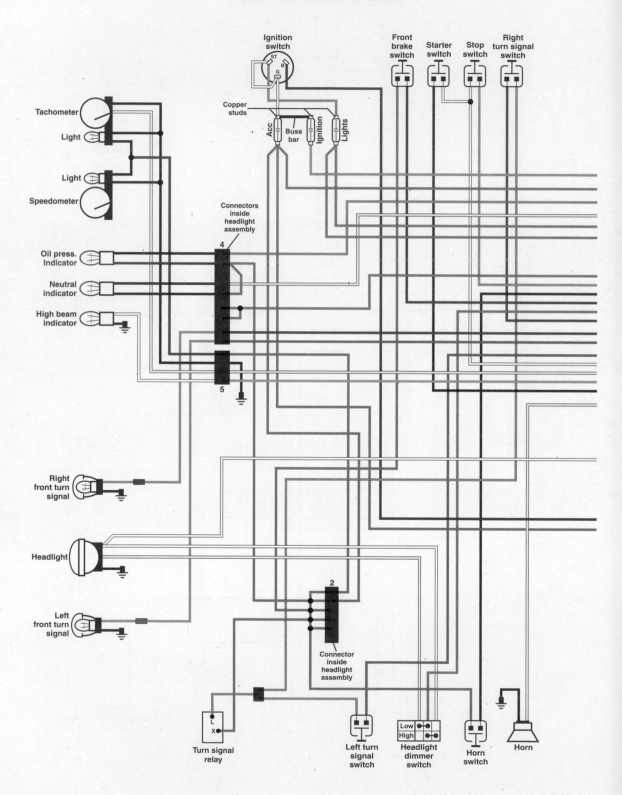

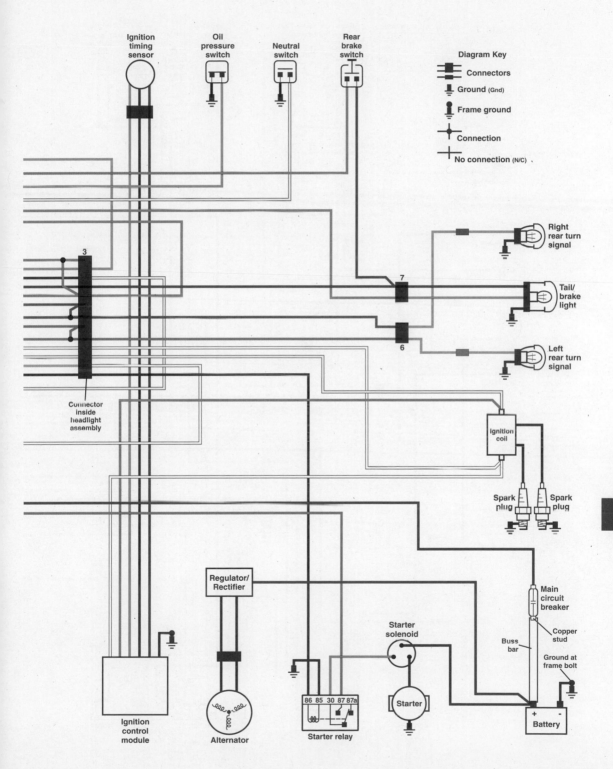

Ignition timing sensor

Oil pressure switch

Neutral switch

Rear brake switch

Diagram Key

Connectors

Ground (Gnd)

Frame ground

Connection

No connection (N/C)

Right rear turn signal

Tail/ brake light

Left rear turn signal

3

7

6

Connector inside headlight assembly

Ignition coil

Spark plug

Spark plug

12

Regulator/ Rectifier

Main circuit breaker

Copper stud

Buss bar

Ground at frame bolt

Starter solenoid

Ignition control module

Alternator

86  85  30  87 87a

Starter relay

Starter

+  Battery  −

## 1980-1984 FXWG

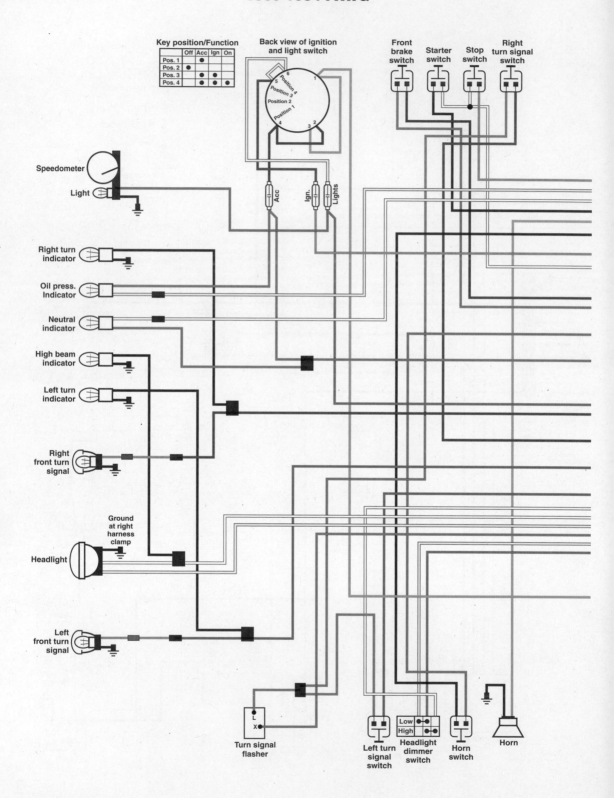

Key position/Function

| | Off | Acc | Ign | On |
|---|---|---|---|---|
| Pos. 1 | | ● | | |
| Pos. 2 | ● | | | |
| Pos. 3 | | ● | ● | |
| Pos. 4 | | ● | ● | ● |

Back view of ignition and light switch

Front brake switch

Starter switch

Stop switch

Right turn signal switch

Speedometer

Light

Acc

Ign.

Lights

Right turn indicator

Oil press. Indicator

Neutral indicator

High beam indicator

Left turn indicator

Right front turn signal

Ground at right harness clamp

Headlight

Left front turn signal

Turn signal flasher

Left turn signal switch

Headlight dimmer switch

Horn switch

Horn

Low

High

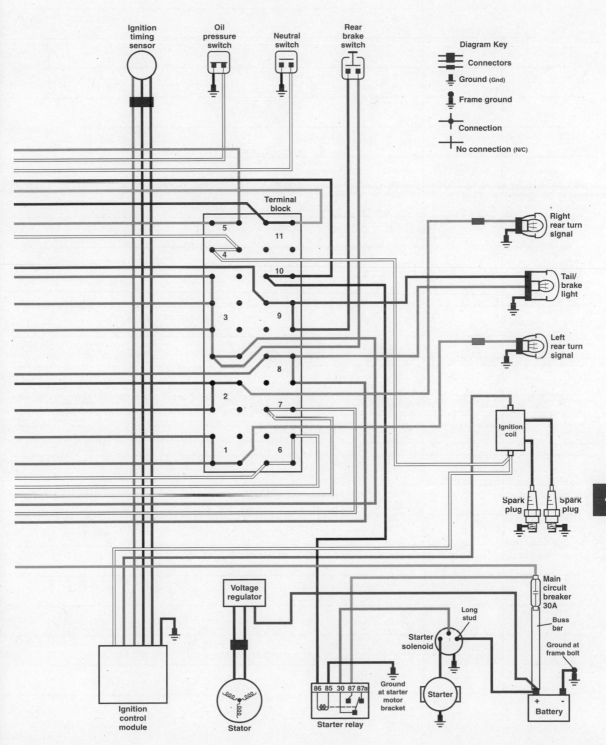

# MAINTENANCE LOG

| Date | Miles | Type of Service |
|------|-------|-----------------|
|      |       |                 |
|      |       |                 |
|      |       |                 |
|      |       |                 |
|      |       |                 |
|      |       |                 |
|      |       |                 |
|      |       |                 |
|      |       |                 |
|      |       |                 |
|      |       |                 |
|      |       |                 |
|      |       |                 |
|      |       |                 |
|      |       |                 |
|      |       |                 |
|      |       |                 |
|      |       |                 |
|      |       |                 |
|      |       |                 |
|      |       |                 |
|      |       |                 |
|      |       |                 |
|      |       |                 |
|      |       |                 |
|      |       |                 |